Microfluidics and Nanotechnology

Biosensing to the Single Molecule Limit

Devices, Circuits, and Systems

Series Editor
Krzysztof Iniewski
CMOS Emerging Technologies Research Inc.,
Vancouver, British Columbia, Canada

PUBLISHED TITLES:

Atomic Nanoscale Technology in the Nuclear Industry
Taeho Woo

Biological and Medical Sensor Technologies
Krzysztof Iniewski

Building Sensor Networks: From Design to Applications
Ioanis Nikolaidis and Krzysztof Iniewski

Circuits at the Nanoscale: Communications, Imaging, and Sensing
Krzysztof Iniewski

Electrical Solitons: Theory, Design, and Applications
David Ricketts and Donhee Ham

Electronics for Radiation Detection
Krzysztof Iniewski

Embedded and Networking Systems: Design, Software, and Implementation
Gul N. Khan and Krzysztof Iniewski

Energy Harvesting with Functional Materials and Microsystems
Madhu Bhaskaran, Sharath Sriram, and Krzysztof Iniewski

Graphene, Carbon Nanotubes, and Nanostuctures: Techniques and Applications
James E. Morris and Krzysztof Iniewski

High-Speed Photonics Interconnects
Lukas Chrostowski and Krzysztof Iniewski

Integrated Microsystems: Electronics, Photonics, and Biotechnology
Krzysztof Iniewski

Integrated Power Devices and TCAD Simulation
Yue Fu, Zhanming Li, Wai Tung Ng, and Johnny K.O. Sin

Internet Networks: Wired, Wireless, and Optical Technologies
Krzysztof Iniewski

Low Power Emerging Wireless Technologies
Reza Mahmoudi and Krzysztof Iniewski

Medical Imaging: Technology and Applications
Troy Farncombe and Krzysztof Iniewski

PUBLISHED TITLES:

MEMS: Fundamental Technology and Applications
Vikas Choudhary and Krzysztof Iniewski

Microfluidics and Nanotechnology: Biosensing to the Single Molecule Limit
Eric Lagally and Krzysztof Iniewski

MIMO Power Line Communications: Narrow and Broadband Standards, EMC, and Advanced Processing
Lars Torsten Berger, Andreas Schwager, Pascal Pagani, and Daniel Schneider

Nano-Semiconductors: Devices and Technology
Krzysztof Iniewski

Nanoelectronic Device Applications Handbook
James E. Morris and Krzysztof Iniewski

Nanoplasmonics: Advanced Device Applications
James W. M. Chon and Krzysztof Iniewski

Nanoscale Semiconductor Memories: Technology and Applications
Santosh K. Kurinec and Krzysztof Iniewski

Novel Advances in Microsystems Technologies and Their Applications
Laurent A. Francis and Krzysztof Iniewski

Optical, Acoustic, Magnetic, and Mechanical Sensor Technologies
Krzysztof Iniewski

Radiation Effects in Semiconductors
Krzysztof Iniewski

Semiconductor Radiation Detection Systems
Krzysztof Iniewski

Smart Sensors for Industrial Applications
Krzysztof Iniewski

Technologies for Smart Sensors and Sensor Fusion
Kevin Yallup and Krzysztof Iniewski

Telecommunication Networks
Eugenio Iannone

Testing for Small-Delay Defects in Nanoscale CMOS Integrated Circuits
Sandeep K. Goel and Krishnendu Chakrabarty

Wireless Technologies: Circuits, Systems, and Devices
Krzysztof Iniewski

FORTHCOMING TITLES:

3D Circuit and System Design: Multicore Architecture, Thermal Management, and Reliability
Rohit Sharma and Krzysztof Iniewski

Circuits and Systems for Security and Privacy
Farhana Sheikh and Leonel Sousa

FORTHCOMING TITLES:

CMOS: Front-End Electronics for Radiation Sensors
Angelo Rivetti

Gallium Nitride (GaN): Physics, Devices, and Technology
Farid Medjdoub and Krzysztof Iniewski

High Frequency Communication and Sensing: Traveling-Wave Techniques
Ahmet Tekin and Ahmed Emira

High-Speed Devices and Circuits with THz Applications
Jung Han Choi and Krzysztof Iniewski

Labs-on-Chip: Physics, Design and Technology
Eugenio Iannone

Laser-Based Optical Detection of Explosives
Paul M. Pellegrino, Ellen L. Holthoff, and Mikella E. Farrell

Metallic Spintronic Devices
Xiaobin Wang

Mobile Point-of-Care Monitors and Diagnostic Device Design
Walter Karlen and Krzysztof Iniewski

Nanoelectronics: Devices, Circuits, and Systems
Nikos Konofaos

Nanomaterials: A Guide to Fabrication and Applications
Gordon Harling and Krzysztof Iniewski

Nanopatterning and Nanoscale Devices for Biological Applications
Krzysztof Iniewski and Seila Selimovic

Optical Fiber Sensors and Applications
Ginu Rajan and Krzysztof Iniewski

Organic Solar Cells: Materials, Devices, Interfaces, and Modeling
Qiquan Qiao and Krzysztof Iniewski

Power Management Integrated Circuits and Technologies
Mona M. Hella and Patrick Mercier

Radio Frequency Integrated Circuit Design
Sebastian Magierowski

Semiconductor Device Technology: Silicon and Materials
Tomasz Brozek and Krzysztof Iniewski

Smart Grids: Clouds, Communications, Open Source, and Automation
David Bakken and Krzysztof Iniewski

Soft Errors: From Particles to Circuits
Jean-Luc Autran and Daniela Munteanu

FORTHCOMING TITLES:

VLSI: Circuits for Emerging Applications
Tomasz Wojcicki and Krzysztof Iniewski

Wireless Transceiver Circuits: System Perspectives and Design Aspects
Woogeun Rhee and Krzysztof Iniewski

Microfluidics and Nanotechnology

Biosensing to the Single Molecule Limit

Edited by
Eric Lagally
Western Govenors University

Managing Editor
Krzysztof Iniewski
CMOS Emerging Technologies Research Inc.
Vancouver, British Columbia, Canada

CRC Press is an imprint of the
Taylor & Francis Group, an **informa** business

CRC Press
Taylor & Francis Group
6000 Broken Sound Parkway NW, Suite 300
Boca Raton, FL 33487-2742

First issued in paperback 2017

Version Date: 20140311

ISBN 13: 978-1-138-07239-8 (pbk)
ISBN 13: 978-1-4665-9490-6 (hbk)

Library of Congress Cataloging-in-Publication Data

Microfluidics and nanotechnology : biosensing to the single molecule limit / editors, Eric Lagally, Krzysztof Iniewski.
pages cm -- (Devices, circuits, and systems)
Includes bibliographical references and index.
ISBN 978-1-4665-9490-6 (hardback)
1. Biosensors. 2. Biomolecules. 3. Cytochemistry. 4. Microfluidics--Industrial applications. 5. Molecular recognition. 6. Nanofluids. I. Lagally, Eric.

R857.B54M548 2014
572--dc23 2014002073

Visit the Taylor & Francis Web site at
http://www.taylorandfrancis.com

and the CRC Press Web site at
http://www.crcpress.com

Contents

Preface xi
Editors xiii
Contributors xv

PART I Detection Technologies

Chapter 1 Toward Unsupervised Smart Chemical Sensor Arrays 3
Leonardo Tomazeli Duarte and Christian Jutten

Chapter 2 Active CMOS Biochip for Electrochemical DNA Assays 19
Peter M. Levine, Ping Gong, Rastislav Levicky, and Kenneth L. Shepard

Chapter 3 Label-Free DNA Sensor Based on a Surface Long-Period Grating 87
Young-Geun Han

Chapter 4 Measuring the Physical Properties of Cells 101
Shirin Mesbah Oskui and William H. Grover

Chapter 5 Technologies for Low-Cost, Hall Effect–Based Magnetic Immunosensors 131
Simone Gambini, Karl Skucha, Jungkyu Kim, and Bernhard E. Boser

PART II Integrated Microfluidic and Nanofluidic Systems

Chapter 6 Two-Dimensional Paper Networks for Automated Multistep Processes in Point-of-Care Diagnostics 155
Elain Fu, Barry Lutz, and Paul Yager

Chapter 7 Droplet-Based Digital Microfluidics for Single-Cell Genetic Analysis 171

Yong Zeng and Richard A. Mathies

Chapter 8 Droplet-Based Microfluidics for Biological Sample Preparation and Analysis 201

Xuefei Sun and Ryan T. Kelly

Chapter 9 A Review of Tubeless Microfluidic Devices 221

Pedro J. Resto, David J. Beebe, and Justin C. Williams

Index 265

Preface

Microtechnology and more recent nanotechnology methods have enabled the fabrication of a wide variety of new chemical and biological sensors. These sensors demonstrate exquisite sensitivity and low limits of detection, enabling exploration of new scientific frontiers. In particular, the novel physics that emerges at small length scales allows parallel, mass-fabricated sensors for detection to both single-cell and single-molecule limits. Stochastic differences between individual cells and molecules have been shown to play important roles in larger biological systems, and these novel sensors have begun to test and elucidate these effects.

This book focuses on the combination of soft materials like elastomers and other polymers with materials like semiconductors, metals, and glass to form integrated detection systems for biological and chemical targets. Microfluidic advances in this arena include systems for forming and analyzing tiny droplets (so-called droplet microfluidics); the combination of electrostatic and dielectrics to manipulate droplets on the microscale; miniaturized separation systems, including electrophoresis, for detecting a wide range of genetic targets from single cells; and novel optical and mechanical detection methods at the single-cell and single-molecule scales.

This book represents a snapshot of the state of the art from the world's leading microfluidics and nanotechnology laboratories. The combination of different materials at both of these length scales is driving a powerful new set of scientific inquiries that have to date been impossible to address using other technologies.

The book is arranged in two major sections. In Part I, the authors discuss a number of unique detection technologies. In Chapter 1, Leonardo Duarte and Christian Jutten describe systems for addressing the blind source separation problem and selectivity in an array of chemical sensors. Their approach of developing signal processing algorithms for monitoring multiple uncalibrated sensors provides a powerful method for unsupervised sensor arrays. In Chapter 2, Peter Levine and coworkers describe their recent work with CMOS sensor arrays for detecting DNA hybridization using electrochemical methods. In Chapter 3, Young-Geun Han describes methods for fiber optic–based long-period gratings for label-free detection of DNA. In Chapter 4, Shirin Oskui and William Grover describe microelectromechanical resonator methods for the detection of the mass, density, and volume of single cells. In Chapter 5, Simone Gambini describes techniques for Hall effect–based magnetic immunosensors and demonstrates the ability of such systems to detect as few as eight magnetic microbeads in an immunoassay.

Part II of the book is dedicated to descriptions of recent work in microfluidic and nanofluidic transport and integration of detection systems into more comprehensive lab-on-a-chip systems. Chapter 6 describes the work of Elaine Fu and coworkers to develop paper microfluidic lab-on-a-chip systems, and their investigations into the flow dynamics and transport properties of fluids in paper materials. Chapters 7 and 8 focus on two approaches for implementing droplet-based microfluidics. In Chapter 7, Yong Zeng and Richard Mathies describe the design, fabrication, and scale-up of membrane microvalve-based systems for controlled droplet generation and demonstrate genetic detection assays to the level of single cells and single gene copies using such systems. Chapter 8 focuses on the work of Xuefei Sun and Ryan T. Kelly, who describe methods of picoliter-volume droplet generation, extraction, and coupling of these droplet manipulation techniques to existing mass spectrometry detection systems. Finally, Chapter 9 presents a review by Pedro Resto, David Beebe, and Justin Williams of the leading methods for generating flow using physical effects including evaporation, concentration gradients, and backflow. They then review recent work to integrate these techniques into larger high-throughput analysis systems.

Our hope is that the book will serve both as a resource on the most recent advances in microfluidics for researchers around the world as well as an inspiration for more junior researchers, so that they might further expand the field in the future.

Editors

Dr. Eric Lagally is currently a faculty member at Western Governors University, where he teaches science to adult students. His research interests are in microfluidics for biological detection, particularly genetic detection of bacterial pathogens and the selection and use of affinity reagents including antibodies and nucleic acid aptamers. He founded and consulted for Lagally Consulting, and before that served as assistant professor at the University of British Columbia from 2006 to 2011, where his research program was responsible for developing multiplexed surface plasmon resonance microfluidics as well as dielectrophoresis chips for whole-cell detection of *Mycobacterium tuberculosis*. He has published 30 papers, both peer-reviewed and in conference proceedings, and is coinventor on a patent for microfluidic valve technologies. He has also authored review papers and chapters in edited books on microfluidic technologies. He received his PhD in bioengineering in 2003 from the Joint UC-Berkeley/UC-San Francisco Bioengineering Graduate Group.

Dr. Krzysztof (Kris) Iniewski is managing R&D at Redlen Technologies, a start-up company in Vancouver, Canada. Redlen's revolutionary production process for advanced semiconductor materials enables a new generation of more accurate, all-digital, radiation-based imaging solutions. Kris is also a president of CMOS Emerging Technologies Research, Inc. (www.cmosetr.com), which organizes high-tech events covering communications, microsystems, optoelectronics, and sensors. Over his career, Dr. Iniewski has held numerous faculty and management positions at the University of Toronto, University of Alberta, Simon Fraser University, and PMC-Sierra. He has published over 100 research papers in international journals and conferences. He holds 18 international patents granted in the United States, Canada, France, Germany, and Japan. He is a frequent invited speaker and has consulted for multiple organizations internationally. He has written and edited several books for CRC Press, Cambridge University Press, IEEE Press, Wiley, McGraw Hill, Artech House, and Springer. His personal goal is to contribute to healthy living and sustainability through innovative engineering solutions. In his leisure time, Kris can be found hiking, sailing, skiing, and biking in beautiful British Columbia. He can be reached at kris.iniewski@gmail.com.

Contributors

David J. Beebe
Department of Biomedical Engineering
University of Wisconsin
Madison, Wisconsin

Bernhard E. Boser
Department of Electrical Engineering and Computer Science
University of California
Berkeley, California

Leonardo Tomazeli Duarte
School of Applied Sciences
University of Campinas
Limeira, Brazil

Elain Fu
Department of Bioengineering
University of Washington
Seattle, Washington

Simone Gambini
Department of Electrical and Electronic Engineering
The University of Melbourne
Melbourne, Australia

Ping Gong
Seventh Sense Biosystems
Cambridge, Massachusetts

William H. Grover
Department of Bioengineering
University of California
Riverside, California

Young-Geun Han
Department of Physics
Hanyang University
Seoul, South Korea

Krzysztof Iniewski
CMOS Emerging Technologies Research Inc.
Vancouver, British Columbia, Canada

Christian Jutten
Gipsa-lab
Joseph Fourier University
Grenoble, France

Ryan T. Kelly
Environmental Molecular Sciences Laboratory
Pacific Northwest National Laboratory
Richland, Washington

Jungkyu Kim
Department of Mechanical Engineering
Texas Tech University
Lubbock, Texas

Eric Lagally
Western Governors University
Salt Lake City, Utah

Rastislav Levicky
Department of Chemical and Biomolecular Engineering
NYU Polytechnic School of Engineering
Brooklyn, New York

Peter M. Levine
Department of Electrical and Computer Engineering
University of Waterloo
Waterloo, Ontario, Canada

Barry Lutz
Department of Bioengineering
University of Washington
Seattle, Washington

Richard A. Mathies
Department of Chemistry
University of California
Berkeley, California

Shirin Mesbah Oskui
Department of Bioengineering
University of California
Riverside, California

Pedro J. Resto
Department of Mechanical Engineering
University of Puerto Rico
Mayagüez, Puerto Rico

Kenneth L. Shepard
Department of Electrical Engineering
Columbia University
New York, New York

Karl Skucha
Department of Electrical Engineering and Computer Science
University of California
Berkeley, California

Xuefei Sun
Biological Sciences Division
Pacific Northwest National Laboratory
Richland, Washington

Justin C. Williams
Department of Biomedical Engineering
University of Wisconsin-Madison
Madison, Wisconsin

Paul Yager
Department of Bioengineering
University of Washington
Seattle, Washington

Yong Zeng
Department of Chemistry and Bioengineering Graduate Program
University of Kansas
Lawrence, Kansas

Part I

Detection Technologies

1 Toward Unsupervised Smart Chemical Sensor Arrays

Leonardo Tomazeli Duarte and Christian Jutten

CONTENTS

1.1 Introduction 3
1.2 Smart Chemical Sensor Arrays 4
1.2.1 Potentiometric Sensors 4
1.2.2 Selectivity Issues in Potentiometric Sensors 5
1.2.3 Chemical Sensor Arrays 6
1.3 Blind Source Separation 7
1.3.1 Problem Description 7
1.3.2 Strategies to Perform BSS 8
1.3.2.1 ICA Methods 8
1.3.2.2 Bayesian Approach 9
1.3.3 Nonlinear Mixtures 10
1.4 Application of BBS Methods to Chemical Sensor Arrays 11
1.4.1 First Results 11
1.4.2 ICA-Based Methods for the Case with Different Valences 11
1.4.3 Use of Prior Information to Estimate the Electrode's Slope 13
1.4.4 Bayesian Source Separation Applied to Chemical Sensor Arrays 14
1.5 Practical Issues 15
1.5.1 Dealing with the Scale Ambiguity 15
1.5.2 ISEA Database 16
1.6 Conclusions 16
References 16

1.1 INTRODUCTION

One of the major challenges in chemical analysis is how to deal with the lack of selectivity that is typical of chemical sensors [1]. Recently, much attention has been given to an approach borrowed from the field of signal processing to deal with the interference problem related to chemical sensors. In this alternative approach, the sensing mechanism is based on an ensemble of sensors, usually referred to as smart sensor arrays (SSAs), that are not necessarily highly selective with respect to a given analyte. The rationale behind such an approach is that, although the sensors within the array may respond to several chemical species, if there is enough diversity between the sensors, then advanced signal processing methods can be applied in order to estimate the concentration—or detect the presence—of the chemical species under analysis.

Most chemical SSAs are based on *supervised* signal processing methods, which require calibrating (or training) samples to adjust the parameters of the adopted signal processing method. The application of supervised methods in both quantitative and qualitative analysis has been proving quite successful in tasks such as odor and taste automatic recognition systems (electronic noses [2] and tongues [3], respectively). However, despite the success of SSAs based on supervised methods, this approach has two important practical problems. First, the acquisition of training samples is usually a costly and time-consuming task. Second, due to the drift in the response of chemical sensors, the calibration procedure must be performed every time the sensor array is used.

In view of the practical limitations associated with supervised SSAs, some researchers have been developing systems based on unsupervised (or blind) methods. The idea here is to adjust the SSA data processing stage by considering only the array responses and possibly some information on the mechanism of the interference phenomenon. As a result, the calibration stage may be eliminated or at least simplified in unsupervised solutions. When unsupervised systems for performing quantitative analysis are considered, the task at hand is to solve what the signal processing community calls the *blind source separation* (BSS) problem [4–6]. The goal of BSS is to retrieve a set of signals (sources) based only on the observation of a set of signals that correspond to mixed versions of the original sources. Therefore, in quantitative analysis using SSAs, the sources would correspond to the time variations of the concentrations of each chemical species under analysis, whereas the mixed signals would be given by the array responses.

In this chapter, we discuss the main aspects underlying the application of BSS methods to the problem of quantitative analysis using SSAs. Our focus will be on sensor arrays composed of potentiometric electrodes. Concerning the chapter's organization, we first provide, in Section 1.2, a brief introduction to potentiometric sensors. In Section 1.3, we introduce the BSS problem. Then, in Section 1.4, we provide an overview of current results concerning the application of BSS methods to quantitative analysis using potentiometric sensor arrays. Section 1.5 is devoted to the discussion of some important practical aspects related to the use of BSS methods for performing chemical analyses. Finally, Section 1.6 concludes the chapter.

1.2 SMART CHEMICAL SENSOR ARRAYS

1.2.1 Potentiometric Sensors

Potentiometric sensors are devices that measure the variation in electrical potential induced by variation in the concentration of a given chemical species. The best-known example of a potentiometric sensor is the ion-selective electrode (ISE) [1,7,8].

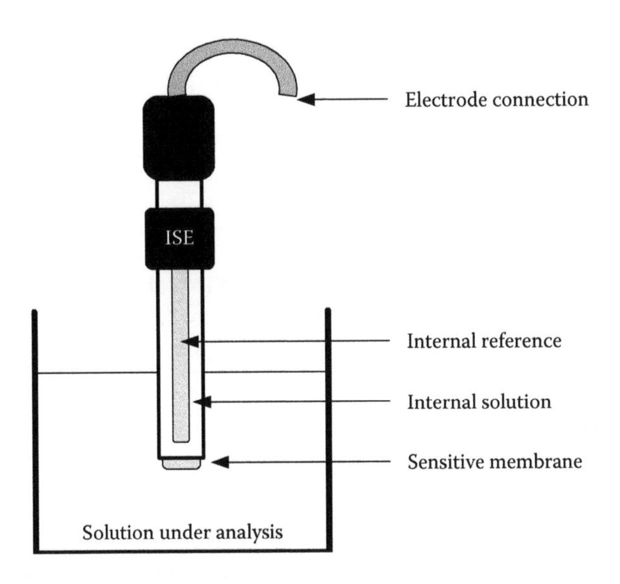

FIGURE 1.1 The ion-selective electrode (ISE).

This device consists of an internal solution, an internal reference electrode, and a sensitive membrane, as depicted in Figure 1.1. The electrical potential in this membrane results from electrochemical equilibrium and is directly related to the activity (which can be seen as the effective concentration) of a given target ion.

A well-known example of an ISE is the glass electrode [1,7] used for measuring the pH of a given solution. In addition, some ISEs are customized for different ions such as ammonium, potassium, and sodium. These devices have been intensively used, for instance, in food and soil inspection, clinical analysis, and water-quality monitoring. One of the explanations for the success of ISEs in such applications is the simplicity of their approach. Analysis using ISEs does not require sophisticated laboratory equipment and procedures and, thus, can be carried out in the field if necessary. Moreover, ISEs are very economical compared with other chemical sensing systems. Another popular example of a potentiometric sensor is the ion-sensitive field-effect transistor (ISFET) [9]. To a certain extent, the ISFET can be seen as a miniaturized version of the ISE as the transduction mechanisms of both devices are essentially the same.

1.2.2 Selectivity Issues in Potentiometric Sensors

Electrodes such as ISEs and ISFETs have an important drawback: these devices are not selective as the generated potential usually depends on a given target ion and other interfering ions. This may become important when the target ion and the interfering ions have similar physical or chemical properties or both [7]. In such cases, the measurements taken by the chemical sensor become uncertain when the concentrations of the interfering ions are high enough.

The interference phenomenon in potentiometric sensors can be modeled with the Nicolsky–Eisenman (NE) equation [1,7,8,10]. Assuming that s_i and s_j correspond, respectively, to the activity of the target ion and the activity of interfering ions, the response of a potentiometric sensor according to the NE equation is given by

$$x = e + d\log\left(s_i + \sum_{j,j \neq i} a_{i,j} s_j^{z_i/z_j} \right), \tag{1.1}$$

where e and d are constants that depend on some physical parameters and z_i denotes the valence of the i-th ion. The parameters a_{ij}, which are called selectivity coefficients, model the interference between the ions under analysis.

1.2.3 Chemical Sensor Arrays

The key concept underlying SSAs is diversity. Indeed, in an SSA, as illustrated in Figure 1.2, there is a data processing block that makes use of the diversity within the array to remove the effects caused by the interference. Since the signal processing block is at the core of the SSA, an interesting aspect of this design is its flexibility: the same SSA can conduct different types of analyses following minor and often automatic changes. To some extent, the term *smart* refers to this adaptive character of SSAs. Besides this advantage, other assets of SSAs include robustness and cost; indeed, SSAs are usually composed of simple, and thus cheap, electrodes. There are also embedded SSAs, which contain a microcontroller.

The signal processing techniques used in SSAs can first be classified with respect to the paradigm adopted to adjust their parameters. If the method makes use of a set of training (or calibration) data, it is called *supervised.* In supervised methods, the calibration data are considered in a stage (training stage) that precedes the effective use of the array. If, on the other hand, no calibration stage is considered, then the signal processing method is referred to as *unsupervised* or *blind.*

However, supervised qualitative chemical analysis creates a pattern classification problem [11]. This problem can be tackled through machine learning techniques, such as multilayer perceptron (MLP) neural networks [12] and support vector machines (SVM) [13]. Supervised quantitative analysis through sensor arrays can be formulated as a multivariate regression problem.

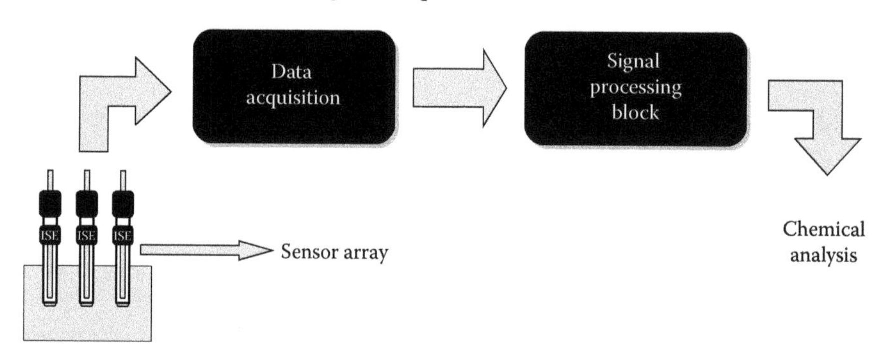

FIGURE 1.2 Smart chemical sensor array.

The majority of SSAs are based on a supervised paradigm. However, there is growing interest in unsupervised methods. For instance, a number of works have considered chemical unsupervised qualitative analysis. In this case, since training points are not available, the classes that should be classified are not known beforehand. The resulting problem here is known as clustering analysis and can be approached via algorithms like k-means and self-organizing maps (SOM) [13].

While the three signal processing tasks discussed so far (supervised and unsupervised qualitative analysis, and supervised quantitative analysis) are now familiar to those working with chemical SSAs, unsupervised quantitative chemical analysis is less developed. As will become clear later, unsupervised analysis can be formulated as a BSS problem.

1.3 BLIND SOURCE SEPARATION

In BSS [4], the goal is to retrieve a set of signals (sources) that were submitted to a mixing process. Since the mixing process is also assumed to be unknown, source separation is conducted by considering only observed signals (mixtures). BSS methods have been intensively used in biosignal processing, audio analysis, and image processing (see the textbooks [4–6] for more details).

Although the formulation of BSS is simple, its resolution became possible only after the introduction of a new learning paradigm in signal processing. Indeed, in a paper published in the 1980s by Hérault et al. [14], it was shown that the standard approach that was typically considered in statistical signal processing at that time—methods based on second-order statistics—could not solve the BSS problem. Below we will provide a mathematical formulation of the BSS problem and present the main strategies to solve this problem.

1.3.1 Problem Description

Let us consider that vectors $\boldsymbol{s}(t) = [s_1(t)\ s_2(t)\ \ldots\ s_N(t)]^T$ and $\boldsymbol{x}(t) = [x_1(t)\ x_2(t)\ \ldots\ x_M(t)]^T$ represent the sources and the mixtures, respectively. Moreover, let the mixing process be represented by a mathematical mapping $\boldsymbol{F}(\cdot)$, so the mixtures can be expressed as follows

$$\boldsymbol{x}(t) = \boldsymbol{F}(\boldsymbol{s}(t)). \tag{1.2}$$

BSS methods aim to estimate the sources $\boldsymbol{s}(t)$ based only on the mixtures $\boldsymbol{x}(t)$, that is, without making use of any precise information about either the mixing mapping $\boldsymbol{F}(\cdot)$ or the sources $\boldsymbol{s}(t)$. It should be stressed here that, in the context of SSAs, the sources $\boldsymbol{s}(t)$ correspond to the concentrations of the chemical species under analysis and the mixtures $\boldsymbol{x}(t)$ to the signals recorded by the array.

The BBS problem and solution are illustrated in Figure 1.3. The basic idea is to define a separating system, which is represented by $\boldsymbol{G}(\cdot)$ so the signals $\boldsymbol{y}(t) = [y_1(t)\ y_2(t)\ \ldots\ y_N(t)]^T$, given by

$$\boldsymbol{y}(t) = \boldsymbol{G}(\boldsymbol{x}(t)), \tag{1.3}$$

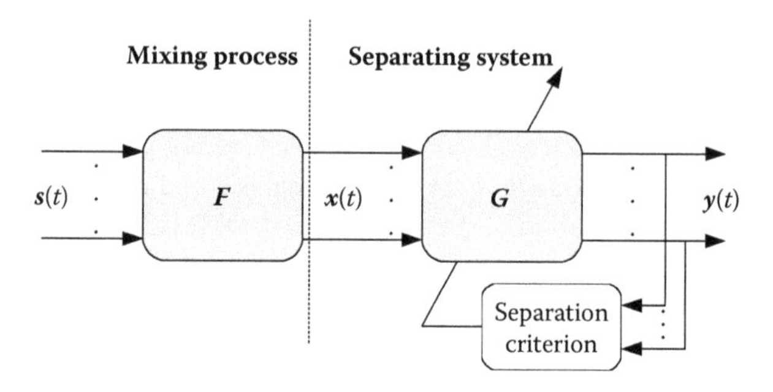

FIGURE 1.3 General scheme of the problem of blind source separation.

are as close as possible to the actual sources $\boldsymbol{s}(t)$. Since the sources are unknown, the fundamental question in this context is how to formulate a separation criterion that will guide adjustment of the parameters of the separating system $\boldsymbol{G}(\cdot)$. In Section 1.3.2, we discuss some of the main strategies applied to accomplish this task. Henceforth, we assume that the number of mixtures and sources are the same ($N=M$).

1.3.2 Strategies to Perform BSS

The BSS problem is badly posed and, thus, cannot be solved if a minimum of prior information about the sources is not available. For example, let us consider that one knows that the sources present a given property. If this property is lost after the mixing process, a possible method to build a separation criterion would be to search for the separating system $\boldsymbol{G}(\cdot)$ that provides estimations $\boldsymbol{y}(t)$ having the same property observed in $\boldsymbol{s}(t)$. The idea is valid when the recovery of such an original characteristic assures that $\boldsymbol{G}(\cdot)$ provides a perfect inversion of the mixing mapping $\boldsymbol{F}(\cdot)$. This is the case, for instance, in the context of independent component analysis (ICA), as discussed in Section 1.3.2.1.

1.3.2.1 ICA Methods

Initially, ICA methods were designed to deal with the situation in which the mixing process is linear and instantaneous, that is, without memory. In this case, it is possible to express the mixing process as follows

$$\boldsymbol{x}(t)=\boldsymbol{A}\boldsymbol{s}(t), \tag{1.4}$$

where $\boldsymbol{A}$ corresponds to the mixing matrix. Typically, it is assumed that the number of sensors is equal to the number of sources, $\boldsymbol{A}$ being a square matrix in this case. Therefore, in this situation, the separating system can also be defined as a square matrix, so the retrieved signals are given by

$$\boldsymbol{y}(t)=\boldsymbol{W}\boldsymbol{s}(t). \tag{1.5}$$

Ideally, the separating matrix should be given by $\boldsymbol{W}=\boldsymbol{A}^{-1}$.

In ICA, the sources are viewed as realizations of independent random variables. Since the independence property is lost after the linear mixing process, the basic idea in ICA is to adjust $\boldsymbol{W}$ so the signals $\boldsymbol{y}(t)$ are again independent. Comon [15] showed that, for the situation in which at most one source follows a Gaussian distribution, the recovery of independent components $\boldsymbol{y}(t)$ is possible if, and only if, $\mathbf{WA}=\mathbf{PD}$, where $\mathbf{P}$ is permutation matrix and $\mathbf{D}$ a diagonal matrix. In other words, recovering independent components assures that one can indeed recover the waveform of the original sources, but it is impossible to determine their exact order and original scales. These limitations are usually referred to as the order and scale ambiguities, and arise because statistical independence is a property that is invariant to order and scales.

ICA can be formulated as an optimization problem in which the cost function is related to a measure of statistical independence. Usually in the literature, this cost function is called contrast function. For instance, a measure that is commonly employed in the context of information theory, *mutual information* [16], can be defined as a contrast function. Indeed, the mutual information of a set of random variables is always greater than or equal to zero, and is zero if and only if these variables are mutually independent, which thus provides a natural measure of statistical independence.

Other approaches that also lead to simple ICA algorithms are described in the literature. Examples include the Infomax approach, cumulant-based methods, nonlinear decorrelation methods, and nonlinear principal component analysis (see [4,6] for an introduction). An interesting point is that all these approaches are somehow connected and can be described through a consistent theoretical framework [17], resulting thus in practical algorithms that are similar.

1.3.2.2 Bayesian Approach

Another approach that has been applied in the context of BSS is based on Bayesian estimation. Among the features found in this framework is the possibility of incorporating prior information that can be described in a probabilistic manner. Moreover, although Bayesian BSS methods usually consider, as prior information, that the sources are independent, this assumption is rather an instrumental one as Bayesian methods do not rely on an independence recovery procedure. The development of Bayesian methods for BSS was addressed in a number of works (see chapter 12 of [4] for an introduction), and one can find applications in, for instance, spectroscopy data analysis [18] and hyperspectral imaging [19].

In Bayesian BSS, instead of defining a separating system aimed at inverting the mixing process, one searches for a generative model that can correctly explain the observed data. To clarify this idea, let the $N\times T$ matrix $\boldsymbol{X}$ represent all the observations (T samples of N mixtures) of the problem. Moreover, let the matrix $\boldsymbol{\Theta}$ represent all the unknown terms of our problem, that is, the sources and the coefficients of the mixing process. The key concept in the Bayesian approach is the posterior probability distribution, that is, the probability distribution of the unknown parameters $\boldsymbol{\Theta}$ conditioned to the observed data $\boldsymbol{X}$. This posterior distribution can be obtained via Bayes' rule, as follows:

$$\mathrm{p}(\Theta|\boldsymbol{X}) = \mathrm{p}(\boldsymbol{X}|\Theta)\frac{\mathrm{p}(\Theta)}{\mathrm{p}(\boldsymbol{X})}. \tag{1.6}$$

In this expression, $p(\boldsymbol{X}|\boldsymbol{\Theta})$ is the likelihood function and is directly related to the assumed mixing model. The term $p(\boldsymbol{\Theta})$ denotes the prior distribution, and should be defined by taking into account the available information. For example, if the sources are nonnegative [20], which is the case in chemical arrays, it is natural to consider prior distributions whose support takes only nonnegative values.

There are several possible ways to estimate $\boldsymbol{\Theta}$ from $\boldsymbol{X}$ in a Bayesian framework. For instance, a possible strategy is to find $\boldsymbol{\Theta}$ that maximizes the posterior distribution. This strategy is known as maximum *a posteriori* (MAP) estimation. Another way to obtain an estimation is based on the Bayesian minimum mean square error (MMSE) estimator. In this situation, the resulting estimator is obtained by taking the expected value of the posterior distribution $p(\boldsymbol{\Theta}|\boldsymbol{X})$.

1.3.3 Nonlinear Mixtures

In some practical cases, the linear approximation does not give a good enough description of the interference model. For example, this is the case when dealing with potentiometric sensors (see Section 1.2.2). In such situations, the mixing process that takes place at the sensor array becomes nonlinear and, thus, must be tackled with nonlinear BSS models.

Nonlinear BSS presents some problems that are not present in a linear context. The most important one is that ICA does not lead to source separation in a general nonlinear context. In other words, there are some mixing models for which retrieving independent components is not enough to retrieve the sources. In view of this difficulty, some researchers have been considering constrained classes of nonlinear models for which ICA is still valid [21]. For instance, as shown in [22,23], ICA-based solutions assure source separation in an important class of constrained models known as post-nonlinear (PNL) models.

The basic structure of a PNL model is shown in Figure 1.4. The mixing process is composed of a mixing matrix $\boldsymbol{A}$, which is followed by component-wise functions, represented by the vector of functions $\boldsymbol{f}(\cdot)$. The mixing model can be represented by $\boldsymbol{x}(t)=\boldsymbol{f}(\boldsymbol{A}\boldsymbol{s}(t))$. As shown in Figure 1.4, the separating system in this case can be defined as $\boldsymbol{y}(t)=\boldsymbol{W}\boldsymbol{g}(\boldsymbol{x}(t))$. Under the assumptions that the sources are independent and that the component-wise functions $\boldsymbol{f}(\cdot)$ and $\boldsymbol{g}(\cdot)$ are monotonic, the separating matrix $\boldsymbol{W}$ and the compensating functions $\boldsymbol{g}(\cdot)$ can be adjusted by an independence recovery procedure [22].

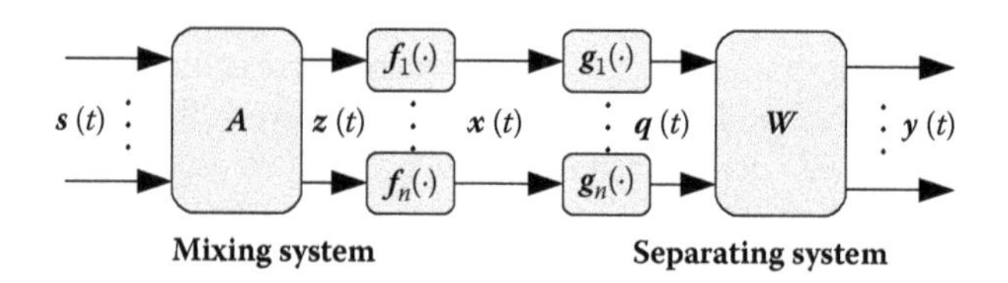

FIGURE 1.4 The post-nonlinear model (mixing and separating structures).

1.4 APPLICATION OF BBS METHODS TO CHEMICAL SENSOR ARRAYS

1.4.1 First Results

The application of BSS methods to chemical sensor arrays is a relatively new topic and has been considered in a small number of papers. The first contributions in this area were made by Sergio Bermejo and collaborators [24]. In their work, linear source separation was employed to process the data obtained by an array of ISFETs, which, as discussed before, can be modeled by the NE equation. Hence, the mixing process that takes place in ISFETs is basically the same as that in ISEs.

As discussed before, one of the problems encountered when applying source separation methods to potentiometric sensor arrays is related to the nonlinear nature of these sensors. In [24], the nonlinear terms present in the mixing process are compensated for by using a small set of calibration points. Therefore, the solution proposed in [24] is not completely blind since only the linear part of the mixing process is processed in a blind fashion.

The work presented in [25] investigated the application of the mutual information-based PNL method proposed in [22] to ISFET arrays. Indeed, if one assumes that the interference problem can be modeled by the NE equation, and when all the ions under analysis have the same valences, the resulting mixing process becomes a special case of the PNL model, where the component-wise functions are given by

$$f_i(t) = e_i + d_i \log(t). \tag{1.7}$$

Ideally, the parameters d_i could be set by simply considering the theoretical Nernstian slope value predicted by the NE equation [1]. However, d_i usually deviates from this theoretical value and therefore should be estimated. The parameter e_i does not need to be estimated in a BSS framework since it only introduces a scale gain, and, as already discussed, BSS methods are not able to detect the original scales of the sources. The approach proposed by Bedoya et al. [25] was assessed by considering a set of synthetic data that were obtained from actual measurements.

1.4.2 ICA-Based Methods for the Case with Different Valences

As pointed out by Bedoya et al. [25], when the valences of all ions under analysis are equal, the ratio z_i/z_j in Equation 1.1 equals one, and, as a consequence, the mixing process in this case is given by a PNL model. When the valences are different, the resulting mixing model becomes tougher because nonlinearity (power term) appears inside the logarithmic term.

Elsewhere, we proposed algorithms suited to the situation in which there are power terms inside the logarithm function [26,27]. In view of the complexity of the resulting mixing model in this case, we consider some simplifying assumptions. First, we assume that, although they are different, the valences of the ions under analysis are known in advance. Second, we consider that the parameters e_i and d_i are previously known. The estimation of d_i can be done, for instance, by the blind methods that will be discussed in Section 1.4.3. Moreover, even if e_i is not exactly known, we can still apply our method, but the best we can do is to retrieve each source up to an unknown multiplicative gain. A last simplification is that we only considered the case with two sources (ions) and two mixtures (electrodes). This assumption is realistic in many practical situations where there is one interfering ion that is dominant while the others can be neglected.

When the simplifications described above are considered, the resulting mixing model can be written as

$$\begin{aligned} x_1 &= s_1 + a_{12}s_2^k, \\ x_2 &= s_2 + a_{21}s_1^{1/k}. \end{aligned} \tag{1.8}$$

The term k corresponds to the valence ratio z_1/z_2. Our analysis assumes that k is always a natural number. This situation arises, for instance, in the analysis of Ca^{2+} and Na^+, since $k=2$ in this case. It is worth noticing that, since the sources are non-negative in our problem, there is no risk of having complex-value numbers from the term $s_1^{1/k}$. Finally, as we are interested in an ICA solution, it is assumed that the sources are statistically independent.

The first problem that arises when dealing with the nonlinear mixing model (Equation 1.8) concerns the definition of a proper separating system. Differently from linear and even PNL systems, the mixing model (Equation 1.8) is not invertible, and, as a consequence, it is not possible to define a direct separating system. In order to overcome this problem, we previously defined [26] a separating structure based on a recurrent system. This strategy, which was first adopted in the context of another class of nonlinear models known as linear-quadratic models [28], is able to perform a sort of implicit inversion of the mixing model, given that, when the mixing parameters are known, the equilibrium points of the defined recurrent system correspond to the actual sources. Of course, these equilibrium points can be attained only if they are stable. As a consequence, for values of ionic activities and selectivity coefficients that lie outside the stability region, the adopted recurrent system cannot be used. In [26], we provide these stability regions.

Concerning the estimation of the parameters of the adopted recurrent separating system, we previously proposed [26] an ICA method based on the minimization of a measure of nonlinear correlation. The idea was implemented by considering a gradient descent approach. In particular, we studied some convergence issues related to the obtained gradient-based learning rule. In order to obtain a more robust learning algorithm than the recurrent separating system, we considered in [27] an algorithm based on mutual information minimization. Our approach was based on the concept of the differential of the mutual information, which was proposed in [29]. More details can be found in [27].

To illustrate the performance of the method proposed in [27], we considered the problem of estimating the activities of Ca^{2+} and Na^+ ($k=2$) by using two ISEs, each tailored to a different ion under analysis. The selectivity coefficients were obtained from [30] and the temporal evolution of the ionic activities (our sources) were artificially generated. In Figure 1.5, we consider the situation in which the activity of

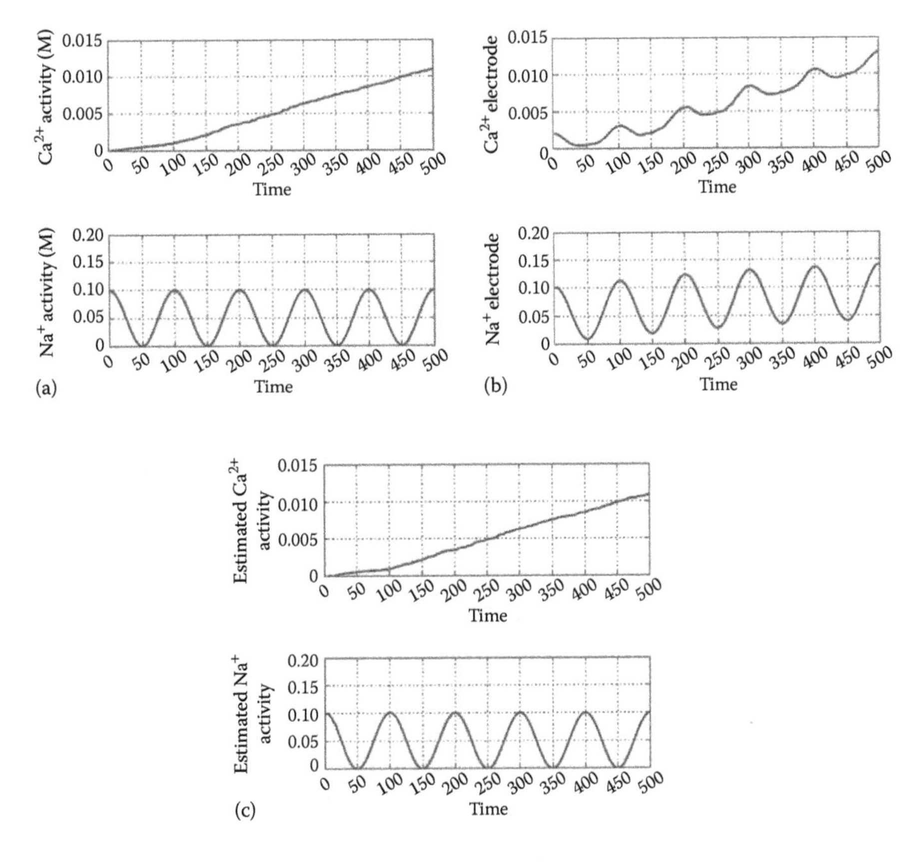

FIGURE 1.5 Example of mutual information minimization algorithm: (a) sources; (b) mixtures; (c) retrieved sources after 600 iterations.

the ion Ca^{2+} lies in the interval $[10^{-4}]$ to $[10^{-3}]$ M, whereas the activity of Na^{+} varies between $[10^{-4}]$ and $[10^{-1}]$ M, and we plot the sources (ionic activities), the mixtures (array responses), and the retrieved sources estimated by the algorithm [27]. The total number of samples of the mixtures was 500. As shown in Figure 1.5, (a) there is indeed a mixing process taking place at the sensor array, and (b) our method was able to provide a very good estimation of the sources.

1.4.3 Use of Prior Information to Estimate the Electrode's Slope

The ICA methods discussed in Section 1.4.2 were designed to deal with the mixing model (Equation 1.8), and, thus, assumed that the slopes d_i were known in advance. In practice, though, it is necessary to estimate these slopes. In Duarte and Jutten [31], we addressed this problem by considering the following additional assumption on the sources: there is at least a period of time where one, and only one, of the sources takes a constant value different from zero. In the context of chemical sources, this hypothesis seems valid in some applications as can be seen, for instance, in the waveforms presented in [32].

1.4.4 Bayesian Source Separation Applied to Chemical Sensor Arrays

As discussed in Section 1.3.2.2, another way to take advantage of prior information on the sources and mixing coefficients is to consider a Bayesian approach. One of the motivations behind this approach is related to the use of some prior information whose incorporation is easier in a probabilistic framework. For instance, in our problem, the fact that the sources are nonnegative can be easily taken into account by considering nonnegative priors.

In [33], we developed a Bayesian source separation method for mixtures generated by the NE equation. The basis of our approach, which can be applied for cases of both equal and different valences, relies on the attribution of a nonnegative prior, the lognormal distribution, to the sources. Moreover, since the selectivity coefficients usually lie in the interval [0,1], we attributed to these parameters a uniform prior with support given by [0,1]. Concerning the implementation of the Bayesian estimation, we considered an approach based on Markov chain Monte Carlo methods. More details can be found in [33].

We tested our Bayesian method considering the set of real data available in the ISE array (ISEA) dataset (see Section 1.5.2). For instance, we show in Figures 1.6 through 1.8, respectively, the sources, the mixtures, and the retrieved sources in a case in which the goal is to estimate the activities of K^+ and NH_4^+ by considering an SSA composed of one K^+-ISE and one NH_4^+-ISE. These results attest that the Bayesian algorithm was able to retrieve a good estimation of the sources even under adverse conditions, namely, a reduced number of samples and sources with a high degree of correlation. Conversely, the application of an ICA-based PNL source separation method could not separate the sources in this case [33]. This limitation arises from the fact that the sources were not independent in this scenario, thus violating the essential assumption made in ICA.

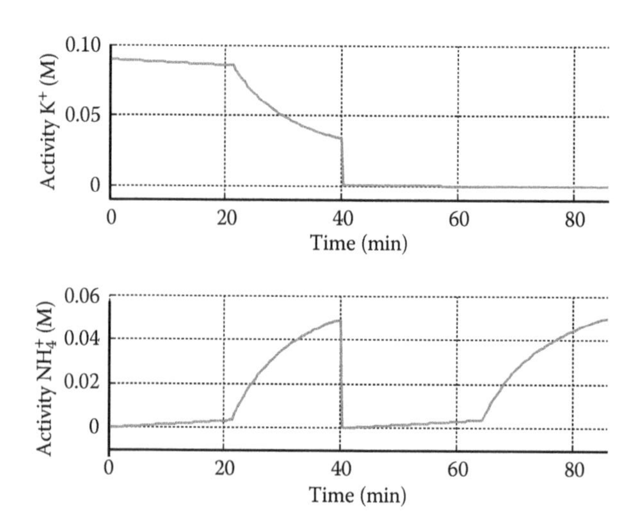

FIGURE 1.6 Application of the proposed Bayesian source separation approach to an array composed of a K^+-ISE and NH_4^+-ISE (sources).

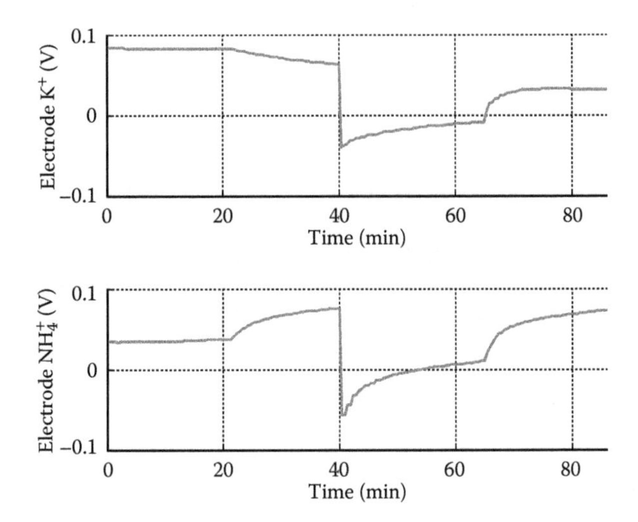

FIGURE 1.7 Application of the proposed Bayesian source separation approach to an array composed of a K^+-ISE and NH_4^+-ISE (mixtures).

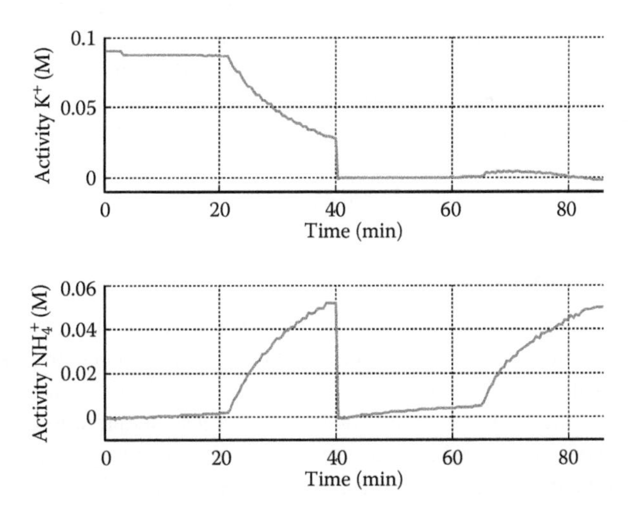

FIGURE 1.8 Application of the proposed Bayesian source separation approach to an array composed of a K^+-ISE and NH_4^+-ISE (estimated sources).

1.5 PRACTICAL ISSUES

1.5.1 Dealing with the Scale Ambiguity

As mentioned before, in BSS there is always a scale ambiguity, since separation criteria usually cannot recover the correct amplitude of the sources. While in many applications the scale ambiguity can be accepted, this limitation poses a problem in chemical sensing applications since the main goal is to retrieve the correct value of the concentration. Therefore, the source separation step must be followed by postprocessing whose goal is to retrieve the correct scale. This additional stage requires at least two calibration points. In view of this requirement, should we not simply use the available calibration points for performing supervised processing? For example, we could define a separating system (the inverse of the NE model) adjusted under supervision. The key point here is that, as shown in [33], in an experiment with one K^+-ISE and one NH_4^+-ISE, the supervised solution required at least 20 calibration points to provide a good estimation of the sources. Conversely, a good estimation was achieved by the Bayesian approach proposed in [33] followed by a postprocessing stage that made use of only three calibration points to retrieve the correct scale of the sources. This difference highlights the benefits of unsupervised methods: although it is impossible to operate with no calibration points, the number of calibration points needed in an unsupervised approach is rather small, which can be quite advantageous in practice.

1.5.2 ISEA Database

In the framework of our research on BSS-based chemical sensor arrays, we had some difficulty finding actual data that could be used to validate the developed methods. Thus, at www.gipsa-lab.inpg.fr/isea, we provide a dataset (publicly available) acquired in a set of experiments with ISEAs. These experiments are described in [34]. Basically, we considered three scenarios: (a) analysis of a solution containing K^+ and NH_4^+; (b) analysis of a solution containing K^+ and Na^+; and (c) analysis of a solution containing Na^+ and Ca^{2+}. These data can be used in the development of unsupervised signal processing methods as well as in the case of supervised solutions since we also provide the original sources.

1.6 CONCLUSIONS

The goal of this chapter was to provide an overview of the application of BBS methods to potentiometric sensor arrays. The results are promising because these methods may work even if only a few calibration points are available. Of course, in a practical context, this feature is quite useful; for instance, a less demanding calibration step may help analysis in the field. However, despite the encouraging results obtained in this research, many questions must still be answered before source separation blocks can be incorporated into commercial chemical analyzers. Measures required include the development of methods for more than two sources and the provision of more precise mixing models, which may eventually increase the estimation quality, thus permitting the use of source separation methods even in very-high-precision applications.

REFERENCES

1. P. Gründler, *Chemical Sensors: An Introduction for Scientists and Engineers*, Springer: Berlin, 2007.
2. H. Nagle, R. Gutierrez-Osuna, and S. S. Schiffman, The how and why of electronic noses, *IEEE Spectrum*, 35(9), 22–31, 1998.

3. Y. G. Vlasov, A. V. Legin, and A. M. Rudnitskaya, Electronic tongue: Chemical sensor systems for analysis of aquatic media, *Russian Journal of General Chemistry*, 78, 2532–2544, 2008.
4. P. Comon and C. Jutten (eds), *Handbook of Blind Source Separation: Independent Component Analysis and Applications*, Academic Press: Oxford, 2010.
5. J. M. T. Romano, R. R. F. Attux, C. C. Cavalcante, and R. Suyama, *Unsupervised Signal Processing: Channel Equalization and Source Separation*, CRC Press: Boca Raton, 2011.
6. A. Hyvärinen, J. Karhunen, and E. Oja, *Independent Component Analysis*, John Wiley & Sons: New York, 2001.
7. P. Fabry and J. Fouletier (eds), *Microcapteurs Chimiques et Biologiques: Application en Milieu Liquide*, Lavoisier: Paris, 2003, in French.
8. E. Bakker, Electrochemical sensors, *Analytical Chemistry*, 76, 3285–3298, 2004.
9. P. Bergveld, Thirty years of ISFETOLOGY. What happened in the past 30 years and what may happen in the next 30 years, *Sensors and Actuators B*, 88, 1–20, 2003.
10. E. Bakker and E. Pretsch, Modern potentiometry, *Angewandte Chemie International Edition*, 46, 5660–5668, 2007.
11. R. Blatt, A. Bonarini, E. Calabro, M. D. Torre, M. Matteucci, and U. Pastorino, Lung cancer identification by an electronic nose based on an array of MOS sensors, in *Proceedings of International Joint Conference on Neural Networks (IJCNN)*, pp. 1423–1428. 12–17 August, 2007, IEEE, Orlando, FL.
12. M. Pardo and G. Sberveglieri, Classification of electronic nose data with support vector machines, *Sensors and Actuators B*, 107, 730–737, 2005.
13. R. O. Duda, P. E. Hart, and D. G. Stork, *Pattern Classification*, 2nd ed., Wiley-Interscience: New York, 2000.
14. J. Hérault, C. Jutten, and B. Ans, Détection de grandeurs primitives dans un message composite par une architecture de calcul neuromimétique en apprentissage non supervisé, in *Proceedings of the GRETSI*, pp. 1017–1022. May 20–24, 1985, Groupement de Recherche en Traitement du Signal et des Images, Nice, France.
15. P. Comon, Independent component analysis, a new concept? *Signal Processing*, 36, 287–314, 1994.
16. T. M. Cover and J. A. Thomas, *Elements of Information Theory*, Wiley-Interscience: New York, 1991.
17. T.-W. Lee, M. Girolami, A. J. Bell, and T. J. Sejnowski, A unifying information-theoretic framework for independent component analysis, *Computers & Mathematics with Applications*, 39, 1–21, 2000.
18. S. Moussaoui, D. Brie, A. Mohammad-Djafari, and C. Carteret, Separation of non-negative mixture of non-negative sources using a Bayesian approach and MCMC sampling, *IEEE Transactions on Signal Processing*, 54, 4133–4145, 2006.
19. N. Dobigeon, S. Moussaoui, M. Coulon, J.-Y. Tourneret, and A. O. Hero, Joint Bayesian endmember extraction and linear unmixing for hyperspectral imagery, *IEEE Transactions on Signal Processing*, 57, 4355–4368, 2009.
20. A. Cichocki, R. Zdunek, A. H. Phan, and S. Amari, *Nonnegative Matrix and Tensor Factorizations: Applications to Exploratory Multiway Data Analysis and Blind Source Separation*, John Wiley & Sons: New York, 2009.
21. C. Jutten and J. Karhunen, Advances in blind source separation (BSS) and independent component analysis (ICA) for nonlinear mixtures, *International Journal of Neural Systems*, 14, 267–292, 2004.
22. A. Taleb and C. Jutten, Source separation in post-nonlinear mixtures, *IEEE Transactions on Signal Processing*, 47(10), 2807–2820, 1999.
23. S. Achard and C. Jutten, Identifiability of post-nonlinear mixtures, *IEEE Signal Processing Letters*, 12(5), 423–426, 2005.

24. S. Bermejo, C. Jutten, and J. Cabestany, ISFET source separation: Foundations and techniques, *Sensors and Actuators B*, 113, 222–233, 2006.
25. G. Bedoya, C. Jutten, S. Bermejo, and J. Cabestany, Improving semiconductor-based chemical sensor arrays using advanced algorithms for blind source separation, in *Proceedings of Sensors for Industry Conference (SIcon)*, pp. 149–154. January 24–27, 2004, IEEE, New Orleans.
26. L. T. Duarte and C. Jutten, Blind source separation of a class of nonlinear mixtures, in *Proceedings of the 7th International Workshop on Independent Component Analysis and Signal Separation (ICA 2007)*, pp. 41–48. September 9–12, 2007, Springer, London, UK.
27. L. T. Duarte and C. Jutten, A mutual information minimization approach for a class of nonlinear recurrent separating systems, in *Proceedings of the IEEE Workshop on Machine Learning for Signal Processing (MLSP)*, pp. 127–129. August 27–29, 2007, IEEE, Thessaloniki, Greece.
28. S. Hosseini and Y. Deville, Blind separation of linear-quadratic mixtures of real sources using a recurrent structure, in *Proceedings of the 7th International Work-Conference on Artificial and Natural Neural Networks (IWANN)*, pp. 241–248. June 3–6, 2003, Springer-Verlag, Menorca, Spain.
29. M. Babaie-Zadeh, C. Jutten, and K. Nayebi, Differential of the mutual information, *IEEE Signal Processing Letters*, 11(1), 48–51, 2004.
30. Y. Umezawa, P. Bühlmann, K. Umezawa, K. Tohda, and S. Amemiya, Potentiometric selectivity coefficients of ion-selective electrodes, *Pure and Applied Chemistry*, 72, 1851–2082, 2000.
31. L. T. Duarte and C. Jutten, A nonlinear source separation approach to the Nicolsky–Eisenman model, in *Proceedings of the 16th European Signal Processing Conference (EUSIPCO 2008)*, August 25–29, 2008, European Association for Signal Processing, Lausanne, Switzerland.
32. M. Gutiérrez, S. Alegret, R. Cáceres, J. Casadesús, and O. M. M. del Valle, Nutrient solution monitoring in greenhouse cultivation employing a potentiometric electronic tongue, *Journal of Agricultural and Food Chemistry*, 56, 1810–1817, 2008.
33. L. T. Duarte, C. Jutten, and S. Moussaoui, A Bayesian nonlinear source separation method for smart ion-selective electrode arrays, *IEEE Sensors Journal*, 9(12), 1763–1771, 2009.
34. L. T. Duarte, C. Jutten, P. Temple-Boyer, A. Benyahia, and J. Launay, A dataset for the design of smart ion-selective electrode arrays for quantitative analysis, *IEEE Sensors Journal*, 10(12), 1891–1892, 2010.

2 Active CMOS Biochip for Electrochemical DNA Assays

Peter M. Levine, Ping Gong, Rastislav Levicky, and Kenneth L. Shepard

CONTENTS

2.1 Introduction ... 21
2.1.1 Electrochemical DNA Sensing ... 22
2.1.2 Active CMOS Biochip for Electrochemical DNA Sensing ... 23
2.1.3 Chapter Overview ... 23
2.1.4 Chapter Organization ... 24
2.2 Background and Review ... 24
2.2.1 Affinity-Based DNA Sensors ... 24
2.2.2 Performance Metrics ... 24
2.2.3 Fluorescence-Based DNA Microarrays ... 25
2.2.3.1 Commercially Available Microarray Systems ... 27
2.2.4 Electrochemical DNA Sensors ... 27
2.2.4.1 Sensing Methodology ... 27
2.2.4.2 Real-Time DNA Detection ... 28
2.2.5 Principles of Electrochemistry ... 29
2.2.5.1 Electrode–Electrolyte Interface ... 29
2.2.5.2 Electrochemical Reactions at Equilibrium ... 30
2.2.5.3 Potentiostat Operation ... 31
2.2.5.4 Electrode Materials ... 32
2.2.5.5 Cyclic Voltammetry ... 32
2.2.6 DNA Immobilization Chemistry ... 34
2.2.7 Review of Electrochemical DNA Sensing Techniques ... 35
2.2.7.1 Redox-Label-Based Electrochemical DNA Sensors ... 35
2.2.8 CMOS Electrochemical DNA Sensors ... 36
2.2.8.1 Prior Work on CMOS Electrochemical DNA Sensors ... 36
2.3 Design of an Active CMOS Biochip for Label-Based Electrochemical DNA Assays ... 38
2.3.1 Label-Based Electrochemical DNA Sensing Method ... 39
2.3.2 Active CMOS Biochip System Requirements ... 40

2.3.2.1 Input Current Range 40
2.3.2.2 Bandwidth 41
2.3.2.3 Working Electrode Area 41
2.3.2.4 Silicon Area of Sensor Interface Electronics 41
2.3.3 System Architecture 42
2.3.3.1 Dual-Slope ADC Architecture and Operation 42
2.3.3.2 Working Electrode Array 45
2.3.3.3 Potentiostat Control Amplifiers 45
2.3.4 Electrochemical Cell Interface Model 45
2.3.5 Design of ADC Circuit Components 46
2.3.5.1 Integration Amplifier 46
2.3.5.2 Thermal and Flicker Noise 47
2.3.5.3 Bandwidth and Stability 48
2.3.5.4 Integration Capacitor 49
2.3.5.5 Comparator 49
2.3.5.6 Current Sources and Switches 50
2.3.5.7 Digital Counter and Control Circuitry 51
2.3.6 Design of Control Amplifier Circuits 51
2.3.7 Sensor Detection Limit and Noise Analysis 53
2.3.7.1 Integrated Noise 55
2.3.7.2 Shot Noise from a Redox Current 55
2.3.7.3 Thermal Noise from Electrolyte Resistance 56
2.3.7.4 Amplifier Thermal Noise 56
2.3.7.5 Amplifier Flicker Noise 57
2.3.7.6 Quantization Noise 57
2.3.7.7 Target Coverage Detection Limit 57
2.3.7.8 Effect of Working Electrode Area on Detection Limit 59
2.3.7.9 Effect of ADC Integration Time on Detection Limit 59
2.3.7.10 Sensor Dynamic Range 60
2.3.8 CMOS Biochip Fabrication and PostProcessing 61
2.3.8.1 Initial Electrode Fabrication Procedure 62
2.3.8.2 Improved Electrode Fabrication Procedure 63
2.4 Experimental Results 65
2.4.1 CMOS Chip Packaging and Experimental Setup 66
2.4.2 Electrical Characterization 66
2.4.3 Integrated Electrochemical Measurements of a Bulk Redox Species 68
2.4.3.1 Biochip Preparation and Experimental Procedure 69
2.4.3.2 Measurement Results 69
2.4.4 Integrated Electrochemical Measurement of DNA Probe Surface Coverage 71
2.4.4.1 Biochip Preparation and Experimental Procedure 71
2.4.4.2 Measurement Results 72
2.4.5 Label-Based Electrochemical DNA Sensing 73
2.4.5.1 ADC Operation 73
2.4.5.2 Chemical Preparation 73

2.4.5.3 Chip Surface Preparation 74
2.4.5.4 Measurement Results 74
2.5 Summary 80
References 80

2.1 INTRODUCTION

The development of methods and technology to perform genome sequencing and genomic assays has enabled significant advances in biomedical research in the past and will have a profound impact on the areas of disease prevention, personalized medicine, and clinical diagnostics in the future [1]. Over the last two decades, enormous effort from the public and private sectors has been spent on "sequencing" the human genome; that is, determining the exact series of the approximately 6 billion nucleotide base pairs contained in the 46 chromosomes found in most human cells. Ever since drafts of the human genome were first published in 2001 by both the publicly funded International Human Genome Sequencing Consortium through the Human Genome Project (HGP) [2] and a team led by J. Craig Venter of the privately owned biotechnology company Celera Genomics [3], the cost of human genome sequencing has declined by many orders of magnitude due to the introduction of increasingly automatable methods and technology. In particular, shotgun sequencing approaches [4] involving dideoxy chain termination techniques and electrophoretic size separation (known as "Sanger sequencing," after the individual who developed the technique [5]) have been replaced by massively parallel optical [6,7] and electronic [8] sequencing-by-synthesis platforms. These innovations have brought the cost of human genome sequencing down from US$3 billion, in the case of the HGP [9], to well under US$1 million [8]. Further technological developments will undoubtedly occur until one's own genome can be sequenced for less than US$1000 [10].

In addition to the numerous breakthroughs in DNA sequencing technology, methods and platforms for executing genomic assays have also undergone many advancements. Perhaps the most significant of these was the introduction of DNA microarrays in the mid-1990s. Presently, DNA microarrays are widely used for rapid, quantitative analysis of gene expression on a large scale [11–13]. Microarrays typically use a passive substrate, such as a glass slide, on which single-stranded DNA (ssDNA) "probe" molecules, representing thousands of individual genes, are arranged in a regular pattern. Probes are then allowed to bind to, or "hybridize" with, complementary, fluorophore-labeled ssDNA "target" molecules in an analyte solution. After nonhybridized targets are removed from the array through a washing step, the locations of hybridized targets on the microarray surface are measured with an optical scanner consisting of one or more laser sources for excitation and a photomultiplier tube (PMT) or charge-coupled device (CCD) camera to detect the emitted light. Relative expression levels of bound targets at different array sites can then be quantified from the resulting image.

While DNA microarrays are used extensively by research laboratories worldwide for applications such as transcriptional profiling, genotyping, and epigenetic studies [1], they have found limited use in clinical diagnostics. It is believed that microarray-based tests are superior to traditional DNA-based assays for clinical applications due to their ability to simultaneously measure the relative expression levels of a large number of clinically relevant genes [14]. In addition, microarrays are well suited for use in single-nucleotide polymorphism (SNP) analysis in which the existence of a single-base mismatch on a DNA strand can be used to identify a genetic disorder. However, DNA microarrays have found only limited use in clinical settings because their results often do not meet the stringent accuracy and reproducibility requirements set forth by the clinical and regulatory communities [15]. Despite this, it is expected that the human and biological factors that give rise to poor repeatability can be mitigated through the development and adoption of standardized methods [16].

The high instrumentation cost associated with microarray platforms is another reason why these systems are not widely used in clinical laboratories [14,15]. For example, the entry-level capital cost of optical scanners is typically well above US$100,000, which is too high for many laboratories. However, platform costs can be reduced using alternative detection methods and technologies other than traditional fluorescence-based DNA assays on passive, glass substrates.

In addition to laboratory-based clinical applications, genomic assays based on microarray technology could have future use in point-of-care (POC) genetic diagnostic applications. The ultimate goal would be to create a portable, handheld device for use in a physician's office, ambulance, or at the hospital bedside that could immediately provide time-critical information about the genetic composition of a patient, bacteria, or virus [17]. Microarrays are well suited to this application because they can screen for numerous illnesses containing many genetic markers and, therefore, could be used for such applications as predicting adverse drug reactions [18] or identifying bacterial or viral infections.

Unfortunately, traditional fluorescence-based DNA microarrays are not well suited for use in POC diagnostics. This is because the bulky and heavy lasers and optical detectors employed in these systems defy portability for POC applications. A platform for genomic assays that does not require light as an intermediary for sensing is, therefore, a valuable alternative.

2.1.1 Electrochemical DNA Sensing

Electrochemical sensing approaches to DNA detection, which have gained interest in recent years but are not yet as well developed as optical techniques [19], have the potential to reduce instrumentation costs and provide portable diagnostic platforms for genomic assays [20]. This is because the need for light as an intermediary for sensing is eliminated and, therefore, no optical equipment is required. Electrochemical DNA sensors instead measure electronic activity that results from hybridization of ssDNA targets with probes immobilized on "working" electrodes (WEs) immersed in an electrolyte. The nature of this electronic activity depends on the sensing methodology used but often involves the introduction of electrochemically active labels on the target DNA. An electronic feedback control system known as a "potentiostat" applies a desired input voltage to the electrochemical cell and measures the resulting current at the WE.

2.1.2 Active CMOS Biochip for Electrochemical DNA Sensing

A complementary metal-oxide-semiconductor (CMOS) platform is well suited to the design of electrochemical DNA sensors for genomic assays because the high level of parallel detection demanded by microarray applications requires active multiplexing. This can only be achieved through integration of the WEs directly onto an active CMOS substrate, known as a "biochip," containing the sensor electronics. CMOS is also an attractive technology for this application because it can be augmented using standard microfabrication techniques to accommodate the noble-metal electrodes most commonly used in electrochemical sensors.

In addition, the cost of fabricating individual CMOS biochips can become very low by leveraging the tremendous economies of scale associated with the semiconductor manufacturing industry. For example, a complete mask set for a mature CMOS process has a fixed cost on the order of US$50,000 [21]. This is used to fabricate multiple chips on a single silicon wafer that is typically 6–12 inches in diameter. A bare wafer and associated production expenses can cost around US$5000 [22]. This would translate to a typical manufacturing cost per chip of around US$40–US$50, depending on the number of chips manufactured. As a result, affordable diagnostic devices for widespread use in the healthcare industry could be produced relatively cheaply.

CMOS biochips also have much potential for use in POC genetic diagnostic applications because they can be integrated with other technologies to construct compact, self-contained sensing platforms. For example, biochemical sample preparation could be performed *in situ* using microfluidic channels with integrated pumps and valves [23]; solid-state transducers could be used to detect the presence and concentration of specific genetic sequences in a DNA analyte; and microelectronic integrated circuits could amplify and condition the electronic signals from the transducer, convert this information to a digital format, process it in order to extract relevant biochemical data, and then transmit or display these data externally.

2.1.3 Chapter Overview

This chapter presents an active CMOS biochip for electrochemical DNA sensing to support genomic assays. A CMOS-integrated sensor array is constructed and is designed to perform label-based sensing of DNA surface hybridization in which the redox molecule ferrocene (an electroactive species that undergoes reduction or oxidation depending on the applied potential) is covalently attached to the target DNA. The system consists of a 4×4 array of WEs, on which ssDNA probes are immobilized, and full potentiostat electronics, including control-loop amplifiers and a current-input analog-to-digital converter (ADC) at each electrode site. Because only labeled target molecules localized immediately at the electrode surface allow facile electron transfer, no washing step is required as in traditional fluorescence-based DNA microarrays and, therefore, DNA hybridization can be monitored in real time.

2.1.4 Chapter Organization

This chapter is organized as follows. Section 2.2 provides background information on platforms for analyzing DNA and electrochemical-based DNA detection techniques in particular, as well as electrochemistry principles. A review of previous work on electrochemical DNA sensors and CMOS-integrated DNA sensors is also included. Section 2.3 discusses the design and implementation of an active CMOS biochip for label-based, electrochemical DNA assays and Section 2.4 presents experimental results from this biochip [24,25]. Section 2.5 concludes the chapter.

2.2 BACKGROUND AND REVIEW

Affinity-based electrochemical DNA sensors detect charge transfer and interfacial charge fluctuations resulting from surface-based DNA probe–target hybridization. This section provides the necessary background for understanding the operation of electrochemical DNA sensors as well as their construction on active CMOS substrates.

2.2.1 Affinity-Based DNA Sensors

Affinity-based sensors are used to detect the concentration of "target" molecules in a complex analyte based on their specific physicochemical interactions with surface-bound "probe" molecules. This interaction often results in a binding reaction between probe and target to create a "duplex" molecule on the surface. Appropriate transducers are used to convert changes in, for example, light intensity, interfacial capacitance, or mass due to probe–target binding, often to an electronic signal that can be readily observed.

Affinity-based DNA sensors detect changes that occur from hybridization between ssDNA probes and targets and are used widely in genomic assays. Perhaps the most popular platform for constructing large-scale genomic assays that rely on affinity-based molecular interactions is the DNA microarray. Microarrays enable highly multiplexed, parallel detection and have traditionally supported probe immobilization on a passive substrate.

2.2.2 Performance Metrics

Important measures of performance for affinity-based DNA sensors include detection limit, specificity, and dynamic range (DR). The detection limit is usually quoted as the lowest density of target that can be reliably detected on the transducer or microarray surface, or the smallest target concentration that can be detected in solution. A relationship exists between these two values and will be discussed in more detail in Section 2.2.3. Often, a specific signal-to-noise ratio (SNR) is used to define the detection limit (an SNR of 3 is commonly used [26]), where the noise level can be determined from the sensor response to a blank solution.

Specificity refers to the sensor's ability to respond only to a DNA target in the analyte having a particular sequence and not to other noncomplementary molecules. The DR of the sensor is the ratio of the largest measurable target concentration to the limit of detection [27]. The former variable is often set by saturation of the probe layer with target molecules.

2.2.3 Fluorescence-Based DNA Microarrays

Fluorescence-based DNA microarrays have arguably become the standard technique for quantifying extents of hybridization in multiplexed genomic assays. Applications include gene expression profiling and functional genomics [28,29], drug development [30,31], mutational analysis [32], pharmacogenomics [33], and evolutionary studies [34]. Figure 2.1 displays the setup of a typical microarray experiment in which up to a million distinct groups of ssDNA probe molecules are immobilized in an array format on an electronically passive substrate such as glass, plastic, or silicon (Si). The 10^6–10^{10} ssDNA molecules at each site all contain the same known nucleotide sequence, but the sequence is intentionally varied from site to site. In "oligonucleotide" microarrays (made popular by Affymetrix GeneChip technology), the number of bases comprising each probe molecule is small (between 15 and 70) and represents subsequences of a gene [35]. Conversely, "complementary DNA" (cDNA) arrays can contain entire genes at each site and, therefore, use probes having lengths in the hundreds to thousands [36]. Probes can be constructed externally using solid-phase chemical synthesis and then immobilized on the microarray through mechanical contact "spotting" or noncontact ink-jet printing to create 10–100 μm reaction zones [28]. In this case, silanization of glass [37] or the treatment of Si substrates with an organic film [38] supports covalent probe attachment to the surface. Alternatively, probes can be constructed *in situ* using photolithographic techniques [39].

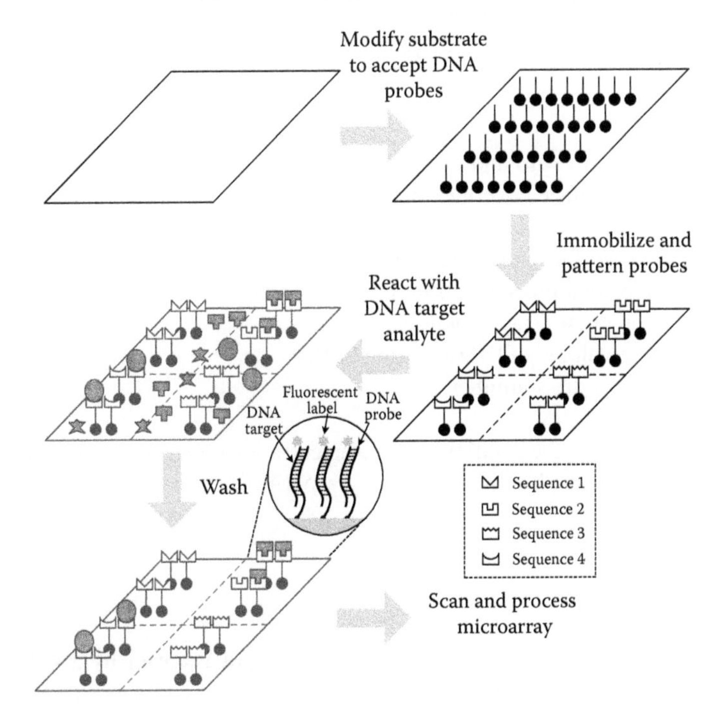

FIGURE 2.1 A typical fluorescence-based DNA microarray for a genomic assay. (Adapted from Patounakis, G., Active CMOS substrates for biomolecular sensing through time-resolved fluorescence detection, PhD. Dissertation, Columbia University, New York, NY, 2005.)

Single-stranded DNA target molecules, having an unknown sequence of bases, are conjugated with chemical labels that, upon absorbing photons at one wavelength, emit photons at a longer wavelength due to energy loss from vibrational or conformational mechanisms. This wavelength translation is expressed as the "Stokes shift." Fluorescent dyes, such as Cy3, Cy5, and fluorescein, are commonly used as labels and usually emit light with wavelengths in the 500–700 nm range.

When the target analyte is introduced to the microarray, probes and targets hybridize to an extent defined by the magnitude of the equilibrium association constant K_a for the binding reaction. The value of K_a is valid only for a particular temperature, hybridization-buffer ionic strength, and probe and target sequence [41]. Probes and targets that are fully complementary exhibit kinetics that give rise to a higher K_a compared to those having fewer complementary bases. Hybridization of surface-bound probe P to target T in solution to form a duplex molecule PT on the surface can be expressed as a single-step, reversible reaction:

$$P+T \rightleftarrows PT. \tag{2.1}$$

Assuming the above reaction reaches equilibrium, K_a is given by

$$K_a = k_f/k_r = \frac{S_T}{\left(S_P - S_T\right)\left[C_T - S_T\left(\frac{A}{N_A V}\right)\right]}, \tag{2.2}$$

where:

k_f and k_r are the forward and reverse rate constants of the reaction, respectively
S_P is the surface coverage (density) of hybridized and nonhybridized probe molecules
S_T is the surface coverage of a particular hybridized target
C_T is the solution concentration of the target
A is the microarray spot area
V is the volume of hybridization buffer
N_A is Avogadro's number

If it is assumed that the solution target concentration changes very little as a result of probe–target surface binding, that is $C_T \gg S_T(A/N_AV)$, then Equation 2.2 can be simplified to

$$K_a = \frac{x}{1-x}\frac{1}{C_T}, \tag{2.3}$$

where x is known as the "fractional occupancy" and is given by $x = S_T/S_P$ [42]. Rearranging the above equation slightly gives:

$$C_T = \frac{x}{1-x}\frac{1}{K_a}, \tag{2.4}$$

which shows that the absolute target concentration in solution can be determined if the fractional occupancy and equilibrium constant are known. However, due to the difficulty in characterizing and measuring these parameters accurately in complex, multitarget analyte, microarray analyses are usually limited to relative, as opposed to absolute, target concentration measurements [43].

The array is usually exposed to the analyte for many hours so that the hybridization process approaches equilibrium. Afterward, the array must be washed so that unbound targets and targets that have nonspecifically attached (adsorbed) to the microarray surface are completely eliminated from the substrate. This helps to reduce the background "noise" level. One implication of this is that hybridization cannot be monitored in real time using conventional microarrays, that is, throughout the entire binding process. This is because the analyte would have to be removed from the microarray with a washing step before equilibrium is reached in order to obtain a meaningful signal.

2.2.3.1 Commercially Available Microarray Systems

The size and weight of commercial fluorescence-based microarray scanners and imagers on the market today make them impractical for use in POC diagnostic applications. Commercial microarray systems are almost always manufactured as monolithic, laboratory-scale instruments in which the light source, optics, and detectors are all self-contained [44,45]. The Affymetrix GeneChip Scanner 3000 7G, for example, has dimensions (W × H × D) of 33 × 46 × 56 cm and weighs approximately 32 kg [44]. Although these instruments are bulky and heavy, their detection limit is very low at around 0.05–0.1 fluors μm^{-2}. This corresponds to approximately 10^6 hybridized targets/cm^2, assuming each DNA target has been conjugated with 10 fluorescent labels on average [36].

2.2.4 Electrochemical DNA Sensors

As discussed in the introduction, electrochemical DNA sensors translate affinity-based chemical reactions involving nucleic acids directly to an electronic signal, completely eliminating the need for light. This not only results in platform size reduction, but also supports real-time monitoring of DNA hybridization, which is difficult, if not impossible, using conventional fluorescence-based microarrays. Furthermore, electrochemical techniques have the potential to eliminate the need for chemical labels in DNA assays.

2.2.4.1 Sensing Methodology

Electrochemical DNA sensing is carried out in this work using a method analogous to fluorescence-based DNA sensing. As shown in Figure 2.2, ssDNA probes are first immobilized on an electrode made from a noble metal, such as gold (Au), immersed in an electrolyte (Figure 2.2a). Next, DNA targets that have been labeled with an electroactive chemical species are introduced to the electrolyte and allowed to interact with the probes (Figure 2.2b). These labels are designed to transfer an electron to the electrode when a potential is applied. An

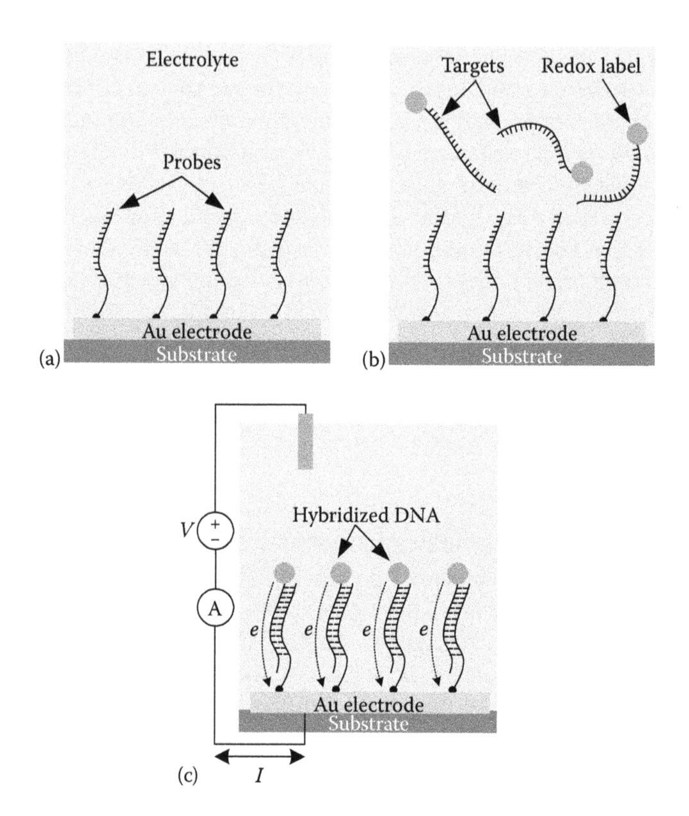

FIGURE 2.2 The electrochemical DNA sensing technique used in this work. (a) DNA probes are immobilized on an Au electrode and immersed in an electrolyte. (b) DNA targets conjugated with electroactive labels are introduced. (c) An applied potential causes the labels on hybridized DNA targets to transfer charge to the electrode. The current measured is proportional to the amount of surface hybridization.

electrochemical cell is then established by applying a potential between the Au electrode and a second electrode using the voltage source V (Figure 2.2c). When this occurs, only those labels attached to surface-hybridized targets are able to transfer electrons to the electrode. A quantitative measure of the amount of surface hybridization is obtained by monitoring the level of dc current I using the ammeter A.

2.2.4.2 Real-Time DNA Detection

Many electrochemical DNA sensors allow for real-time detection of the hybridization process because only chemical changes resulting from an affinity-based reaction localized near the electrode surface give a positive signal. As a result, a washing step to remove unbound and nonspecifically bound targets is unnecessary.

Observation of the entire hybridization process in real time has several advantages. First, the point at which equilibrium has been reached can be determined unambiguously so that assay reproducibility and sensitivity can be improved [46]. Second, kinetic studies of DNA binding can be carried out in order to provide insight into the physical properties governing affinity-based reactions [47]. Third, the additional data provided by real-time sensing allow temporal averaging of the measured signal. This could improve the SNR by reducing the effect of independent noise sources. Some of these include nonideal instrumentation and interfering biochemical processes like cross-hybridization, which become more noticeable in assays exhibiting low expression levels [48,49].

2.2.5 Principles of Electrochemistry

Electrochemistry is a branch of chemistry that deals with the processes and factors that influence the charge distribution around, and transport across, an electrode–electrolyte interface. Electrochemical DNA sensing involves relating this charge to the extent of DNA hybridization at an electrode surface. In this section, basic principles of electroanalytical chemistry and electrochemical instrumentation relevant to the design and validation of an active CMOS biochip for electrochemical DNA assays are presented.

2.2.5.1 Electrode–Electrolyte Interface

When an electrode is immersed in an electrolyte containing water molecules and solvated and unsolvated anions and cations, "layers" of charge form at the interface in order to maintain electrical neutrality, as shown in Figure 2.3. These layers can be modeled as two series capacitors having an overall capacitance per unit area of C'_d, usually in the 10–40 μF cm^{-2} range [50].

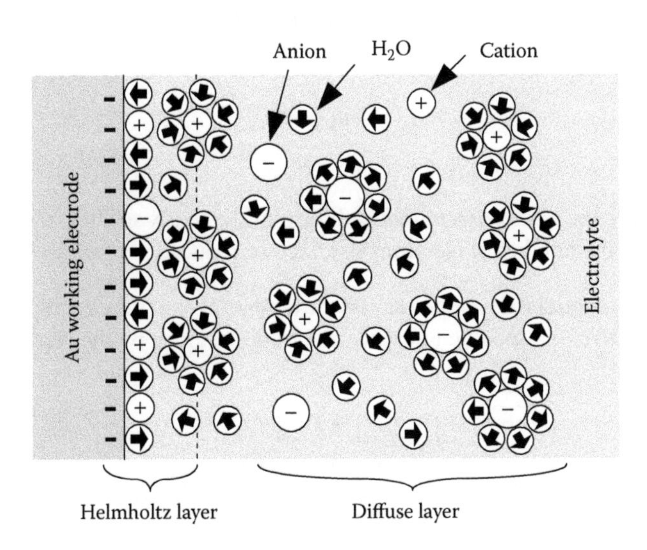

FIGURE 2.3 The distribution of charges at an electrode–electrolyte interface.

The first capacitor can be modeled as a parallel-plate capacitor having a value that is independent of the electrolyte concentration. This "Helmholtz" layer is composed of solvated positive ions physically separated from the negatively charged electrode surface by a monolayer of water molecules and ions attached to the surface. The capacitance per unit area C'_H (in units of F cm^{-2}) is therefore given by

$$C'_H = \frac{\varepsilon\varepsilon_0}{d}, \tag{2.5}$$

where:

- ϵ is the dielectric constant of the medium
- ϵ_0 is the permittivity of free space
- d is the distance between the two layers of charge

The second capacitor (proposed by Gouy and Chapman) arises from the Boltzmann distribution of ions in a "diffuse layer" beyond the Helmholtz layer. Unlike C_H, the capacitance per unit area of the diffuse layer, C'_{GC}, depends on electrolyte concentration.

The overall capacitance per unit area of the interfacial region C'_d can be expressed as

$$\frac{1}{C'_d} = \frac{1}{C'_H} + \frac{1}{C'_{GC}}, \tag{2.6}$$

which arises from the Gouy–Chapman–Stern model of the interface. For simplicity, this structure is often referred to entirely as the "double-layer capacitance."

2.2.5.2 Electrochemical Reactions at Equilibrium

Assume the following reaction occurs at an electrode immersed in an electrolyte containing a bulk reduction–oxidation (redox) species:

$$O + ne \rightleftarrows R, \tag{2.7}$$

where:

- O and R are the oxidized and reduced form of the redox species, respectively
- n is the number of electrons oxidized or reduced per molecule

Then, the potential E developed across the double layer of the electrode–electrolyte interface (relative to a specific reference electrode) is given by the Nernst equation:

$$E = E^0 + \frac{RT}{nF}\ln\left(\frac{C_O}{C_R}\right), \tag{2.8}$$

where:

- E^0 is the standard potential at which the reaction in Equation 2.7 proceeds
- R is the molar gas constant

F is the Faraday constant

C_O and C_R are the concentrations of the oxidized and reduced forms of the redox species, respectively

The above equation holds only under equilibrium conditions, where no current flows across the interface.

2.2.5.3 Potentiostat Operation

A potentiostat can be used to investigate the interface under nonequilibrium conditions. A potentiostat is a feedback control system used to apply a desired potential to an electrode–electrolyte interface and to measure simultaneously the movement of charge through the interface. A standard potentiostat used in a typical electrochemical cell, shown conceptually in Figure 2.4a, consists of three electrodes immersed in an electrolyte: a WE at which the reaction of interest occurs and where DNA probes would be immobilized when performing DNA sensing, a "reference" electrode (RE) to hold the electrolyte at a known potential, and a "counter" electrode (CE) that behaves like an anode or cathode depending on the direction of charge flow.

The voltage source V between WE and CE is adjusted to establish a desired cell input voltage V_{in} between the WE and RE. Direct current I can flow through the external circuit (as measured by the ammeter A) as redox species in the electrolyte, for example, donate electrons to the WE and accept electrons from the CE. In addition, a displacement (charging) current can flow as ions segregate to the WE and CE to form space-charge regions. The high-impedance voltmeter (VM) attached to the RE ensures that very low current flows through this interface, maintaining it at equilibrium.

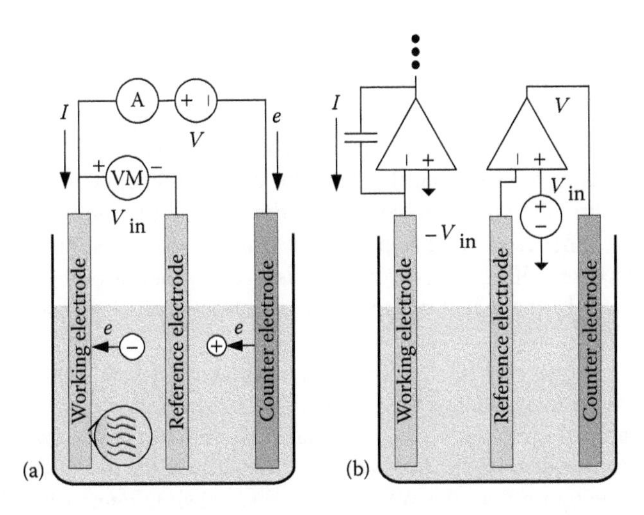

FIGURE 2.4 Description of potentiostat operation. (a) Conceptual diagram. (b) Implementation using standard electronic components.

The basic potentiostat circuitry in Figure 2.4a can be implemented using standard electronic components, as shown in Figure 2.4b. The operational amplifier (op amp) on the right establishes the control loop, while the integrator on the left converts the current flowing through the WE to a voltage for digitization and readout. Only a very small current flows through the RE due to the high input impedance of the control amplifier.

2.2.5.4 Electrode Materials

A wide range of materials are generally used to construct the WEs, including metals, oxides, and semiconductors. These are normally selected to provide a highly polarizable (capacitive) interface when immersed in an electrolyte. Gold is the WE material used in the detection system implemented in this work because it is relatively electrochemically inactive (i.e., it will not react with the electrolyte under normal operating potentials and interfere with the chemical processes being studied) and enables straightforward DNA attachment chemistry.

The RE usually exhibits a nonpolarizable interface that is in chemical equilibrium with the electrolyte, thus providing a stable potential to which the WE potential is referenced. One of the most common RE materials is a silver (Ag) wire coated with a layer of silver chloride (AgCl) formed by anodization in a chlorine-containing solution or through exposure to, for example, common household bleach. When the Ag/AgCl wire is placed in a glass reservoir having a salt bridge containing a 3 M sodium chloride (NaCl) filling solution, the following redox reaction provides a stable electrochemical potential:

$$\mathrm{AgCl} + e \rightleftarrows \mathrm{Ag} + \mathrm{Cl}^{-}. \tag{2.9}$$

Platinum (Pt) is commonly used for the CE because of its inertness. Other noble metals, like Au, can also be used.

2.2.5.5 Cyclic Voltammetry

The active CMOS biochip described in Section 2.3 applies cyclic voltammetry (CV), a large-signal electrochemical measurement technique. CV is one of the most popular techniques for studying the behavior of electrochemical systems because it can rapidly provide considerable information on the thermodynamics of redox processes, the kinetics of multistep electron transfer reactions, and surface-based chemical processes [50]. In this work, CV is used to measure bulk redox species as well as quantify the extent of DNA hybridization at an electrode surface using redox labels.

Figure 2.5 shows how CV is typically executed. In CV, voltage V_{in} (always expressed relative to a standard RE potential, denoted here as V_{re}) is set to the initial voltage V_i, ramped up to the vertex potential V_ν at the scan rate ν, and then ramped down at the same rate until the final voltage V_f (usually equal to V_i) is reached, while the current flowing through the WE, I_{we}, is measured simultaneously. When detecting redox species present in the bulk solution, I_{we} as a function of time appears as shown in Figure 2.5. However, I_{we} is normally viewed as a function of V_{in} so that the

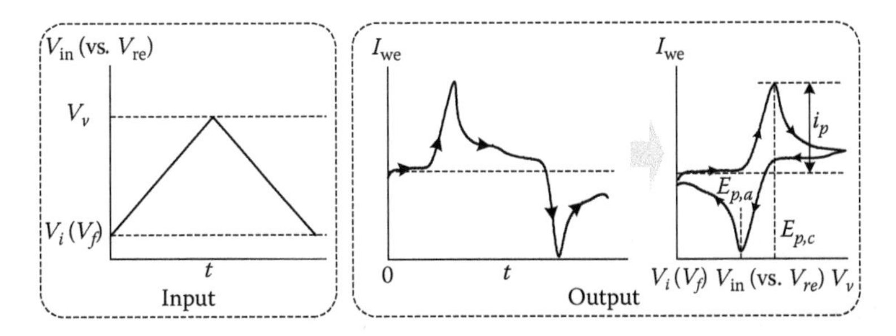

FIGURE 2.5 Cyclic voltammetry of an electrochemical cell containing a bulk redox species.

potentials at which reduction, $E_{p,c}$, and oxidation, $E_{p,a}$, occur (indicated by the forward and reverse current peaks, respectively) may be easily discerned. These potentials, as well as the value of I_{we} at each peak, indicate the degree of reversibility of an electrochemical reaction as well as the amount of chemical product generated or reactant consumed [51]. The peak current i_p for a reversible redox couple (at 25°C) such as that in Equation 2.7, is given by the Randles–Sevcik equation [50]:

$$i_p = \left(2.69\times10^5\right) n^{3/2} A C_O^* D_O^{1/2} \nu^{1/2}, \tag{2.10}$$

where C_O^* is the initial concentration of O in the bulk of the electrolyte (i.e., far from the electrode surface) and D_O is the diffusion coefficient of species O.

The separation ΔE_p between $E_{p,c}$ and $E_{p,a}$ for a reversible couple is given by

$$\Delta E_p = \left|E_{p,a} - E_{p,c}\right| = \frac{0.059}{n}. \tag{2.11}$$

A surface-bound (adsorbed) redox species is not affected by mass-transfer limitations and, therefore, exhibits a current response during CV that differs from that of a bulk redox species. The current response from redox-labeled ssDNA targets follows the same behavior as an adsorbed species because these molecules effectively become "attached" to the electrode surface upon hybridization. Assuming only the adsorbed species O in Equation 2.7 is electroactive, while dissolved O out in the electrolyte is inactive, the redox current i_{ad} is given by [52]

$$i_{ad} = \frac{n^2F^2}{RT}\frac{\nu A\left(S_O/N_A\right)\left(b_O/b_R\right)\exp\left[\left(nF/RT\right)\left(V_{in}-E^0\right)\right]}{\left\{1+\left(b_O/b_R\right)\exp\left[\left(nF/RT\right)\left(V_{in}-E^0\right)\right]\right\}^2}, \tag{2.12}$$

where:

S_O is the surface density of adsorbed O

b_O and b_R are the adsorption coefficients of O and R, respectively

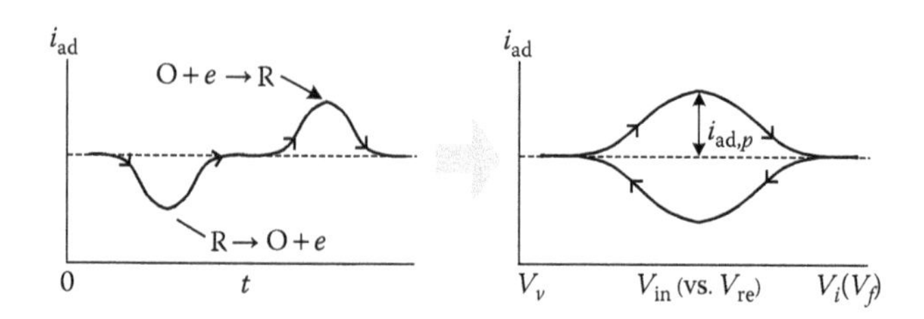

FIGURE 2.6 Cyclic voltammetry measurement of an electrochemical cell containing an adsorbed redox species.

Figure 2.6 displays the characteristic shape of this response. Assuming $b_O = b_R$, the peak current $i_{ad,p}$ can be calculated using

$$i_{ad,p} = \frac{n^2F^2}{4RT}\nu A\left(\frac{S_O}{N_A}\right). \tag{2.13}$$

2.2.6 DNA Immobilization Chemistry

Thiol-metal chemisorption is used in this work to immobilize ssDNA probes on the surface of a WE. Oligonucleotide probes are modified with a thiol (–SH) group at one end. The strong affinity of this group for a noble metal surface, such as Au, enables the formation of bonds between the sulfur (S) and Au atoms as follows [53,54]:

$$\text{ssDNA} - \text{R} - \text{SH} + \text{Au} \rightarrow \text{ssDNA} - \text{R} - \text{S} - \text{Au} + e + \text{H}^+, \tag{2.14}$$

where R is a linker such as $(CH_2)_6$ [55]. Typical surface coverage of probe molecules using this technique is between 10^{11} and 10^{13} molecules cm^{-2}. This depends largely on the immobilization buffer concentration and exposure time.

"Backfilling" the electrode surface with an alkanethiol self-assembled monolayer (SAM) such as mercaptohexanol (MCH) after immobilization of thiolated ssDNA probes is necessary to reduce nonspecific adsorption of probe on the surface. This is because the thiol group on the MCH has a higher affinity for Au than the less strongly adsorbed DNA bases have. This leaves the DNA attached to the Au by the end-point thiol group only, effectively lifting the probe off the surface [56], as displayed in Figure 2.7. In addition, the net negative dipole of the alcohol

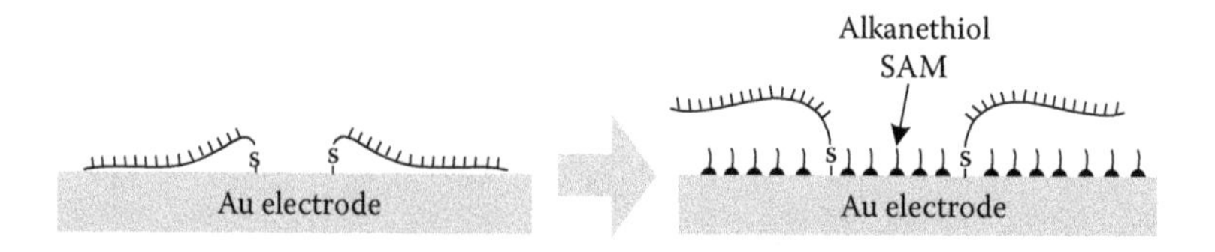

FIGURE 2.7 Backfilling the electrode surface with an alkanethiol self-assembled monolayer lifts the DNA probes off the surface, enhancing hybridization efficiency.

end group repels the DNA backbone, helping to project the probe strand out into solution [57]. This helps to increase hybridization efficiency because the probes have more configurational freedom to form a duplex molecule by coiling around the analyte DNA targets.

2.2.7 Review of Electrochemical DNA Sensing Techniques

Affinity-based electrochemical DNA sensors can be divided into two broad categories: label-based and label-free. In the former, an electroactive chemical label or enzyme attached to the target DNA or a secondary DNA support provides the positive electrochemical signal. Alternatively, this signal can be derived from redox indicator molecules, which associate differently with DNA duplexes than with ssDNA molecules.* Conversely, label-free DNA sensors require no supporting redox species, enzymes, or indicator molecules to provide the electrochemical signal. Instead they rely on, for example, oxidation of individual DNA bases or changes in interfacial capacitance or surface charge due to DNA hybridization. Electrochemical DNA sensors used in biological applications are often referred to as "biosensors" because the reactions of interest occur directly on the transducer surface [54]. Label-based sensing with electroactive redox labels is now discussed in more detail as the active CMOS biochip discussed in this chapter makes use of this approach.

2.2.7.1 Redox-Label-Based Electrochemical DNA Sensors

Chemical modification of DNA targets with an electroactive redox label, enzyme [58,59], or metal nanoparticle [60,61] is used to enhance the detection limit of DNA sensors. Ferrocene is a commonly used redox label because it is easily functionalized and is electrochemically reversible. In addition, it possesses a known redox potential that does not coincide with that of the DNA bases [62]. Ferrocene tags have been used to label phosphoric-, amino-, and thiol-modified ssDNA targets [63]. When these are hybridized to complementary DNA probes on a graphite electrode, the peak anodic current from a differential pulse voltammetry measurement is proportional to the concentration of surface-hybridized probes [64].

Ferrocene is the redox label used in GenMark Diagnostics' eSensor DNA chip [65] (which was originally developed by Motorola's Clinical Micro Sensors Division). In this "sandwich" assay, the ferrocene label is covalently attached to the ribose backbone of a short oligonucleotide signaling probe that selectively binds to a DNA duplex on an Au electrode after hybridization has occurred. The peak Faradaic current measured using ac voltammetry is directly proportional to the number of ferrocene moieties present, which is also proportional to the number of surface-hybridized DNA targets. This system has been used for various clinical applications, including the detection of hemochromatosis [66], human papillomavirus [67], and cystic fibrosis [68].

* DNA detection systems relying on indicator molecules are sometimes categorized as "label-free" in the literature because no chemical modification of the probes or targets is made.

2.2.8 CMOS Electrochemical DNA Sensors

There are currently only a few commercially available electrochemical sensor platforms for clinical genomic assays and none for POC genetic diagnostic applications. This is partly due to the numerous technical challenges encountered when designing a compact device that must incorporate a highly multiplexed transducer interfaced with electronic readout circuitry. Specifically, electrochemical readout normally requires a wired mechanical connection between the electrode and potentiostat electronics. Therefore, the fabrication and interfacing of arrays of microelectrodes on a passive substrate to permit detection of hundreds to thousands of individual sequences for gene-expression and mutation diagnostics presents a significant engineering challenge [19].

Integration of electrode transducer arrays, potentiostat electronics, and digital readout together on an active CMOS substrate enables the construction of portable, highly multiplexed electrochemical DNA sensing platforms for clinical and POC genetic diagnostic applications. This system therefore requires immobilizing DNA probes at electrode sites on the surface of the CMOS chip and carrying out all DNA hybridization reactions there as well. In addition to increased multiplexing and enhanced portability, integrated electronic instrumentation reduces susceptibility to electromagnetic interference that can result from transmitting nonamplified analog signals off chip.

2.2.8.1 Prior Work on CMOS Electrochemical DNA Sensors

Previous work on CMOS-integrated affinity-based electrochemical DNA sensors has involved both label-based and label-free detection. Much of this prior work has focussed on specific modalities of potentiostatic sensing, such as redox cycling and capacitance-to-frequency conversion, to simplify the electronics. In addition, field-effect DNA sensing using special ion-sensitive MOS transistors has been reported. These are all described in more detail below.

2.2.8.1.1 Field-Effect Transistor–Based DNA Sensing

Integrated MOSFETs have been used to perform direct label-based and label-free electronic detection of DNA sequences. DNA probes are usually immobilized on the surface of the oxide above the transistor channel [69,70], directly on an aluminum (Al) electrode above the polysilicon gate of the transistor [71], or on a postprocessed Au thin-film electrode covering the Al [72]. Detection has been performed by monitoring changes in the channel current that result from fluctuations in the electrode surface charge due to DNA hybridization [72]. Alternatively, the transistor current–voltage characteristics can be measured before and after hybridization to determine the resulting threshold-voltage shift [71]. In addition, Ion Torrent (Guilford, CT) has developed and marketed massively parallel CMOS-integrated ion-sensitive FET (ISFET) arrays for detecting hydrogen ion release resulting from nucleotide base incorporation through template-directed DNA polymerase synthesis [8].

2.2.8.1.2 CombiMatrix CMOS Electrochemical DNA Microarray

An electrochemical DNA microarray platform once developed and marketed by CombiMatrix (Mukilteo, WA) consisted of a CMOS microarray chip having more than 12,000 individually addressable, 44 μm diameter, Pt microelectrode sites, and an external (off-chip) reader [73]. The reader contained electronic actuation and measurement circuitry and interfaced with the CMOS microarray chip.

Detection involved a sandwich assay in which biotinylated target DNA first hybridized with immobilized probes. After this, horseradish peroxidase (HRP) was attached to the bound targets using biotin–streptavidin chemistry, which helped catalyze the oxidation of the added species tetramethylbenzidine (TMB). Current generated from the reduction of oxidized TMB from an applied potential was then used to quantify the target concentration. This system was used in influenza A genotyping applications and reportedly provided nearly identical results in certain gene expression studies using standard fluorescence-based DNA microarrays [73].

2.2.8.1.3 Infineon CMOS Electrochemical DNA Microarrays

Schienle et al. at Infineon Technologies (Neubiberg, Germany) have developed CMOS microarrays for detecting DNA oligonucleotides using a redox cycling method in which electronic measurement circuitry is integrated on the same chip as the electrodes. DNA probes are immobilized on an array of 128 interdigitated Au generator and collector electrodes in a 0.5 μm CMOS process [74]. Targets are labeled with the enzyme alkaline phosphatase in the presence of an added para-aminophenylphosphate chemical substrate that is cleaved following hybridization. Upon cleavage, the electrochemically active compound para-aminophenol (p-AP) is produced, which can then be oxidized and reduced by applying a potential of +200 mV and –200 mV to the generator and collector electrodes, respectively. The current produced is measured using an in-pixel current-to-frequency ADC and can be related to the DNA surface coverage. Potentiostat electronics including a control amplifier and CE and REs are also integrated. This system offers highly multiplexed DNA detection with relatively simple and compact readout circuitry, but requires a complex full-back-end CMOS process to create an array of Au electrodes on the surface [75]. Furthermore, real-time monitoring of DNA hybridization is not possible with this platform.

2.2.8.1.4 Toshiba CMOS Electrochemical DNA Sensor Array

Gemma et al. of Toshiba reported a CMOS DNA chip having 40 on-chip Au WEs ranging in diameter from 2 to 200 μm [76]. The redox "groove binder" Hoechst 33258 is introduced to the assay following DNA hybridization. The peak current level produced from Hoechst oxidation during linear-sweep voltammetry is used to provide the hybridization signal. The system includes an integrated potentiostat as well as an individual mirror-based current amplifier and continuous-time current-to-voltage converter connected to each electrode. DNA target concentrations down to approximately 1 nM are observed using this platform. Despite the high overall performance of this system, real-time sensing is not possible.

2.2.8.1.5 Capacitance-Based and Other Label-Free CMOS Electrochemical DNA Sensors

Previous work on label-free CMOS DNA sensors has involved detecting capacitance changes and electropolymer redox reactions in response to DNA surface hybridization. Hassibi and Lee reported on a 10 × 5 CMOS electrochemical array designed to sense biomolecular reactions occurring at on-chip 60 × 60 μm^2 Al WEs [77,78]. Each electrode is connected to a pseudodifferential amplifier network that can be programmed to implement either a transimpedance amplifier, field-effect (ISFET) sensor, or current integrator. The amplifiers also incorporate switched biasing to reduce flicker noise.

Stagni et al. have reported a label-free CMOS DNA sensor for detecting changes in capacitance due to DNA hybridization occurring on an integrated 8 × 16 electrode array [79]. The measurement technique involves periodically switching the current flowing through two interdigitated Au electrodes (with a similar structure and deposition procedure as described in [74]) functionalized with DNA probes, and detecting the transient charging and discharging voltage waveform produced between the electrodes. Since the time constant of this transient signal is proportional to the electrode interfacial capacitance, the frequency of this waveform over many switching cycles can be related to capacitance changes caused by DNA hybridization. Detectable capacitance changes of more than 1 nF due to specific DNA hybridization at the on-chip electrodes have been reported.

A CMOS electrochemical DNA detection system from Heer et al. was used to detect DNA hybridization occurring at a 24 × 24 array of integrated polypyrrole-coated platinum electrodes having diameters between 10 and 40 μm [80]. Changes in the total charge transferred by the surface layer are measured before and after hybridization using CV. These fluctuations are caused by changes in the interaction between chloride ions in the electrolyte and the polymer as negatively charged DNA is brought to the surface. The sensor chip is fabricated in a 0.6 μm process and contains 24 acquisition channels, each consisting of a delta-sigma ADC that uses the intrinsic capacitance of the electrode–electrolyte interface to integrate the charge transferred from the electrode during CV.

2.2.8.1.6 CMOS Microarray for DNA Polymerization Detection

Anderson et al. developed a label-free, 5 × 5, CMOS-integrated microarray to detect DNA using a sequencing-by-synthesis approach [81]. The individual deoxynucleotide triphosphates (dNTPs) are added to the analyte one after another and can bind to the individual bases in ssDNA targets immobilized at planar electrodes on the SiO_2 surface. The enzyme DNA polymerase must be present in solution for base incorporation to occur. When a single nucleotide is incorporated, integrators at each array site measure the induced charge arising from the diffusion of protons away from the electrode.

2.3 DESIGN OF AN ACTIVE CMOS BIOCHIP FOR LABEL-BASED ELECTROCHEMICAL DNA ASSAYS

This section describes the development of an active CMOS biochip for multiplexed, real-time, electrochemical DNA assays using a label-based sensing approach. Unlike previously reported CMOS electrochemical sensors, the proposed system uses CV and redox-labeled DNA targets to provide a quantitative measure of DNA hybridization in real time. The proposed biochip contains integrated potentiostat control amplifiers to establish an on-chip electrochemical cell, and an array of thin-film, Au WEs on which DNA probes are immobilized. Each WE is connected to an integrated dual-slope ADC to sense the current produced by surface redox reactions resulting from electron transfer during CV measurements. Postprocessing is necessary to construct a biologically compatible surface-electrode array on the fabricated standard-CMOS biochip that can withstand operation in a harsh electrochemical environment.

2.3.1 Label-Based Electrochemical DNA Sensing Method

Quantitative, real-time electrochemical measurement of surface hybridization on the active CMOS biochip is carried out using DNA targets covalently modified with N-(2-ferrocene-ethyl) maleimide (simply referred to as "ferrocene") redox labels. Ferrocene (Fc) undergoes the redox reaction

$$\mathrm{Fc} \rightleftarrows \mathrm{Fc}^{+} + e \tag{2.15}$$

over a potential range between 0 and +0.35 V (vs. a standard Ag/AgCl/3 M NaCl RE). In this work, each DNA target is conjugated with a single Fc redox label. A quantitative measure of the extent of surface hybridization is obtained by integrating the amount of charge transferred during Fc oxidation and then relating this value to a surface target coverage.

The classic electrochemical measurement technique, CV, discussed in Section 2.2, is used to drive the redox reaction associated with the Fc label. Figure 2.8 shows the usual arrangement of the three-electrode potentiostatic cell, the input voltage waveform, and the expected output current from the Fc redox reaction. Using CV, V_{in} applied to the control amplifier input starts at 0 V, is ramped up to +0.35 V at the scan rate V, and is then ramped back down to zero at the same rate. Fc labels attached to hybridized DNA targets are oxidized during the forward scan and reduced in the reverse scan. As discussed in Section 2.2.5.5, the overall shape of i_{we} in Figure 2.8 is attributed to the fact that Fc is a surface (rather than a bulk) redox species. The non-Faradaic baseline (charging) current i_{ch} is due to the WE capacitance $C_{\mathrm{we}} = A_{\mathrm{we}}C'_{\mathrm{we}}$, where C'_{we} is the WE capacitance per unit area and A_{we} is the WE area. This is given by

$$i_{\mathrm{ch}} = \pm \nu C_{\mathrm{we}}. \tag{2.16}$$

The density of hybridized probe–target pairs on the WE surface can be determined by integrating the area enclosed by the Fc redox (Faradaic) current after subtraction of background charging contributions, and then dividing the result (in coulombs) by the magnitude of the electronic charge and by the WE area. The charge passed from the forward or reverse scans, or the average of the two, may be used in this calculation.

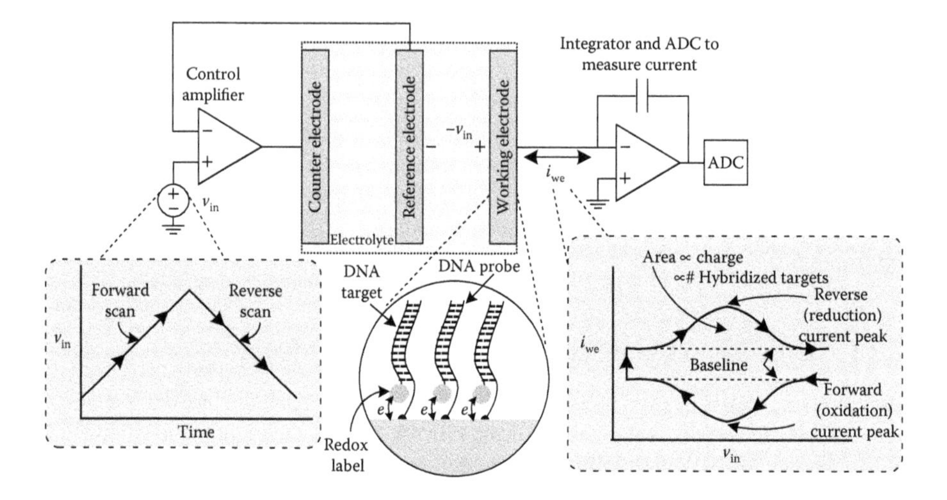

FIGURE 2.8 Electrochemical quantitation of Fc-labeled DNA targets hybridized to probes on the surface of a WE using CV. A potentiostat applies v_{in} to the WE interface and the redox current i_{we} is measured using an integrator and an ADC.

2.3.2 Active CMOS Biochip System Requirements

The range of sensor input current, sensor bandwidth, and chip area constraints are the most important factors in selecting a suitable architecture for the CMOS biochip. Each of these is described in more detail below. The sensor detection limit, which depends on the noise contributed from the electronics and the electrochemical and biological processes occurring at the chip surface, is also an important factor and will be analyzed for the chosen architecture in Section 2.3.7.

2.3.2.1 Input Current Range

Measurement of the small currents produced by surface redox reactions in an electrochemical cell requires amplification prior to readout. Determining the range of input currents produced by Faradaic and non-Faradaic processes is necessary when designing the amplification circuitry. The maximum total current $i_{we,max}$ produced during CV measurements is given by

$$i_{we,max} = i_{ch} + i_{ad,p} = \nu C_{we} + \frac{n^2 F^2}{4RT} \nu A \left(\frac{S_T}{N_A} \right), \tag{2.17}$$

where $i_{ad,p}$, defined previously in Equation 2.13, is the peak current produced by an adsorbed redox species and S_T is the surface target coverage.

The graph in Figure 2.9 shows $i_{we,max}$ as a function of coverage of Fc-labeled targets on a $100 \times 100\ \mu m^2$ WE for a C_{we} of $10\ \mu F\ cm^{-2}$ and CV scan rates of 0.01 and $60\ V\ s^{-1}$. These values are in the expected range for the given parameters. The value of $i_{we,max}$ ranges from 10 pA up to 1 μA, depending on the scan rate and

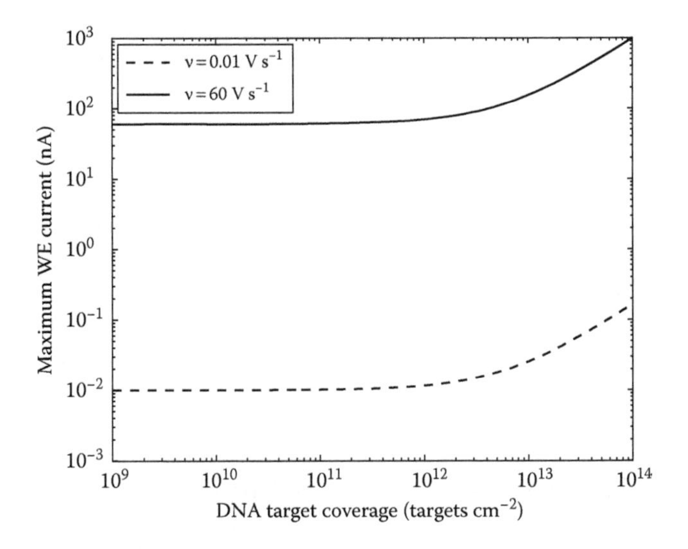

FIGURE 2.9 Expected maximum input current as a function of surface density of hybridized DNA targets for two CV scan rates.

surface target coverage. For coverage below approximately 10^{11} cm^{-2}, the total current is dominated by the concentration-independent charging current. Hybridized target coverage above about 2×10^{13} cm^{-2} is difficult to achieve in practice due to the significant electrostatic repulsion caused by the correspondingly high probe coverage [82].

2.3.2.2 Bandwidth

The required bandwidth of the sensor interface circuitry and ADC sampling rate depends on the bandwidth of the input WE current signal. CV measurements are usually performed at relatively low frequencies. Applying a periodic triangular voltage signal to the control amplifier at a scan rate of 60 V s^{-1} over a 0.35 V range corresponds to an effective frequency of approximately 85 Hz. Assuming the sensor electronics must have a bandwidth 10 times the input voltage frequency to accurately reconstruct the measured current, a 1 kHz sensor input bandwidth is sufficient. Additionally, a 10 kHz ADC would provide over 100 samples of the current waveform produced by the CV measurement above.

2.3.2.3 Working Electrode Area

The density of WEs on the CMOS surface can be increased by reducing the WE area. However, shrinking the WEs also reduces the signal current derived from the redox labels. As a result, the demands on the noise performance of the sensor electronics might become too great to achieve a tolerable SNR. This is discussed in more detail in Section 2.3.7.

2.3.2.4 Silicon Area of Sensor Interface Electronics

Dedicating an individual current sensor and ADC to each WE in the array enables real-time observation of DNA hybridization occurring at all WEs simultaneously.

However, allocating one measurement device per WE could consume a great deal of silicon area and would ultimately limit the maximum density of on-chip WEs. Therefore, a compact interface is desirable.

2.3.3 System Architecture

The overall architecture of the active CMOS biochip is displayed in Figure 2.10. It is composed of a 4 × 4 array of sensor sites where each site contains a square Au WE and a current-input dual-slope (integrating) ADC to measure bidirectional redox and charging currents. Integration-based sensor interfaces are often used in low-current potentiostat applications [83–87]. Potentiostat control amplifiers, connected to integrated CEs and an external RE, establish the on-chip electrochemical cell. Each of these components is discussed in more detail below.

2.3.3.1 Dual-Slope ADC Architecture and Operation

The dual-slope ADC, shown in Figure 2.11, is composed of an integration op amp with the feedback capacitor C_f, two current sources having opposite polarity,

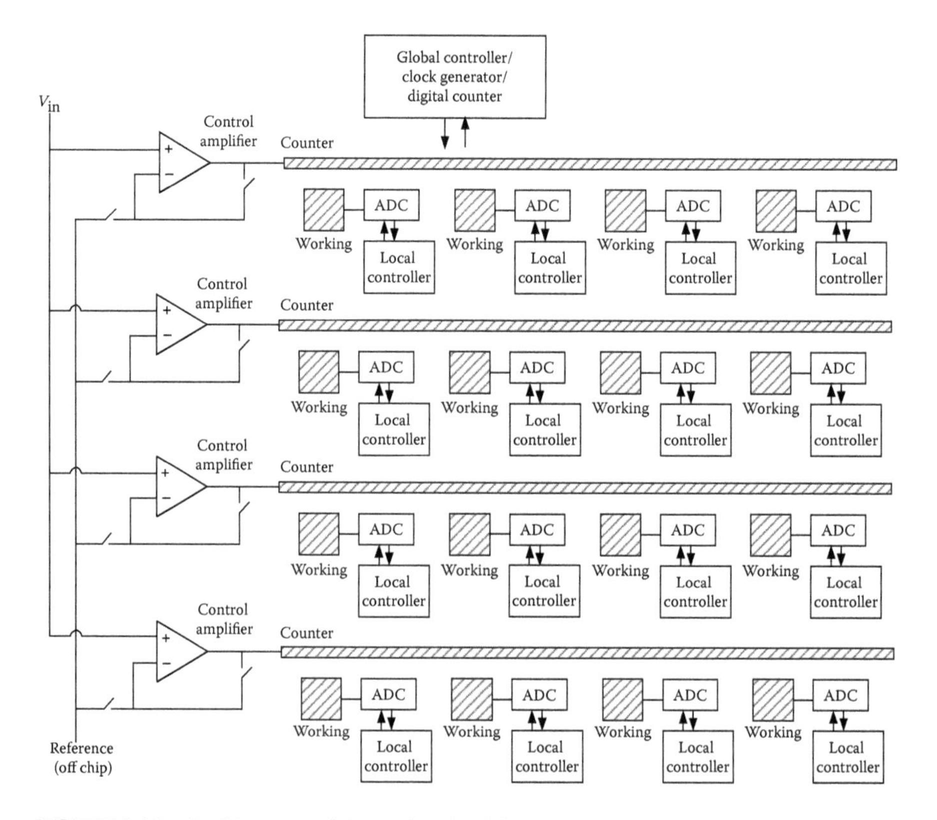

FIGURE 2.10 Architecture of the active CMOS biochip for label-based electrochemical DNA sensing.

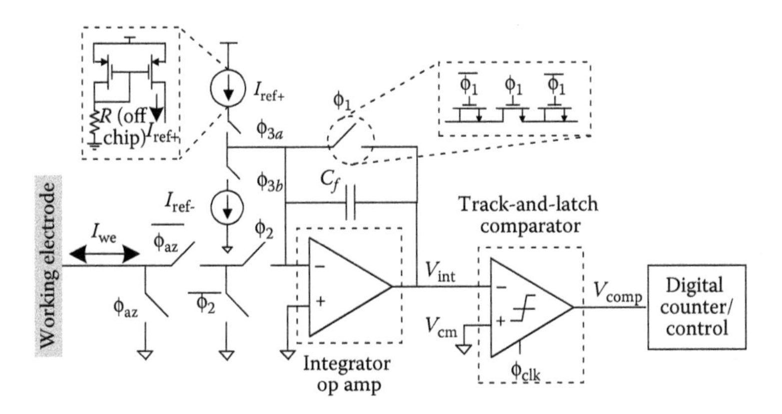

FIGURE 2.11 Architecture of the dual-slope ADC.

a track-and-latch comparator, switches to activate the various components according to the dual-slope algorithm, and digital control circuitry. The dual-slope ADC is well-suited to low-current CV measurements and can provide moderate resolution (9–12 bits) at kilohertz sampling rates. In addition, the current input range can be easily programmed by digitally varying the duration of the temporal integration and discharge intervals. Furthermore, the dual-slope ADC provides an accurate measure of the input current over a wide range that does not depend on the absolute value of C_f. This is advantageous because the absolute value of CMOS-integrated passive components can vary by more than ±20% due to inaccuracies inherent in CMOS manufacturing [88].

Figure 2.12 displays the three steps of the dual-slope algorithm. In the first step in Figure 2.12a, the capacitor is reset when the switch controlled by φ_1 is closed and the integrator output V_{int} is set to the virtual ground voltage V_{cm}. Next, current flowing through the WE I_{in} is integrated onto C_f for a fixed time interval t_1 by asserting the signal φ_2, as shown in Figure 2.12b. The value of V_{int} rises when I_{in} flows in the direction shown. The following expression can be written for the integration capacitor:

$$C_f = \frac{I_{in} t_1}{\Delta V_{int}}. \tag{2.18}$$

The comparator connected to the clock signal φ_{clk} indicates the direction (sign) of I_{in} by comparing V_{cm} to V_{int}. In the final step, digital control circuitry detects the comparator's binary output and selects the current source having the opposite polarity as I_{in}. As shown in Figure 2.12c, the known current I_{in+} is integrated onto C_f when the switch connected to φ_{3a} is closed. The time t_2 required to change the sign of the comparator output is determined when V_{int} crosses V_{cm}. A second expression can be written for the integration capacitor:

$$C_f = \frac{I_{ref+} t_2}{\Delta V_{int}}. \tag{2.19}$$

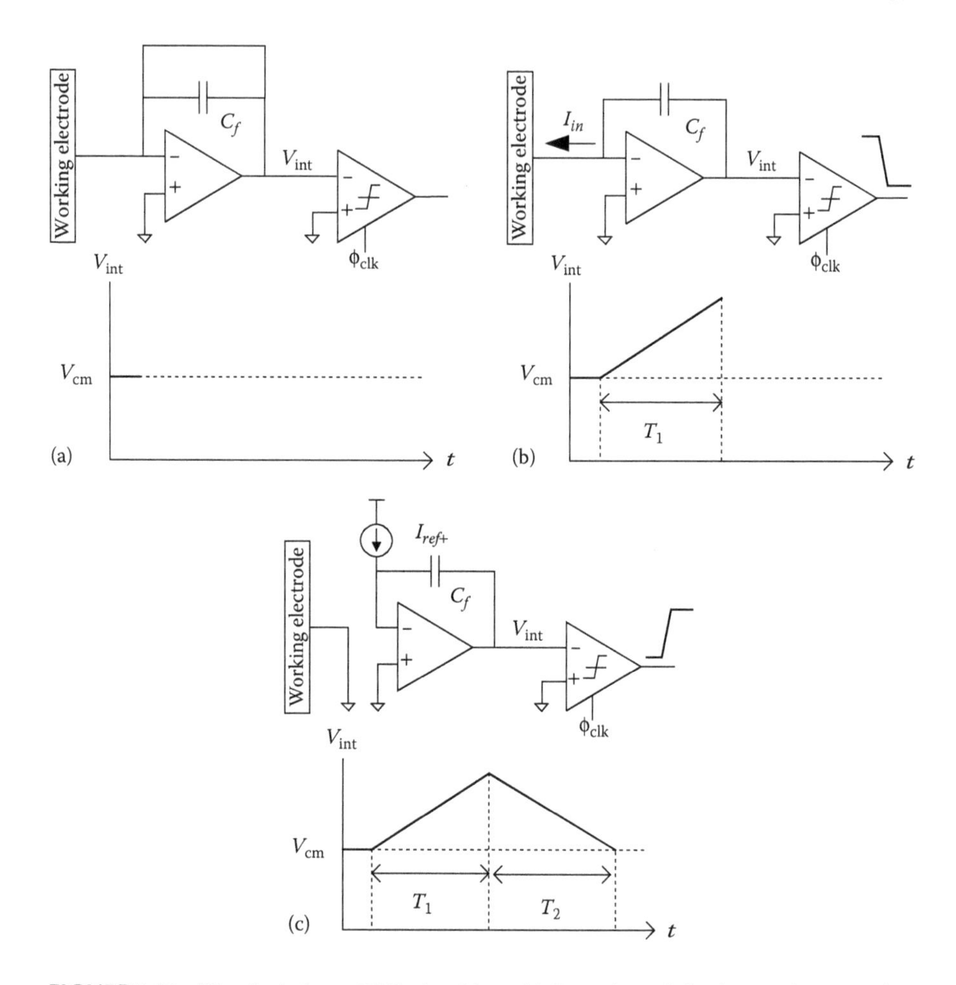

FIGURE 2.12 The dual-slope ADC algorithm. (a) Resetting of the integration capacitor during the first phase. (b) First integration period during which the WE current is integrated onto C_f. (c) Second integration period in which the appropriate current source is activated.

Equating equations 2.18 and 2.19, and solving for I_{in} gives:

$$I_{in} = \frac{t_2}{t_1} I_{ref+}. \tag{2.20}$$

Therefore, an accurate measure of I_{in} is obtained that is independent of the absolute value of C_f. Time t_1 is set and t_2 measured using a digital counter operating at the frequency of φ_{clk}. This rate is much higher than the overall ADC sample rate. The number of bits in the counter sets the nominal resolution of the ADC and effectively quantizes t_2. The minimum sampling period is the sum of the reset time, t_1, and the maximum value of t_2 allowed, $t_{2,max}$. A short time interval t_d is also necessary to allow for the selection of the appropriate current source before the second integration period. The ADC is also designed such that the WE is always connected to V_{cm}, whether it be directly through a switch or to the virtual ground of the integration amplifier. This helps to maintain the desired potential between the WE and RE that is necessary for accurate electrochemical measurements.

The dual-slope ADC also has an "autozeroing" mechanism to reduce the effect of amplifier and comparator offsets on the digital output. Offset removal mitigates the effect of flicker noise on the output and helps to reduce mismatch (i.e., "fixed-pattern" noise) among sensor sites. When the autozeroing signal φ_{az} is asserted in Figure 2.11, only current due to amplifier offset is integrated onto C_f. The digital value representing this current, which is also affected by comparator input offset, can then be subtracted from the digitized WE current. Autozeroing can be performed before each ADC sampling cycle to reduce temperature-dependent offset that occurs as the circuitry heats up over time. This, however, reduces the maximum sample rate.

2.3.3.2 Working Electrode Array

The WEs in the top row of the chip architecture in Figure 2.10 have a side length of 100 μm, while WEs in subsequent rows have side lengths of 90, 80, and 70 μm. This enables studies of the effect of electrode area on the measured cell current for different redox and biomolecular reactions. These sizes are chosen so that the general input current range shown in Figure 2.9 is obtained. WEs that have side lengths less than about 25 μm may operate in the "ultramicroelectrode" regime and thus experience mass-transport effects that deviate from macroscopic behavior [51].

2.3.3.3 Potentiostat Control Amplifiers

Each row of four WEs shares a 2500 × 15 μm^2 CE driven by a control amplifier. The four amplifiers can be operated all in parallel or individually disabled. The inverting amplifier input is connected to an off-chip Ag/AgCl/3 M NaCl RE. The RE is not integrated in order to reduce the amount of CMOS postprocessing necessary. The noninverting control amplifier input is connected to an external waveform generator to produce the CV stimulus.

2.3.4 Electrochemical Cell Interface Model

Figure 2.13a shows a small-signal circuit model of the electrode–electrolyte interfaces in an electrochemical cell, driven by a standard three-electrode potentiostat, that is used in circuit simulations to study system behavior and test potentiostat closed-loop stability. This model has been augmented compared with that discussed in Section 2.2.5.5, to include the CE–electrolyte interface. In Figure 2.13a, R_{s1} and R_{s2} represent the solution resistances of the WE–electrolyte and

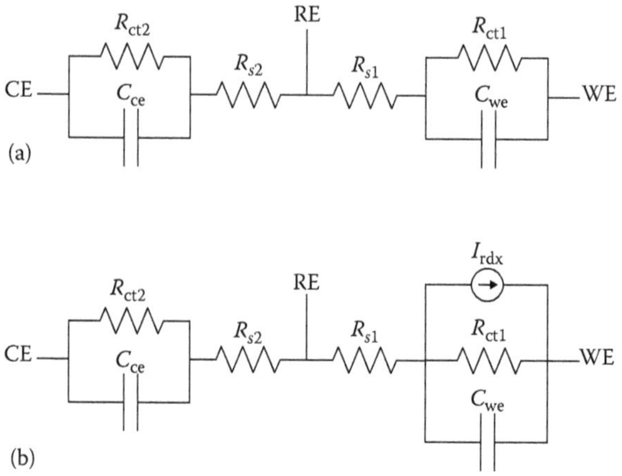

FIGURE 2.13 Electrochemical cell interface models. (a) Small-signal model of the WE–electrolyte and CE–electrolyte interfaces. (b) Interface model that includes a parallel current source to indicate the presence of a surface redox species.

CE–electrolyte interfaces, respectively, while R_{ct1} and R_{ct2} represent the charge-transfer resistances at these interfaces. Also, C_{we} and C_{ce} model the interfacial capacitances, composed of the double- and diffuse-layer capacitances, at the WE and CE, respectively.

Figure 2.13b shows the circuit in Figure 2.13a augmented with the parallel current source I_{rdx} to model the presence of a surface redox species. It is assumed in this case that the capacitors take on their incremental values.

2.3.5 Design of ADC Circuit Components

In this section, design details of the major ADC building blocks, including the integration amplifier and comparator, are discussed.

2.3.5.1 Integration Amplifier

This integration amplifier is implemented using the two-stage operational amplifier (op amp) shown in Figure 2.14. This topology provides a higher dc gain than what could be achieved using a single-stage amplifier, such as a folded cascode. A high dc gain helps minimize voltage fluctuations at the virtual ground, which, in this case, is connected to the WE through switches. A simulated dc gain of 87 dB ($\approx$22.4 kV V^{-1}) is achieved using the two-stage op amp. In addition, the common-source second stage gives a linear output range extending from 0.5 to 2.0 V. The dc gain at these output voltage levels are 85 and 86 dB, respectively. This allows a total range of 1.5 V over which the feedback integration capacitor can be charged. Since the noninverting input of the op amp is normally tied to the mid-rail voltage 1.25 V, the output can swing approximately $\pm$0.75 V in either direction. Other important design considerations, including noise performance, which contributes to the sensor detection limit, and stability are discussed next.

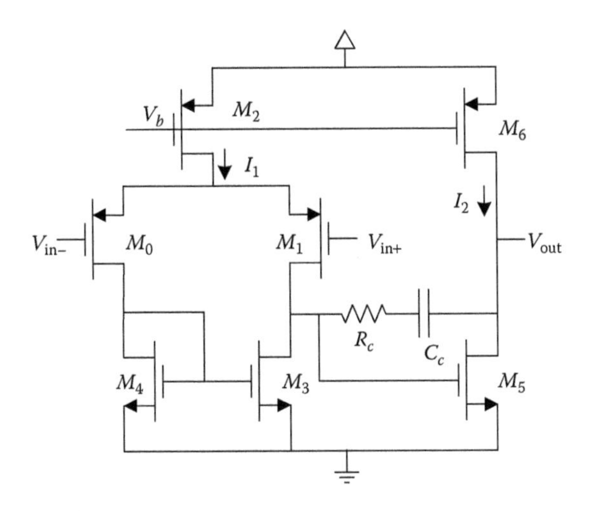

FIGURE 2.14 Two-stage op amp used in dual-slope ADC.

2.3.5.2 Thermal and Flicker Noise

Op amp transistor sizes and bias currents, shown in Table 2.1, are mainly chosen to minimize thermal and flicker noise. Reduction of the latter noise component is especially important in CV measurements because these are carried out at relatively low frequencies. The input-referred flicker noise voltage of the op amp with power spectral density (PSD) $\overline{V_{a,f}^2(f)}$ (with dimension $V^2\ Hz^{-1}$) is given by [21]

$$\overline{V_{a,f}^2(f)} = \frac{2}{C_{ox}f}\left[\frac{K_{fp}}{W_1L_1} + \left(\frac{\mu_n}{\mu_p}\right)\left(\frac{K_{fn}L_1}{W_1L_3^2}\right)\right], \tag{2.21}$$

where:

C_{ox}	is the transistor gate capacitance per unit area
f	is frequency
K_{fp} and K_{fn}	are the technology-dependent flicker-noise constants of the PMOS and NMOS transistors, respectively
W and L	are the transistor width and channel length, respectively
μ_n and μ_p	are the carrier mobilities of the NMOS and PMOS transistors, respectively.

TABLE 2.1
Component Values and Bias Currents for Two-Stage Op Amp Used in ADC Integrator

Transistor	Width/ Length (μm)	Component	Value	Current	Value (mA)
M_0/M_1	4000/1	R_c	150 Ω	I_1	2
M_2/M_6	1500/0.75	C_c	25 pF	I_2	2
M_3/M_4	1000/2				
M_5	1000/0.75				

As shown in the above equation, reducing the flicker noise necessarily involves using large-area input devices. As a result, the input devices M_0 and M_1 in the op amp have a width and length of 4 mm and 1 μm, respectively. In addition, PMOS transistors are used as the input devices because these are known to exhibit less flicker noise than NMOS transistors [89].

Ignoring the effect of the compensation resistor and capacitor R_c and C_c, respectively, the input-referred thermal noise voltage of the op amp with PSD $\overline{V_{a,\text{th}}^2(f)}$ can be calculated as follows [88]:

$$\overline{V_{a,\text{th}}^2(f)} = \frac{16kT}{3}\frac{1}{g_{m1}^2}\left[g_{m1} + g_{m3} + \frac{g_{m5} + g_{m6}}{g_{m5}^2\left(r_{o1} \parallel r_{o3}\right)^2}\right], \tag{2.22}$$

where:

g_m and r_o are the small-signal transconductance and output resistance of each transistor, respectively

the $\parallel$ symbol denotes parallel connection

As the equation suggests, reducing the thermal noise necessarily involves maximizing the value of g_{m1}. To this end, a bias current of 1 mA is designed to flow through M_1, providing a g_{m1} of about 14 mS.

Simulation of the op amp input-referred noise using typical process parameters indicates an overall flicker-noise constant K_f of 1.4×10^{-12} V^2 and a flicker-noise corner at around 700 kHz. The thermal noise power spectral density (PSD) is about 4.0×10^{-18} V^2 Hz^{-1}. The total integrated noise of the op amp over a bandwidth extending from 1 Hz to 1 MHz is approximately 4.5 μV$_{\text{rms}}$. The effect of this on the sensor detection limit will be studied in Section 2.3.7.

2.3.5.3 Bandwidth and Stability

A 25 pF metal-insulator-metal (MIM) capacitor is used to establish the dominant pole of the op amp at 4 kHz. A 150 Ω polysilicon resistor in series with C_c establishes a left-half-plane zero at about 50 MHz. The simulated unity-gain frequency f_0 of the op amp with a load capacitance of 5 pF is 109 MHz. This is roughly equal to $g_{m1}/(2\pi C_c)$.

Phase margin (PM) is an important parameter in assessing the stability of a feedback loop, such as that present in the integration amplifier. A PM less than zero means that the magnitude of the loop gain is greater than unity at frequencies where feedback signals experience a phase shift of more than 180° around the loop. As a result, input and feedback signals effectively add together and are amplified, causing the output to become unstable (grow without bound). Although a PM greater than zero is necessary to ensure stability in a linear feedback system, PMs close to zero are undesirable as these lead to long settling times and ringing in the step response.

The op amp used in the integrator exhibits the worst-case PM when operated in a unity-gain configuration. This occurs during the reset phase of the dual-slope conversion cycle, during which the integration capacitor is shorted out by closing the parallel switch connected to φ_1. Assuming the switch controlled by φ_2, connected directly to the inverting input of the op amp, is open during the reset period, the load capacitance C_L of the op amp is dominated by its own input capacitance plus the input capacitance of the comparator. The simulated PM in this case is approximately 67°.

Alternatively, if the φ_2 switch is closed during reset, the impedance of the electrode–electrolyte interface directly influences the PM. As a result, a C_{we} of around 1 nF in series with an R_{s1} ranging from 100 Ω to 100 kΩ appears in parallel with the usual C_L. Despite this, the PM remains around 67° because the additional pole and zero provided by the interfacial impedance tend to cancel one another for values of R_{s1} that are comparable or above the output resistance of the second stage of the op amp. This behavior is observed in simulation for values of R_{s1} greater than about 1 kΩ.

Table 2.2 summarizes some of the important simulated performance metrics of the op amp used in the ADC integrator.

2.3.5.4 Integration Capacitor

A 5 pF MIM integration capacitor is used in the two-stage op amp feedback loop. The choice of this value is mainly based on area considerations. Alternatively, a nonlinear MOS capacitor could have been used to reduce the occupied area even more. This nonlinearity would effectively be canceled by the dual-slope algorithm.

2.3.5.5 Comparator

The track-and-latch comparator following the integrator in the dual-slope ADC is shown in Figure 2.15. Transistor sizes are included in Table 2.3. In this design, transistors M_5 and M_6 separate the cross-coupled switching transistors at the output from the drains of input transistors M_0 and M_1 to reduce kickback interference. Although autozeroing can mitigate the effect of comparator offset on the digital output, M_0 and M_1 have a width and length of 200 and 1 μm, respectively, to reduce offset caused by mismatch between these transistors. The comparator provides more than 12 bits of resolution at an input clock frequency of 50 MHz in simulation.

TABLE 2.2
Simulated Performance of Two-Stage Op Amp Used in ADC Integrator

Parameter	Conditions	Simulated Value
Supply voltage		2.5 V
DC gain	$V_{out} = 1.65$ V	87 dB
	$V_{out} = 0.5$ V	85 dB
	$V_{out} = 2.0$ V	86 dB
3 dB Bandwidth	$C_L = 5$ pF	4 kHz
Flicker-noise constant K_f		1.4×10^{-12} V^2
Thermal noise PSD		4.0×10^{-18} V^2 Hz^{-1}
Total integrated noise	1 Hz–1 MHz	4.5 μV$_{rms}$

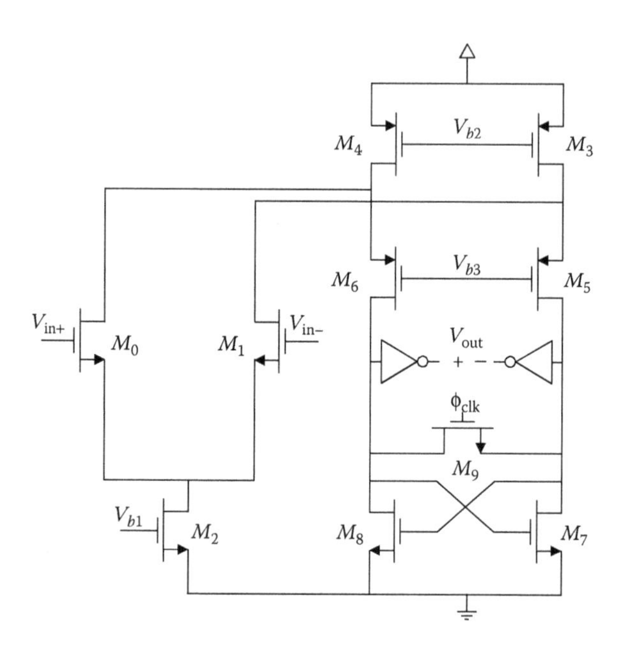

FIGURE 2.15 Track-and-latch comparator used in dual-slope ADC.

TABLE 2.3
Transistor Sizes for Track-and-Latch Comparator Used in Dual-Slope ADC

Transistor	Width/Length (μm)
M_0/M_1	200/1
M_2	300/0.75
M_3/M_4	300/0.24
M_5/M_6	200/0.24
M_7/M_8	5/0.5
M_9	20/0.24

2.3.5.6 Current Sources and Switches

The devices used to source and sink current during the second integration phase of the dual-slope ADC algorithm are composed of PMOS and NMOS current mirrors, respectively, biased using off-chip resistors. The channel length of the PMOS and NMOS transistors in the switches controlled by φ_{3a} and φ_{3b}, respectively, are both 15 μm, while the widths are 5 and 2.5 μm, respectively. Depending on the size of the external tuning resistor, currents ranging from approximately 100 pA to 1 μA can be established.

NMOS pass transistors are used to implement the switches tied to φ_1, φ_2, and φ_4, as these only need to pass mid-rail voltages. The latter two minimum-length switches both have a 50 μm width and exhibit a simulated on-resistance of approximately 50 Ω. The switch controlled by φ_1 has a 20 μm width and an on-resistance of about 120 Ω. In addition, the switches controlled by φ_1 and φ_{2a} are connected to NMOS "dummy" switches on each side, as shown in Figure 2.11, and are clocked on alternate edges. These help reduce conversion errors caused by charge injection [21]. A 100 μm NMOS switch (not shown in Figure 2.11) is directly attached to the WE and can be used to disconnect the electrode from the sensor electronics.

2.3.5.7 Digital Counter and Control Circuitry

A digital counter is required to set t_1 and measure t_2 during dual-slope ADC operation. To allow sufficient range, a 20-bit ripple-carry counter is synthesized using gates from a standard-cell library. This counter can operate at frequencies exceeding 50 MHz, as verified through timing simulation.

Local digital control circuitry, implemented as a finite state machine (FSM), contains comparators to set the duration of t_1 and $t_{2,max}$, detects the analog comparator output, and sets the appropriate switches in the ADC to implement the dual-slope algorithm. The FSM can be bypassed and the ADC completely controlled via an external field-programmable gate array (FPGA). All analog comparator outputs are also connected to pins on the CMOS chip so that these can be observed using external benchtop measurement equipment.

2.3.6 Design of Control Amplifier Circuits

Sufficient current drive, low noise, and stability are the most important factors in the design of the control amplifier. Therefore, this component is constructed using the two-stage op amp in Figure 2.14 with the same transistor sizes as in Table 2.1. The maximum total Faradaic current expected to flow through all 16 on-chip electrodes is about 20 μA. The two-stage op amp can deliver this current easily because it is only 1% of the 2 mA second-stage bias current.

Maintaining stability over a wide range of electrolyte impedances is one of the main goals in the design of the control amplifier. The model shown in Figure 2.16 is used to study the closed-loop stability of the control amplifier when it is driving 16 identical WE–electrolyte interfaces in parallel. In circuit simulation, the feedback loop is cut and the PM is extracted from the open-loop small-signal response. To simplify the analysis, the value of R_{ct} is assumed to be infinite, the value of C_{ce} is assumed to be so large that its impedance is negligible over the operating frequency range, and all WEs are taken to be equal in area. In addition, the resistor R_{sre} has been included to model the series resistance of the highly concentrated RE filling solution (i.e., 3 M NaCl) and the RE is assumed to be placed at an equal distance from the CE and WEs so that all values of R_s are equal.

To investigate the stability range, the values of R_s and C_{we} are swept from 100 Ω to 100 kΩ, and from 100 pF to 10 nF, respectively, using a parametric circuit simulation. These values are in the expected range for a modified Au electrode exposed to various concentrations of buffer. Figure 2.17 shows the PM of

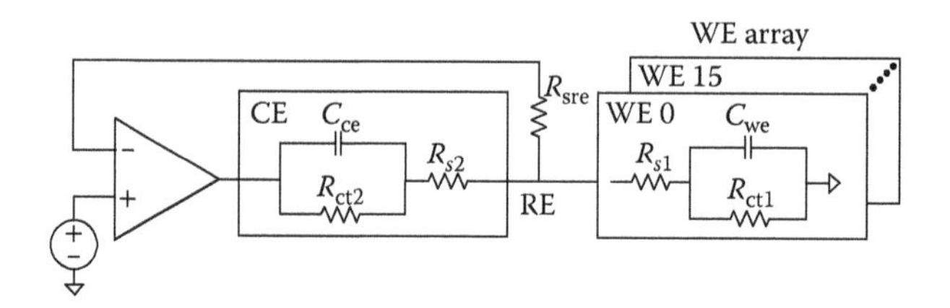

FIGURE 2.16 Model used to simulate closed-loop stability of the control amplifier feedback loop.

the open-loop ac response as a function of R_s and C_{we}. As the graph in Figure 2.17 shows, the PM is above zero for all combinations of R_s and C_{we} considered. It is apparent, however, that different regions of stability exist. To understand these regions better, it is useful to examine the pole and zero locations in the open-loop transfer function. For small values of R_s, the dominant f_{p1} and nondominant f_{p2} pole frequencies of the open-loop transfer function can be expressed approximately as

$$f_{p1} \approx \frac{1}{2\pi} \frac{1}{R_1\left[C_1 + C_c\left(1 + g_{m7}R_2\right)\right] + R_2\left(16C_{we} + C_2\right)} \tag{2.23}$$

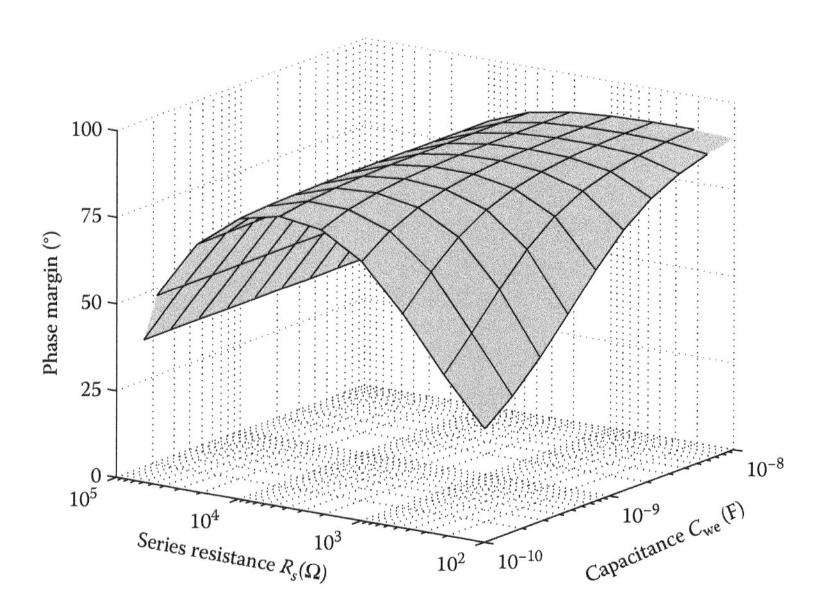

FIGURE 2.17 Phase margin of the control amplifier feedback loop as a function of the WE capacitance and solution resistance when 16 sensors are operating in parallel.

and

$$f_{p2} \approx \frac{1}{2\pi}\left(\frac{g_{m1}}{16C_{\text{we}}+C_2}\right)\left(\frac{1}{1+\frac{C_1}{C_c}}\right), \tag{2.24}$$

where:

R_1 and R_2 represent the equivalent output resistances of the first and second op amp stages, respectively

C_1 and C_2 are the total transistor capacitances at the output of the first and second stages, respectively

The WE impedances also add a zero f_z to the transfer function at

$$f_z = \frac{1}{2\pi R_s C_{\text{we}}}, \tag{2.25}$$

in which the contribution from R_s has been included. In Figure 2.17, values of C_{we} greater than about 500 pF, coupled with small R_s values, give PMs between 60° and 90°. This is because the zero at f_z helps bring the phase shift up at higher frequencies. On the other hand, the PM declines rapidly and approaches about 35° for values of R_s greater than approximately 10 kΩ since f_z is translated to lower frequencies. The rapid decline in PM for the smallest values of R_s and C_{we} occurs because f_z now exists well above f_{p2}.

Not all combinations of R_s and C_{we} in Figure 2.17 can be obtained experimentally. For example, it has been shown that the interfacial capacitance at an Au electrode modified with ssDNA probes changes very little even when the electrolyte concentration is varied by many orders of magnitude [90].

2.3.7 Sensor Detection Limit and Noise Analysis

The smallest DNA target coverage that can be measured by the CMOS sensor array is limited by electrochemical noise processes [91], noise from the integrated electronics, ADC quantization noise, and uncertainties arising from cross-hybridization due to nonspecific binding during operation in a multitarget analyte [92]. Since the last noise source is beyond the scope of this work, only the first three will be analyzed.

For purposes of noise modeling, the simplified circuit model of the dual-slope ADC and electrochemical interface shown in Figure 2.18 are used. The parallel resistance R_{ct} is assumed to be infinite since the WE is normally passivated with an alkanethiol SAM, as discussed in Section 2.2.6. Also, to simplify the analysis, only the current integration period of the ADC for the duration t_1 is considered. It is also assumed that all noise sources are uncorrelated and that the small-signal parameters of the electrochemical cell remain fixed over the input voltage range.

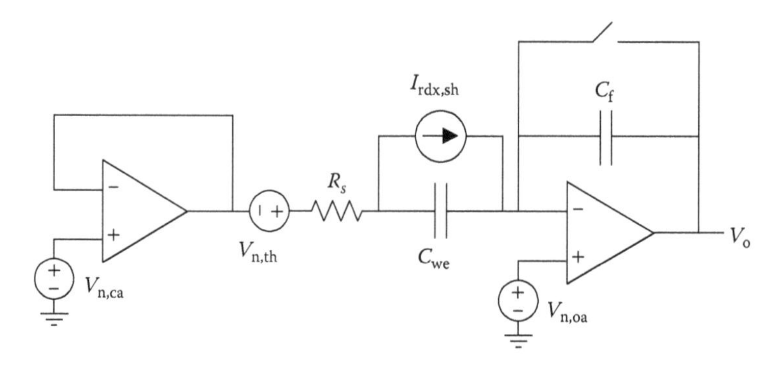

FIGURE 2.18 Simplified circuit model of the dual-slope ADC and electrochemical interface for noise analysis.

The detection limit of the active CMOS biochip can be determined by finding the target coverage S_T in the redox current term $i_{\text{ad},p}$ in Equation 2.17 that produces an SNR of 3 at the output of the integrator. The analysis is simplified by considering only the maximum value of the redox current. The SNR of the integrator output is given by

$$\text{SNR} = \frac{V_{o,\text{max}}^2}{\overline{V_{o,\text{th,rms}}^2} + \overline{V_{o,\text{sh,rms}}^2} + \overline{V_{\text{rst,rms}}^2} + \overline{V_{\text{oa,th,rms}}^2} + \overline{V_{\text{ca,th,rms}}^2} + \overline{V_{\text{oa},f,\text{rms}}^2} + \overline{V_{q,\text{rms}}^2}}. \quad (2.26)$$

The variables in the above expression are now described. $V_{o,\text{max}}^2$ is the square voltage (power) at the integrator output from the signal current $i_{\text{ad},p}$ and is given by

$$V_{o,\text{max}}^2 = \left(\frac{i_{\text{ad},p}t_1}{C_f}\right)^2. \quad (2.27)$$

$\overline{V_{o,\text{th,rms}}^2}$ is the mean square voltage with dimension V_{rms}^2 (also referred to as the average noise power) at the integrator output due to thermal noise produced by R_s in the electrochemical cell noise model in Figure 2.18. Only noise sources associated with the WE–electrolyte interface are included because of the virtual ground provided by the potentiostat control loop at the RE. $\overline{V_{o,\text{sh,rms}}^2}$ is the average power at the integrator output due to shot noise when $i_{\text{ad},p}$ flows. $\overline{V_{\text{rst.rms}}^2}$ arises from kT/C_f noise when the integration capacitor is reset. $\overline{V_{\text{oa,th,rms}}^2}$ and $\overline{V_{\text{oa,th,rms}}^2}$ are the average powers at the integrator output due to thermal noise produced by the integrator op amp and control amplifier, respectively. $\overline{V_{\text{oa},f,\text{rms}}^2}$ is the average flicker noise power at the integrator output produced by the integration op amp. To simplify the analysis, the effect of flicker noise from the control amplifier is not included. Finally, $\overline{V_{q,\text{rms}}^2}$ is the ADC quantization noise power. The charging current contribution is not included in Equation 2.26 because it is a deterministic signal and, therefore, does not affect the SNR. However, charging current does influence the upper limit of the DR of the sensor. Expressions for each of the noise terms in Equation 2.26 are derived below.

2.3.7.1 Integrated Noise

The integrator output voltage $V_{o,n}$ produced after integration period t_1 with an input noise current $I_n(t)$ is given in the time domain by

$$V_{o,n}(t_1) = \frac{1}{C_f}\int_0^{t_1} I_n(\tau)\mathrm{d}\tau = \frac{1}{C_f} I_n(t) * \left[u(t) - u(t - t_1)\right], \tag{2.28}$$

where $u(t)$ is the unit-step function and * represents the convolution operator.

The result can be converted to the frequency domain for noise computation as follows:

$$\overline{V_{o,n}^2}(f) = \left|\frac{1}{j2\pi f C_f}\right|^2 \overline{I_n^2}(f)\left|1 - e^{-j2\pi f t_1}\right|^2 = \left(\frac{t_1}{C_f}\right)^2 \overline{I_n^2}(f)\left[\frac{\sin(\pi f t_1)}{\pi f t_1}\right]^2, \tag{2.29}$$

where $\overline{V_{o,n}^2}(f)$ is the output noise voltage PSD after time t_1 and $\overline{I_n^2}(f)$ is the PSD of the noise current.

Integrating the above equation over frequency gives the average noise power after t_1 as

$$\overline{V_{o,n,\mathrm{rms}}^2} = \int_0^{\infty} \overline{V_{o,n}^2}(f)\mathrm{d}f = \frac{\overline{I_n^2}(f)}{2}\left(\frac{t_1}{C_f^2}\right). \tag{2.30}$$

2.3.7.2 Shot Noise from a Redox Current

Faradaic processes produce shot noise [91] that can be modeled as a current source, as shown in Figure 2.18. This source represents a zero-mean, wide-sense stationary (WSS), white Gaussian noise process with PSD $\overline{I_{\mathrm{rdx,sh}}^2}(f) = 2q \cdot i_{\mathrm{ad},p}$. The average output noise power $V_{o,\mathrm{sh,rms}}^2$ is given by

$$\overline{V_{o,\mathrm{sh,rms}}^2} = \left(\frac{t_1}{C_f}\right)^2 \int_0^{\infty} \overline{I_{\mathrm{rdx,sh}}^2}(f)\left[\frac{\sin(\pi f t_1)}{\pi f t_1}\right]^2 \left[\frac{1}{1 + (2\pi f R_{s1} C_{\mathrm{we}})^2}\right]\mathrm{d}f. \tag{2.31}$$

2.3.7.3 Thermal Noise from Electrolyte Resistance

Series resistance R_{s1} at the WE interface generates thermal noise current with PSD $\overline{I_{n,\text{th}}^2}(f) = 4kT/R_{s1}$, which produces a corresponding voltage noise at the integrator output in Figure 2.18. The average output noise power $\overline{V_{o,\text{th,rms}}^2}$ is given by

$$\overline{V_{o,\text{th,rms}}^2} = \left(\frac{t_1}{C_f}\right)^2 \int_0^{\infty} \overline{I_{n,\text{th}}^2}(f) \left[\frac{(2\pi f R_{s1} C_{\text{we}})^2}{1+(2\pi f R_{s1} C_{\text{we}})^2}\right] \left[\frac{\sin(\pi f t_1)}{\pi f t_1}\right]^2 \left[\frac{1}{1+(2\pi f R_{s1} C_{\text{we}})^2}\right] df. \tag{2.32}$$

For an R_{s1} of 275 Ω (experimentally measured using a 100 × 100 μm² Au WE in a 1 M potassium phosphate buffer [PPB]), C_{we} of 2 nF, T of 298 K, and t_1 set to 15 μs, the rms noise voltage at the integrator output due to the electrolyte resistance is 810 μV_{rms}.

2.3.7.4 Amplifier Thermal Noise

The contribution to total output noise from thermal noise produced by the integrator op amp is found using the circuit in Figure 2.18. Assuming the op amp has a single pole at the frequency f_0 and a dc gain A_0, its open-loop frequency response $A(f)$ is given by

$$A(f) = \frac{A_0}{1 + j\dfrac{f}{f_0}}. \tag{2.33}$$

The average output noise power at the end of the integration period $\overline{V_{\text{oa,th,rms}}^2}$ due to op amp input-referred thermal noise with PSD $\overline{V_{\text{oa,th}}^2}(f)$ is given by

$$\overline{V_{\text{oa,th,rms}}^2} = \int_0^{\infty} \overline{V_{\text{oa,th}}^2}(f) \left|\frac{1-e^{-j2\pi f t_1}}{1+(1+1/A(f))(1-e^{-j2\pi f t_1})}\right|^2 df. \tag{2.34}$$

Using the simulated gain, bandwidth, and noise PSD parameters for the integrator op amp in Table 2.2, the rms thermal noise voltage at the integrator output is approximately 450 μV_{rms}.

Thermal noise produced by the control amplifier with PSD $\overline{V_{\text{oa,th}}^2}(f)$ also contributes to the total noise at the integrator output. The output-referred noise due to the control amplifier $\overline{V_{\text{ca,th,rms}}^2}$ can be computed using

$$\overline{V_{\text{ca,th,rms}}^2} = \left(\frac{t_1}{C_f}\right)^2 \int_0^{\infty} \overline{V_{\text{oa,th}}^2}(f) \left[\frac{(2\pi f C_{\text{we}})^2}{1+(2\pi f R_{s1} C_{\text{we}})^2}\right] \left[\frac{\sin(\pi f t_1)}{\pi f t_1}\right]^2 df. \tag{2.35}$$

Using the op amp thermal noise PSD in Table 2.2 (which is equivalent to the output-referred noise because the control amplifier operates in a unity-feedback configuration), the thermal noise contributed by the control amplifier after an integration period t_1 is about 760 μV_{rms}.

2.3.7.5 Amplifier Flicker Noise

The input-referred flicker noise of the integration amplifier has PSD $\overline{V_{oa,f}^2}(f)$ given by

$$\overline{V_{oa,f}^2}(f) = \frac{K_f}{f}, \quad \text{for } f \geq f_{min}, \tag{2.36}$$

where K_f (with dimensions V^2) is a constant related to the CMOS fabrication parameters and $f_{min} = 1/t_1$ is the frequency at which the integrator is reset.

The flicker noise power at the integrator output after duration t_1, $\overline{V_{oa,f,rms}^2}$ can be found using Equation 2.34, except that $\overline{V_{oa,th}^2}(f)$ must be replaced by $\overline{V_{oa,f}^2}(f)$ and the lower frequency bound on the integral changed to f_{min}. With a K_f of 1.4×10^{-12} V^2 and f_{min} equal to 66.7 kHz, the rms output-referred noise due to op amp flicker noise is approximately 800 μV_{rms}.

2.3.7.6 Quantization Noise

The ADC quantization noise power $\overline{V_{q,rms}^2}$ is given by

$$\overline{V_{q,rms}^2} = \frac{V_\Delta^2}{12}, \tag{2.37}$$

where V_Δ is the voltage difference between two adjacent quantization levels of the ADC.

Assuming V_R is the linear output voltage range of the integrator op amp in the dual-slope ADC in Figure 2.11, and N is the number of bits in the counter used to measure t_2 in Equation 2.20, V_Δ is given by

$$V_\Delta = \frac{V_R}{N}. \tag{2.38}$$

As an example, V_Δ is approximately 0.73 mV assuming a V_R of 0.75 V and a 10-bit nominal resolution. This produces a quantization noise voltage of about 210 μV_{rms}.

2.3.7.7 Target Coverage Detection Limit

Figure 2.19 displays the contributions from each of the noise sources discussed above (except for quantization noise) as a function of DNA surface target coverage. The total noise at the integrator output for an integration period lasting 15 μs stays approximately constant at 1.3 mV_{rms}. As can be seen, the thermal noise generated by the electrochemical cell resistance, integration op amp, and control amplifier dominates the total sensor noise.

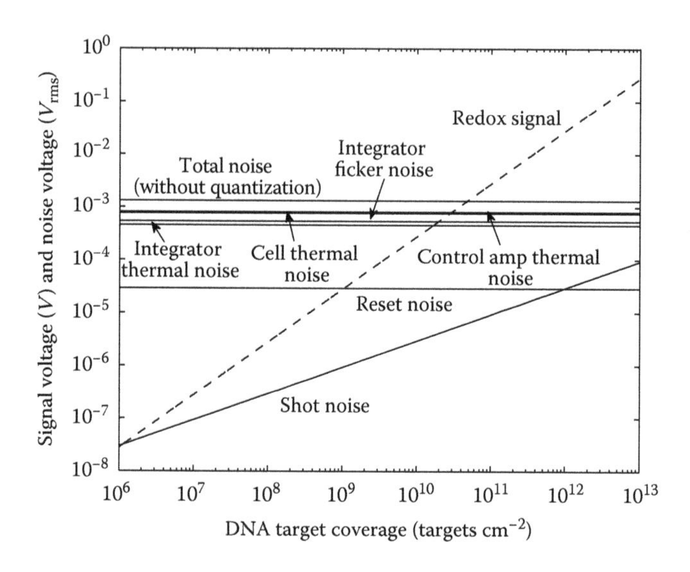

FIGURE 2.19 Signal voltage and noise voltages at the integrator output as a function of DNA surface target coverage.

The quantization-noise-free SNR given by Equation 2.26 (but excluding the $\overline{V_{q,rms}^2}$ term) as a function of surface target coverage is shown in Figure 2.20. The horizontal line indicating an SNR of 3 is drawn to indicate the detection limit. In this case, the quantization-noise-free detection limit is approximately 8.0×10^{10} targets cm^{-2}. The corresponding maximum redox current measured at this limit is 750 pA.

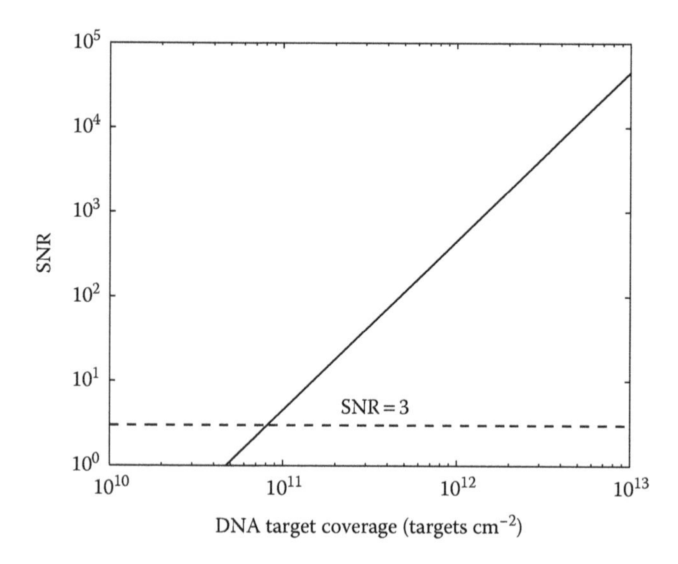

FIGURE 2.20 Quantization-noise-free SNR as a function of DNA surface target coverage.

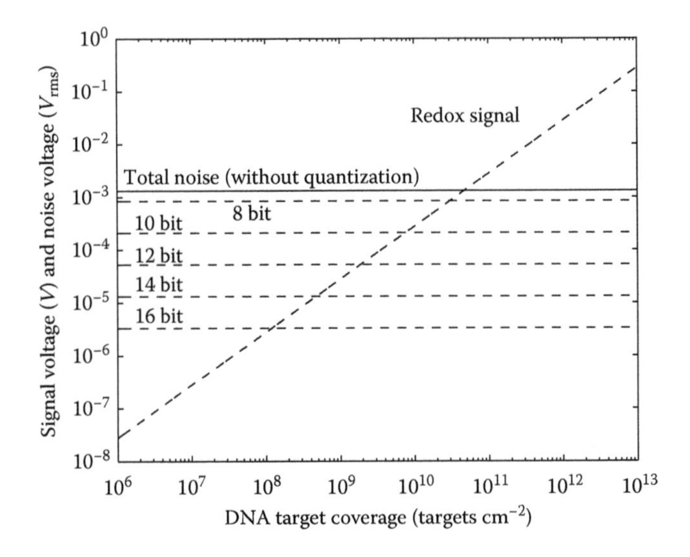

FIGURE 2.21 Quantization noise voltage for different ADC resolutions along with the quantization-free noise level at the integrator output.

Figure 2.21 shows the quantization-free noise voltage determined above as well as the quantization-noise level for different ADC resolutions, assuming a V_R of 0.75 V. It is evident that quantization noise at levels of 10 bits or more does not contribute significantly to the total noise at the integrator output. The minimum target coverage that can be detected with an SNR of 3 using an 8-, 10-, and 12-bit ADC is 9.6×10^{10}, 8.2×10^{10}, and 8.0×10^{10} targets cm^{-2}, respectively.

It is also useful to study the effect of WE area and ADC integration time on the DNA target detection limit of the sensor. Analysis of these is carried out below.

2.3.7.8 Effect of Working Electrode Area on Detection Limit

Figure 2.22 shows the sensor detection limit as a function of the WE area. The detection limit decreases as the WE area increases because the redox signal power goes up with the square of the area, while the thermal noise power produced by R_s and the control amplifier increases by a smaller factor.

2.3.7.9 Effect of ADC Integration Time on Detection Limit

Increasing the duration t_1 over which the ADC integrates the WE current has the effect of reducing the sensor detection limit, as displayed in Figure 2.23a. This is again due to the power-of-two relationship between the maximum redox current level and overall signal power. However, this analysis assumes that the CV scan rate remains unchanged. Using a long t_1 means that the ADC sample rate must be reduced, assuming a constant I_{ref}. Therefore, fewer points on the CV curve can be collected when both t_1 and V are large. This, in turn, reduces the accuracy associated with integrating the WE current to obtain the total charge transferred. In addition, as the scan rate increases, so does the charging current in proportion. As a result, saturation occurs when the sum of the

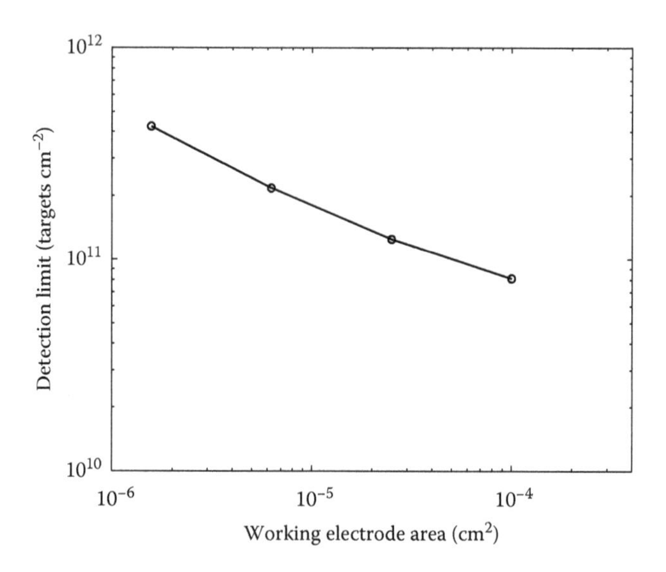

FIGURE 2.22 Effect of working electrode area on sensor target coverage detection limit.

redox signal and charging currents produces a voltage outside the linear range of the integrator op amp. This is evident when t_1 is equal to 30 μs in Figure 2.23b. For longer integration periods, saturation can occur before the SNR even reaches 3.

2.3.7.10 Sensor Dynamic Range

The sensor DR is defined as the ratio of the largest nonsaturating input current I_{max} to the minimum detectable input current I_{min}. These are given by the following expressions:

$$I_{\text{max}} = \frac{C_f V_{o,\text{max}}}{t_1} \tag{2.39}$$

$$I_{\text{min}} = \frac{C_f}{t_1}\left(\overline{V_{o,\text{th}}^2} + \overline{V_{o,\text{sh}}^2} + \overline{V_{\text{rst}}^2} + \overline{V_{\text{oa,th}}^2} + \overline{V_{\text{ca,th}}^2} + \overline{V_{\text{oa},f}^2}\right)^{1/2}, \tag{2.40}$$

where $V_{o,\text{max}}$ is the maximum linear output voltage of the integrator op amp. The DR can therefore be expressed as

$$\begin{aligned} \text{DR} &= \frac{I_{\text{max}}}{I_{\text{min}}} \\ &= \frac{V_{o,\text{max}}}{\left(\overline{V_{o,\text{th}}^2} + \overline{V_{o,\text{sh}}^2} + \overline{V_{\text{rst}}^2} + \overline{V_{\text{oa,th}}^2} + \overline{V_{\text{ca,th}}^2} + \overline{V_{\text{oa},f}^2}\right)^{1/2}}. \end{aligned} \tag{2.41}$$

Assuming a V_R of 0.75 V and a t_1 of 15 μs, the DR of the sensor as a function of the surface target coverage stays fairly constant at around 520, which is equivalent to about 2.8 decades.

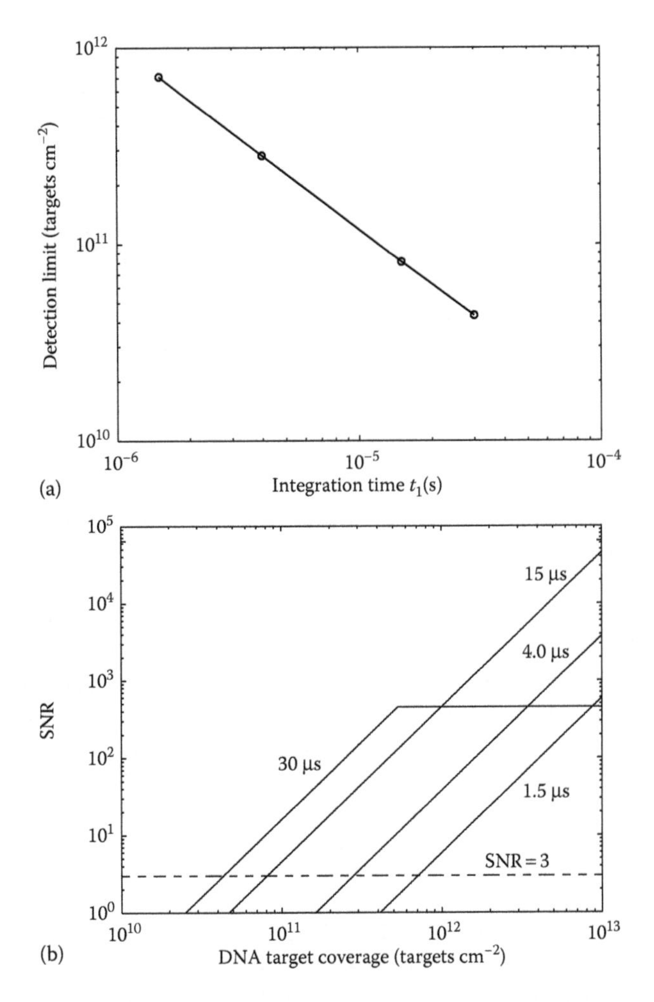

FIGURE 2.23 Effect of ADC integration time t_1 on sensor detection limit. (a) Sensor limit of detection as a function of integration time. (b) SNR for various integration times. Saturation occurs when the sum of the redox signal and charging currents produce a voltage outside the linear range of the integrator op amp.

2.3.8 CMOS Biochip Fabrication and PostProcessing

The electrochemical DNA biochip is fabricated in a Taiwan Semiconductor Manufacturing Company (TSMC, Hsinchu, Taiwan) 2.5 V, five-metal, 0.25 μm, mixed-signal CMOS process. All five metal layers in this process are made of Al. A photograph of the 5×3 mm^2 die is shown in Figure 2.24. Each of the 16 sensor sites on the chip, comprised of a WE and dual-slope ADC, measures approximately 650×350 μm^2. About two-thirds of this area is occupied by the two-stage integrator op amp and compensation capacitor. Each of the control amplifiers occupies an area of about 400×400 μm^2, while the amplifiers in the diagnostic circuits use less than 350×350 μm^2 of silicon area.

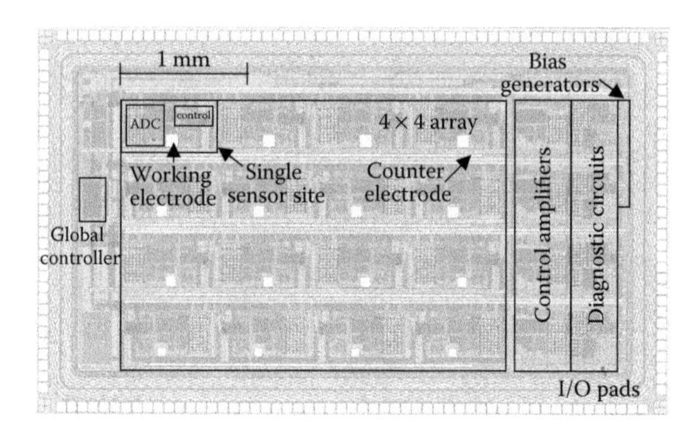

FIGURE 2.24 Die photograph of the active CMOS sensor array for label-based electrochemical DNA assays.

At the time of chip physical design, pads in the top metal layer (M5) are laid out to form the integrated WEs and CEs. Tungsten (W) vias vertically connect M5 to the other Al metal layers below. Openings are then specified in the 1–2 μm thick silicon nitride (Si_3N_4) and SiO_2 passivation layers above the WE and CE metal pads, as shown in Figure 2.25, exposing the top Al metal at these locations. This process is no different from that used to construct bond pads along the chip perimeter.

Postprocessing of the fabricated CMOS chip is necessary to create an array of Au electrodes on the surface at the predefined locations discussed previously. As was described in Section 2.2.6, Au has the advantages of being relatively electrochemically inactive in the presence of strong electrolytes and is easily modified by self-assembly of well-ordered monolayers of thiol, sulfide, or disulfide compounds through Au–S bonding.

2.3.8.1 Initial Electrode Fabrication Procedure

Gold electrodes are constructed on the CMOS surface using standard microfabrication techniques in a clean-room environment at Columbia University. Gold is used for both the WEs and CEs in order to reduce the number of processing steps. Originally, a 20 nm thick layer of titanium (Ti) was deposited using electron-beam (e-beam) evaporation under vacuum directly onto the Al metal pads to act as an adhesion layer, followed by a 200–300 nm thick layer of Au using the same technique. Standard photolithographic techniques were employed to selectively cover the chip with photoresist in order to lift off deposited metal at undesired locations following evaporation. After postprocessing, the chip layer stack appears as shown in Figure 2.26a, while Figure 2.26b shows an optical microscope image of a WE and CE.

Unfortunately, depositing Ti/Au directly onto Al pads causes corrosion problems when the chip is operated in a wet electrochemical environment. This is because Al, unlike Au, is highly electrochemically active and corrodes easily when exposed to an electrolyte. During experiments in which the CMOS chip was operated in an analyte solution (to be discussed in greater detail in Section 2.4), it is thought that dissolved ions, such as chloride, penetrate the Ti/Au through grain boundaries in the thin-film layers. Following penetration, it is believed that the Al pad below reacts

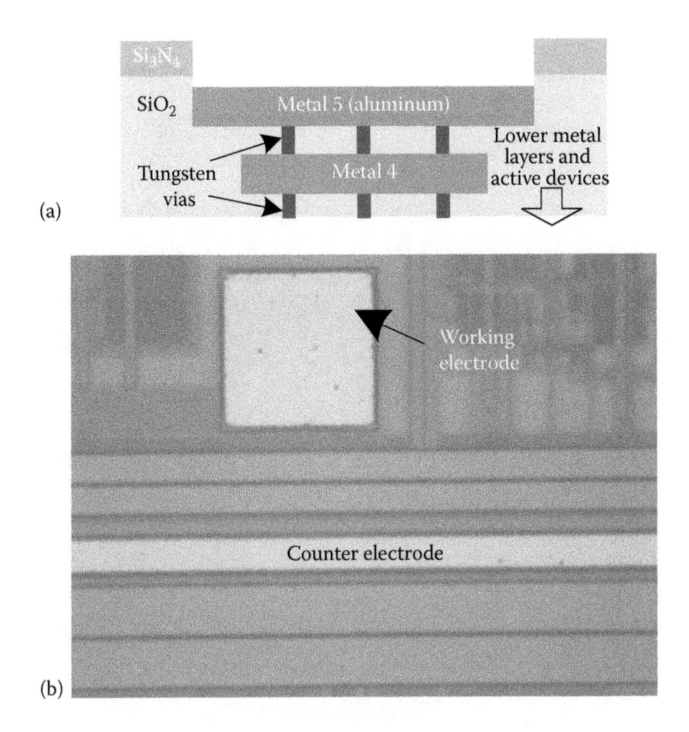

FIGURE 2.25 CMOS biochip Al metal layers with the passivation removed. (a) Layer stack-up showing the two highest Al metal layers (not to scale). (b) Optical photograph of the top metal layer of the CMOS biochip at a WE site with the passivation removed.

with the ions, corroding after only a short time. When this occurs, the Ti/Au film is destroyed as well, as can be seen in Figure 2.27.

2.3.8.2 Improved Electrode Fabrication Procedure

To eliminate the corrosion problem and subsequent destruction of the WEs and CEs, the Al is selectively removed at the electrode locations prior to Ti/Au deposition, as shown in Figure 2.28a. This is done using a phosphoric acid–based wet etchant from Transene (Aluminum Etchant Type A). After Al removal, the Ti/Au layers are e-beam evaporated onto the electrode locations, as displayed in Figure 2.28b. The added metal electrically connects to the lower metal layers through the existing W vias.

Figure 2.29 shows an optical microscope image of a WE site following Al etching. The dense grid of W vias is clearly visible, along with the fourth Al metal layer. Therefore, evaporated metal can easily make a solid electrical connection to the via during deposition.

The advantages of the above electrode fabrication method are that it is both simple and inexpensive. For instance, it does not require an entire back-end process as in [75]. In addition, it needs fewer processing steps than when constructing a "stepped" electrode structure in which the Au electrode is built adjacent to, rather than directly above, an Al pad [93]. Furthermore, only a few inexpensive ingredients are required to prepare the Al etchant, which is fully compatible with standard photolithographic processes.

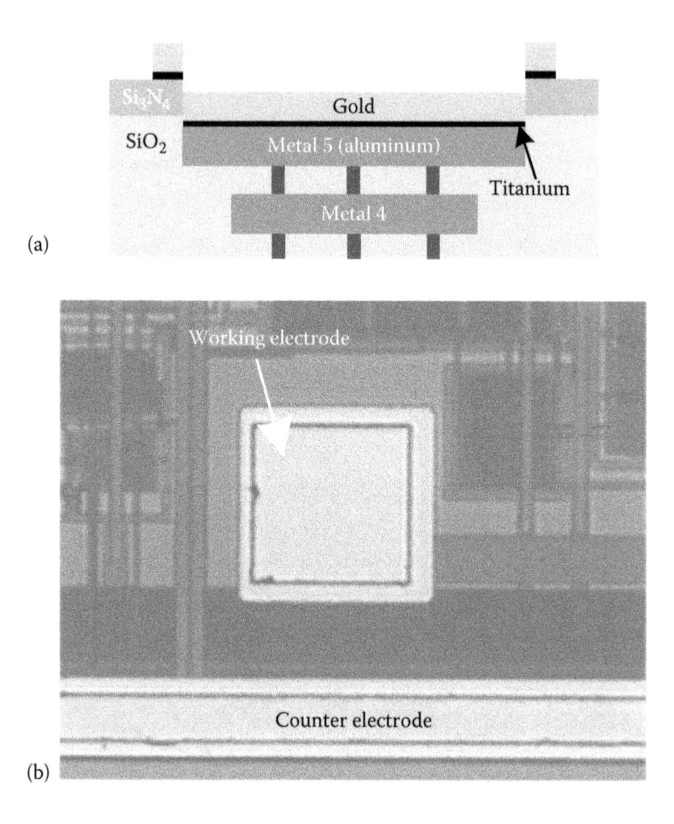

FIGURE 2.26 Original postprocessing method used to construct Au electrodes on the surface of the CMOS biochip. (a) Chip layer stack-up after Ti/Au is deposited directly on the native Al pads. (b) Optical microscope image of a WE and CE after postprocessing.

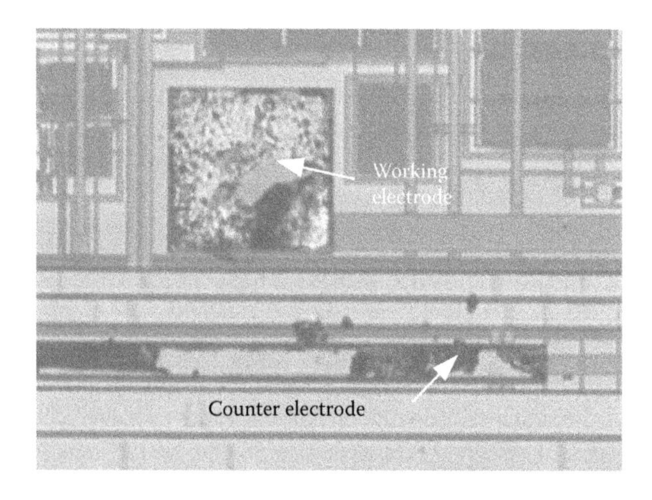

FIGURE 2.27 Visible destruction of an integrated WE and CE after the CMOS biochip is operated in an electrolyte.

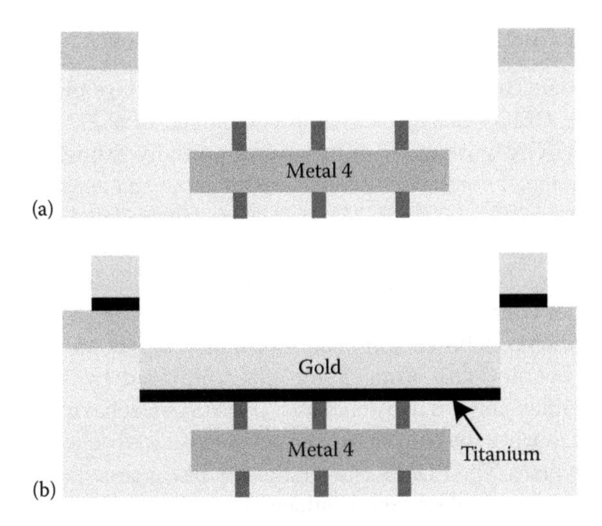

FIGURE 2.28 Improved electrode fabrication procedure. (a) Aluminum at electrode sites is selectively etched away using standard photolithographic techniques. (b) Titanium and Au are deposited onto the electrode sites and electrically connect to the metal layers below through the existing vias.

FIGURE 2.29 A working electrode site following Al removal using a wet etch.

2.4 EXPERIMENTAL RESULTS

Experimental results are presented in this section that validate the operation of the active CMOS biochip for label-based electrochemical DNA assays. Biochip packaging issues, results from electrical characterization, and integrated electrochemical measurements of redox species and ferrocene-labeled DNA targets are described in detail.

2.4.1 CMOS Chip Packaging and Experimental Setup

To enable electronic characterization, electrochemical testing, and external interfacing of the active CMOS biochip, the chip is mounted in a 272-pin, 27×27 mm^2 ball grid array (BGA) package with its surface partially exposed, as displayed in Figure 2.30. The metal bond wires connecting the input and output (I/O) pads along the chip perimeter to pads on the BGA package are encapsulated using a heat-cured, chemical-resistant epoxy (Hysol, Henkel, Düsseldorf, Germany) to protect them from electrolyte exposure. Bonding and packaging of the CMOS chips are performed by an external vendor (Corwil, Milpitas, CA).

The packaged chip is fastened in a surface-mount socket on a custom-designed printed circuit board (PCB) using a top plate provided by the socket vendor. A square, 1 mm thick poly(dimethylsiloxane) (PDMS) sheet having a square hole cut out in the center to leave the chip surface uncovered is sandwiched between the chip package and top plate. The PDMS essentially acts like a gasket, preventing analyte leakage onto the PCB. A glass tube is attached to the top plate using an ultraviolet (UV)-cured epoxy to form a 12 mL analyte reservoir over the chip. An external Ag/AgCl/3 M NaCl RE is held in the reservoir with a Teflon cap.

2.4.2 Electrical Characterization

Electrical characterization of the active CMOS biochip is carried out prior to performing integrated electrochemical measurements. Figure 2.31 shows the output noise spectrum of the control amplifier over a bandwidth from 10 Hz to 21 kHz when this device is operated in a unity-gain configuration. The $1/f$ corner frequency is located above the maximum input frequency of the spectrum analyzer. The measured output noise voltage is 21.2 μV_{rms} over the 10 Hz–21 kHz band when the effect of 60 Hz line interference and other interfering tones is removed.

Characterization of the dual-slope ADC at each sensor site is carried out using the on-chip diagnostic circuits. To allow sufficient bandwidth for CV experiments, the

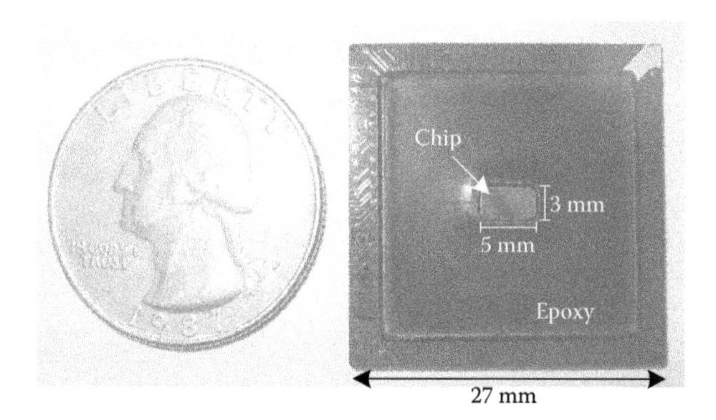

FIGURE 2.30 Packaged CMOS biochip in which the bond wires are covered using a chemical-resistant epoxy to protect them from analyte exposure. The US quarter dollar is included for size comparison.

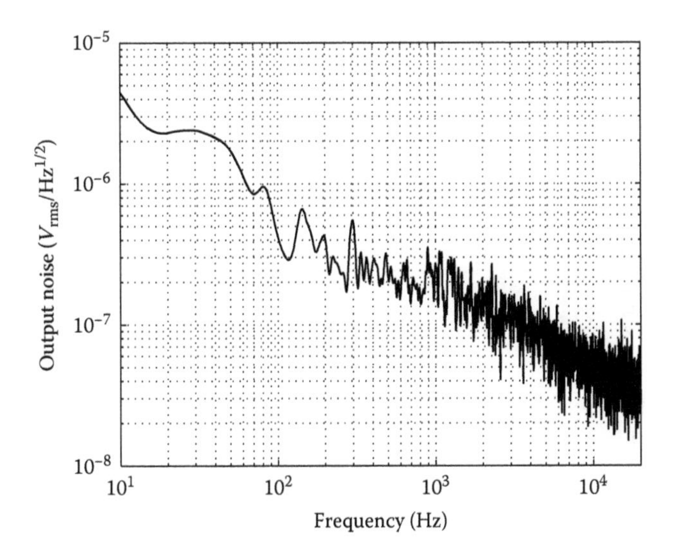

FIGURE 2.31 Output noise spectrum of the control amplifier from 10 Hz to 21 kHz with interference tones removed.

ADCs are operated at a sampling rate f_s of 2.5 kHz with φ_{clk} set to 3.5 MHz. The first integration time t_1 is set to 23 μs and $t_{2,max}$ is 315 μs, providing a nominal resolution of 10 bits (plus an additional sign bit). Although dual-slope ADCs generally feature resolutions of 16 bits or more, these are normally operated at very low sampling rates (a few hertz) to avoid use of an impractically high value of φ_{clk}. For the current application, resolution has been traded off so that a higher sampling rate could be employed. In addition, use of a lower-speed φ_{clk} reduces the level of on-chip switching interference that can couple into the analog circuit blocks.

The remaining time during each conversion cycle is required to reset C_f and select the appropriate reference current source. The maximum I_{we} before integrator saturation using these settings is about 110 nA. Reference currents I_{ref+} and I_{ref-} are set to 15 and 18 nA, respectively. Typical differential nonlinearity (DNL) and integral nonlinearity (INL) values for the ADCs are −0.25 and +0.38 LSB, respectively, with an LSB current of approximately 240 pA.

The DR of the ADC is experimentally verified using a 103 Hz sinusoidal input current. Figure 2.32a displays the typical SNR and signal-to-noise-and-distortion ratio (SNDR) of the ADC as a function of input current level. The lower end of the DR curve is fitted due to the difficulty in providing a sufficiently small ac voltage signal to the on-chip transconductance amplifiers used for ADC testing. The DR is limited at the upper end by integrator saturation. A maximum effective number of bits (ENOB) of 9 is achieved and is limited by the linearity of the test circuits. A DR greater than 10 bits is achieved from circuit simulations of the dual-slope ADC alone. Figure 2.32b shows the result of an 8192-point fast Fourier transform (FFT) of the measured ADC output when a full-scale input is applied. The strong second harmonic is due to the single-ended architecture of the ADC.

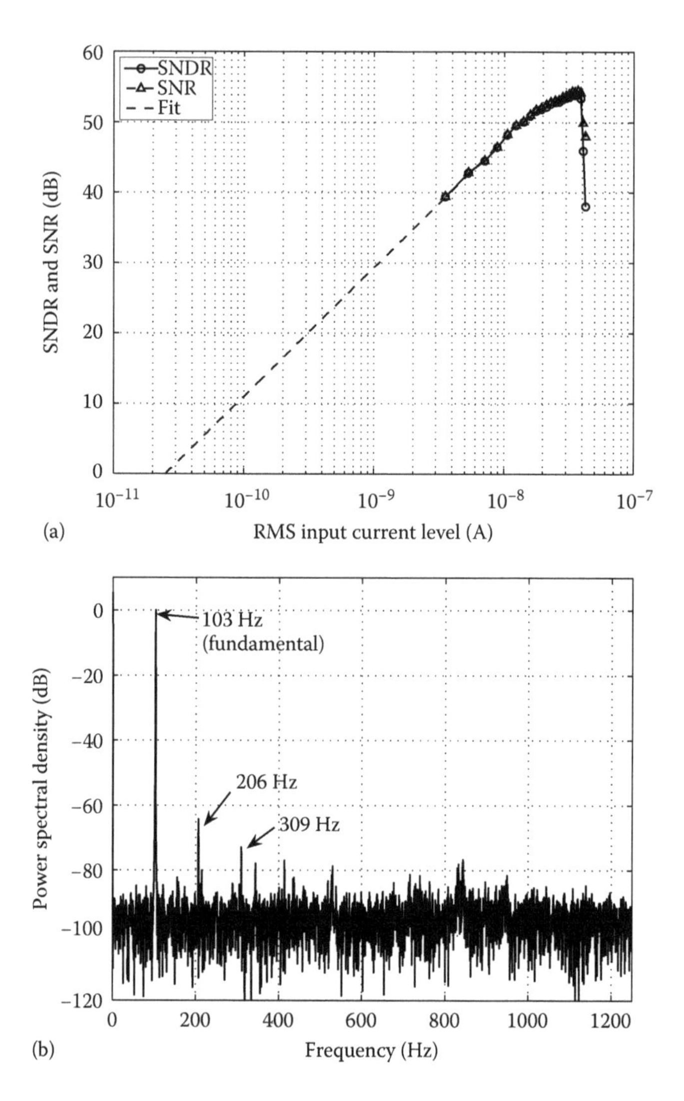

FIGURE 2.32 Typical measured results from ac linearity testing of the dual-slope ADCs. (a) Dynamic range. (b) Output spectrum with a full-scale input at 103 Hz. The resolution bandwidth is 0.31 Hz.

2.4.3 Integrated Electrochemical Measurements of a Bulk Redox Species

As a first example of the use of the active CMOS biochip for basic electrochemical sensing, CV measurements of the bulk redox species potassium ferricyanide, $K_3[Fe(CN)_6]$, are carried out. At the appropriate potential, ferricyanide ions are reduced to ferrocyanide ions in the reaction

$$\mathrm{Fe}\left(\mathrm{CN}\right)_6^{3-} + e \rightarrow \mathrm{Fe}\left(\mathrm{CN}\right)_6^{4-}. \tag{2.42}$$

Potassium ferricyanide is often used by electrochemists to study interfacial properties because of its highly reversible behavior. It is for this reason also that potassium ferricyanide is often used as a diagnostic measurement tool in electrochemical systems.

2.4.3.1 Biochip Preparation and Experimental Procedure

In order to demonstrate proper functioning of the active CMOS biochip, CV measurements of 2 mM potassium ferricyanide in 1 M PPB (pH 7.4), made by combining appropriate amounts of K_2HPO_4 and KH_2PO_4 in water, are carried out. In these experiments, the potential between all the WEs in the array (held at 1.25 V relative to ground) and RE is scanned from +0.75 to −0.5 V and back at various rates while the cell current is observed at one WE. To extend the potential range of the electrochemical cell beyond the 2.5 V power supply limit, the on-chip control amplifiers are bypassed and a discrete, off-chip op amp (AD8628, Analog Devices, Norwood, MA), operating at 3 V, is used to drive an external Pt-wire CE rather than the on-chip Au WEs. A standard Ag/AgCl/3 M NaCl RE (RE-5B, Bioanalytical Systems, West Lafayette, IN) is used in all experiments.

Cleaning of the WE surface is necessary to avoid contamination from one experiment to the next. This is particularly important when performing DNA sensing because probe layers left over from previous experiments affect the hybridization thermodynamics of subsequent studies. Thick, macroscopic Au electrodes can be easily cleaned through mechanical polishing or controlled dissolution through voltage cycling in a strong acid and oxidizing agent. However, such techniques are too invasive and impractical to be applied to the thin-film Au WEs on the CMOS biochip. Therefore, before each new experiment, the chip is placed in a UV/ozone cleaner for 5 min to break down organic contaminants on the Au WE surfaces [94]. These can then be eliminated from the surface by thorough rinsing in deionized water.

2.4.3.2 Measurement Results

Figure 2.33a shows the cell current at one of the 100 × 100 μm^2 WEs when an input scan rate v of 72 mV s^{-1} is used. A zero-phase, low-pass FIR filter is used to post-process the raw data in MATLAB®. The locations of the forward (reduction) and reverse (oxidation) current peaks at +0.22 and +0.30 V, respectively, match those obtained when the same experiment is run on a commercial potentiostat using a 125 μm diameter Au WE. In addition, the 80 mV difference in the potentials at which the maximum forward and reverse currents occur (also known as the "peak potentials") is relatively close to the theoretical value of 59 mV for a fully reversible, single-electron redox process given by Equation 2.11. The magnitude of the current falls after each peak due to mass-transport limitations of the redox species to the WE surface.

In Figure 2.33b, v is increased from 24 to 480 mV s^{-1} and the peak reduction current is measured. As indicated by the Randles–Sevcik equation (Equation 2.10), the peak current i_p at a planar electrode for a reversible reaction under diffusive control (measured after subtracting the charging current background) increases linearly with $v^{1/2}$. This behavior is observed in Figure 2.33b.

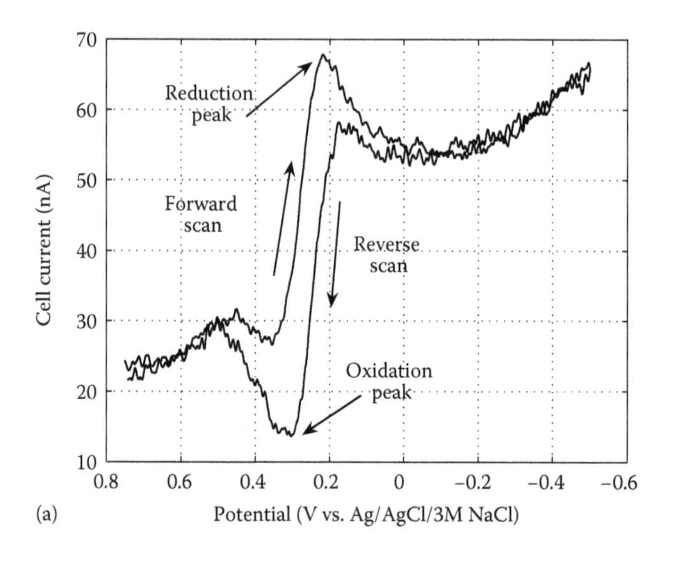

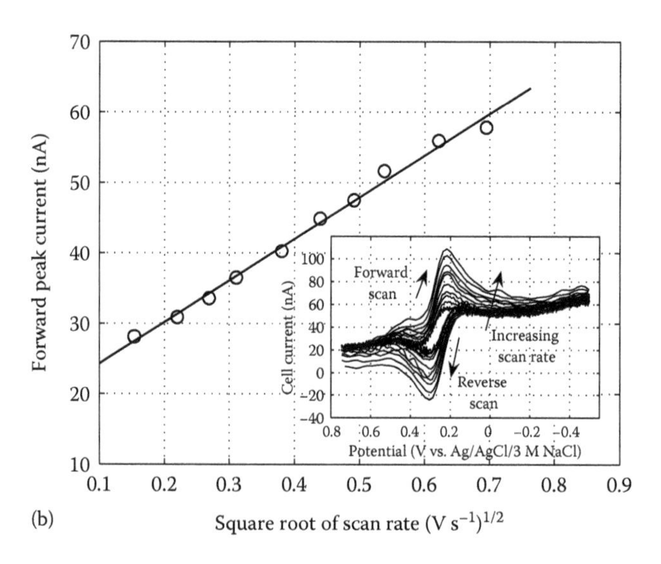

FIGURE 2.33 CV measurements of the bulk redox species potassium ferricyanide using the active CMOS biochip. (a) Measured result at a 100 × 100 μm^2 WE at a scan rate of 72 mV s^{-1}. (b) Peak current dependence of ferricyanide reduction on scan rate.

The linear dependence of i_p on WE area (also reflected in Equation 2.10) is also verified by running a CV scan at 290 mV s^{-1} and observing the current flowing through the 100 × 100, 90 × 90, and 80 × 80 μm^2 WEs. Figure 2.34 shows the results from this measurement. The actual WE area is assumed to be 1.5 times greater than the geometric (drawn) area in order to account for surface roughness of the e-beam-deposited Ti/Au layer. This value is in the range measured in independent, off-chip experiments on Au electrodes and has been verified by others for evaporated thin-film metals [95].

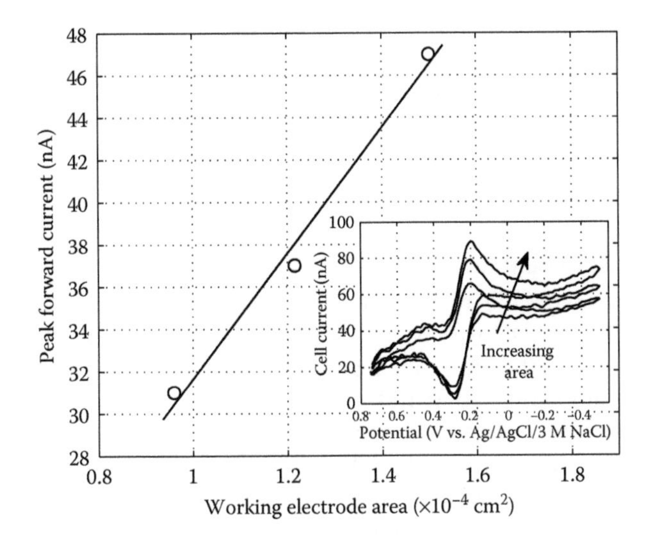

FIGURE 2.34 Dependence of peak current from ferricyanide reduction on the WE area.

2.4.4 Integrated Electrochemical Measurement of DNA Probe Surface Coverage

This section describes the use of the active CMOS biochip in determining the coverage of ssDNA probes on the WE surface. As outlined in Section 2.2.3, knowledge of the ssDNA probe surface coverage is necessary in order to calculate thermodynamic parameters in affinity-based DNA sensing assays.

In these experiments, Au WEs are functionalized with a monolayer of ssDNA probes and CV measurements are then carried out in the presence of the redox intercalator hexaamineruthenium(III) chloride to determine probe surface density. The redox-active counterion $RuHex^{3+}$ associates with the surface-immobilized DNA, causing the thermodynamics of the redox processes to be altered. Prior work has shown that as probe coverage increases, the reduction potential for the reaction

$$RuHex^{3+} + e \rightarrow RuHex^{2+} \quad (2.43)$$

shifts toward more negative values. This phenomenon is due, in part, to changes in local dielectric constant, spatial distribution, and solvation [90]. This technique can provide an absolute measure of the probe coverage once calibration using a surface-analysis technique such as x-ray photoelectron spectroscopy (XPS) has been carried out.

2.4.4.1 Biochip Preparation and Experimental Procedure

The chip is cleaned as described previously and is then incubated for 30 min in a 1 M $MgCl_2$ solution containing a known concentration of thiolated, 20-mer ssDNA probe having the sequence 5′-TTT TTT TCC TTC CTT TTT TT-3′. Next, the chip is incubated in 1 mM mercaptopropanol (MCP) solution for 90 min, forming a SAM that helps to passivate the WE surface and prevent nonspecific interactions between the immobilized DNA and WE, as was discussed in Section 2.2.6. CV at a scan rate of 4 V s^{-1} is then carried out in 7 mL of 10 mM Tris acetate buffer (pH 7.4) with 1 μM $RuHex^{3+}$.

2.4.4.2 Measurement Results

Figure 2.35 displays the results from two different CV experiments at one 90 × 90 μm^2 WE for DNA probe coverages of 1×10^{13} and 4×10^{12} cm^{-2}. These probe densities are obtained by incubating the chip in different concentrations of DNA probe solution (75 and 25 μM, respectively) and are verified using a set of calibration measurements on a commercial potentiostat and a 125 μm diameter Au WE. The overall shape of the CV curves is different from those obtained in the previous section because $RuHex^{3+}$ electrostatically associates with the ssDNA probes at the WE surface and is, therefore, not subject to mass-transport limitations.

The forward peaks occur at –277.2 and –294.8 mV for the lower and higher probe coverage, respectively. This indicates a shift of 17.6 mV toward more negative potentials with the higher coverage, confirming previous observations. Parallel experiments on a commercial potentiostat show similar peak potentials and a closely matching shift of 15.4 mV. In addition, the quantity of $RuHex^{3+}$ near the WE increases with higher probe coverage [90], as is evident in the observed peak-current increase for the 1×10^{13} cm^{-2} measurement.

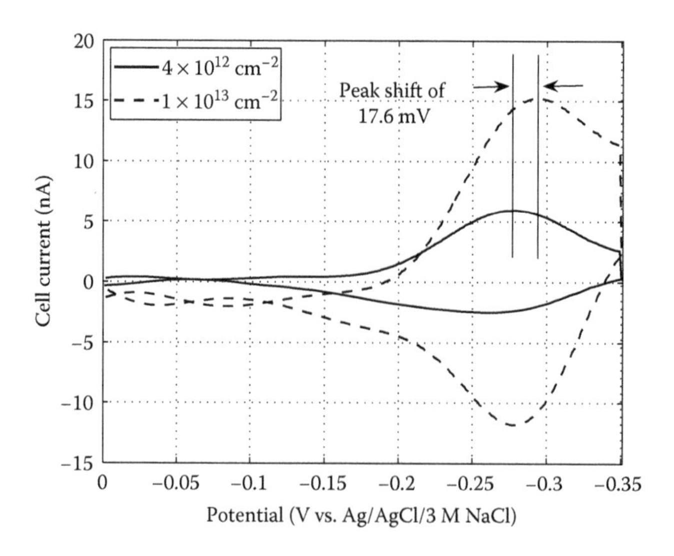

FIGURE 2.35 DNA probe surface coverage measurements using the redox intercalator $RuHex^{3+}$ and the active CMOS biochip.

2.4.5 Label-Based Electrochemical DNA Sensing

Results from quantitative, multiplexed, and real-time electrochemical DNA sensing using the active CMOS biochip are now presented. The measurement technique described in Section 2.3.1, in which ssDNA targets conjugated with Fc redox labels hybridize with surface-immobilized ssDNA probes, is employed.

2.4.5.1 ADC Operation

Due to the high CV scan rate used in the current experiments (60 V s^{-2}), the sampling rates of the integrated dual-slope ADCs are increased to 10 kHz with a φ_{clk} of 3.5 MHz. The values of t_1 and $t_{2,max}$ are 15 and 63 μs, respectively, providing a nominal resolution of almost 8 bits (plus a sign bit). Both reference current sources are set to 60 nA. With these settings, the typical measured SNDR of the ADCs is 43.7 dB at an input current level of 38 nA_{rms} (corresponding to −6 dBFS). The typical maximum DNL and INL are +0.22 and +0.15 LSB, respectively.

2.4.5.2 Chemical Preparation

2.4.5.2.1 DNA Oligonucleotide Sequences

The sequences of the 3′-end thiolated 20-mer DNA oligonucleotide probes (MWG-Biotech, High Point, NC) and 18-mer Fc-conjugated DNA targets used in the active CMOS biochip experiments are shown in Table 2.4. The sequences for probe P1 (from *Homo sapiens* retinoblastoma 1 mRNA, which gives rise to a rare form of eye cancer in humans) and target T1, and P2 and T2, are pairwise complementary, respectively. The sequence of target T1s differs from that of T1 by a single base. These model sequences are selected in order to demonstrate the functionality of the active CMOS biochip and to allow comparison with off-chip electrochemical experiments with a commercial benchtop potentiostat using the same DNA sequences. Methods to prepare the N-(2-ferrocene-ethyl) maleimide redox label and perform target DNA conjugation can be found in [25].

TABLE 2.4
DNA Oligonucleotide Sequences Used in Label-Based DNA Sensing Experiments

Name	Sequence
P1	5′-TTT TAA ATT CTG CAA GTG ATJ-3′
P2	5′-TTT TTT TCC TTC CTT TTT TTJ-3′
T1	5′-FcCAC TTG CAG AAT TTA AAA-3′
T1s	5′-FcCAC TTG CTG AAT TTA AAA-3′
T2	5′-FcAAA AAG GAA GGA AAA AAA-3′

Note: "J" represents a thiol (–SH) group.

2.4.5.3 Chip Surface Preparation

The surface of the CMOS chip is cleaned as described previously and then incubated in a 1 $MgCl_2$ solution containing 500 nM ssDNA probe for 30 min. This provides a probe surface density of approximately 8×10^{12} cm^{-2}, which is determined from a set of calibration measurements performed off chip [82]. After probe immobilization, the chip is incubated in a 1 mM MCP solution for 90 min. All CV experiments are run using 7 mL of 1 M PPB (pH 7.4).

2.4.5.4 Measurement Results

2.4.5.4.1 Basic DNA Hybridization Detection

Figure 2.36 shows a typical output from the ADC measured at one of the 100×100 μm^2 WEs in the array, functionalized with probe P1, approximately 50 min after 50 nM of target T1 is introduced to the system. The current peaks due to the Fc redox reactions are evident above the charging current level, indicating hybridization between P1 and T1. Hysteresis causes the forward and reverse charging currents to differ somewhat. Based on the average charging current, the WE interfacial capacitance is measured to be approximately 7 μF cm^{-2}, which is in the expected range for an MCP-modified Au electrode at the DNA coverage and buffer ionic strength used [90].

Figure 2.37 indicates that the peak current level (average of forward and reverse peaks after background subtraction) increases linearly with scan rate, as predicted by theory for a surface redox species (see Equation 2.12). This confirms that the signal current originates from surface-hybridized DNA targets and not diffusing species in solution. Additionally, the measured linear dependence of peak current on WE area, as predicted by theory, is shown in Figure 2.38.

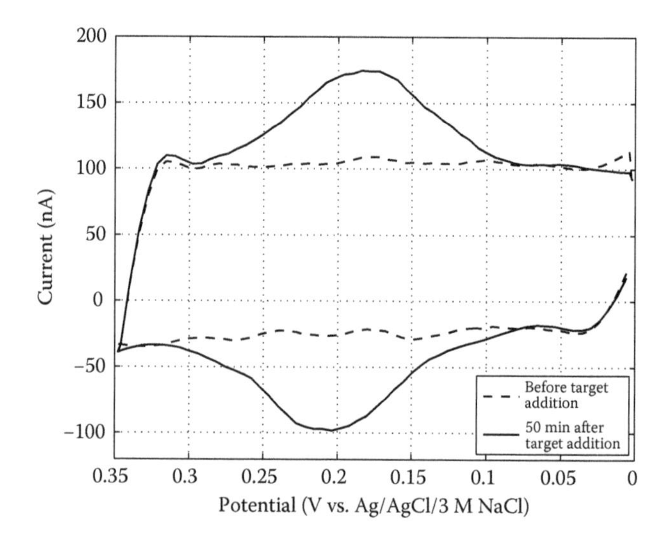

FIGURE 2.36 A typical CV measurement using the active CMOS biochip in which the charging current, observed before complementary DNA target is added to the analyte, is shown along with the sensor response from hybridization, 50 min after target addition, at one 100×100 μm^2 WE.

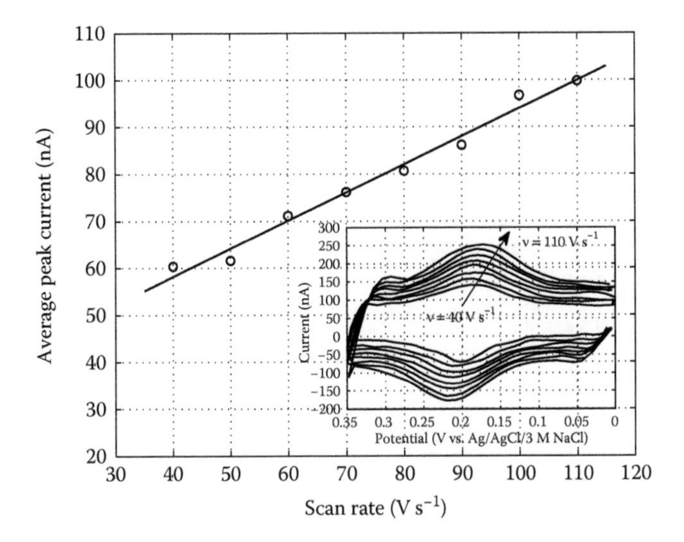

FIGURE 2.37 Linear dependence of the average peak current (after background subtraction) on the CV scan rate. Inset shows the resulting waveform from each scan.

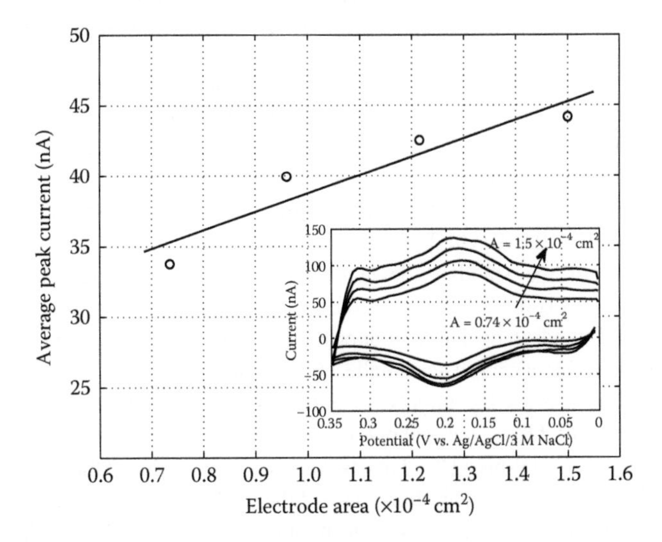

FIGURE 2.38 Linear dependence of the average peak current (after background subtraction) on the WE area. Inset shows the resulting waveform from each scan.

2.4.5.4.2 DNA Target Concentration Series

The active CMOS biochip enables real-time quantitation of surface-hybridized targets in a multiplexed fashion. This, in turn, allows large-scale optimization of parameters affecting hybridization in diagnostic assays including probe coverage, target concentration, probe and target sequence, buffer ionic strength, and temperature [41]. As a first step in this direction, it is important to examine the relationship between the concentration of DNA target in solution and the magnitude of the sensor output signal.

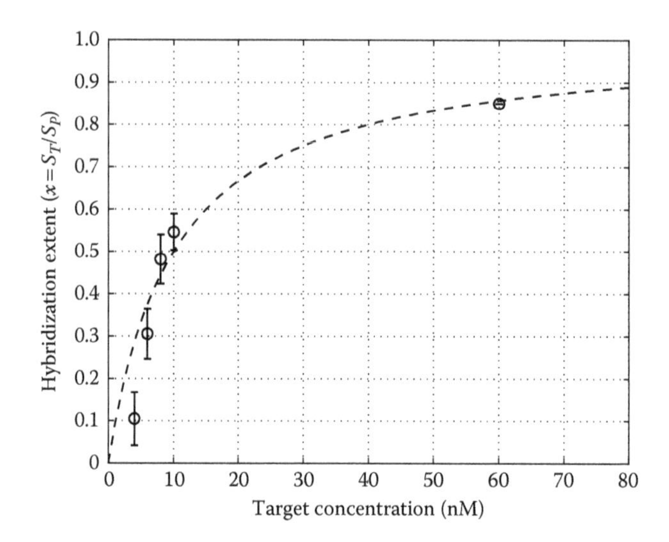

FIGURE 2.39 Results of a DNA target concentration, measured at one $100 \times 100\ \mu m^2$ WE, in which the hybridization extent x is plotted as a function of the solution target concentration. The dashed line is the Langmuir fit to the measured equilibrium isotherm. Errors bars for the four lowest target concentrations indicate the standard deviation from three separate experiments.

Figure 2.39 displays the results of a target concentration series, measured at one $100 \times 100\ \mu m^2$ WE, in which the extent of hybridization $x = S_T/S_P$, is plotted as a function of the solution target concentration. By fitting the measured data in the figure according to Equation 2.3, K_a is found to be approximately $1 \times 10^8\ M^{-1}$. This value of K_a falls in the range determined in previous studies of surface-based DNA sensing assays (10^7–$10^9\ M^{-1}$) [41].

Measurement of the standard deviation of the charging current background in Figure 2.36 provides some indication of the experimental detection limit of the sensor. With a measured standard deviation of approximately 2.1 nA (or, equivalently, a current noise variance of $4.4 \times 10^{-18}\ A^2$), the smallest target coverage that could be measured with an SNR of 3 is about 3.9×10^{11} targets cm^{-2}, based on solving for S_T in the redox current term in Equation 2.17. This value is about five times larger than the theoretical limit found in Section 2.3.7.7 using a 9-bit ADC. This discrepancy may be due to the fact that the additional noise from the second integration phase of the dual-slope ADC was neglected in the analysis in addition to the flicker noise contribution from the control amplifier.

2.4.5.4.3 Real-Time DNA Sensing

Figure 2.40 demonstrates observation of DNA probe–target hybridization in real time when 60 nM of T1 is hybridized to complementary P1 at one of the $100 \times 100\ \mu m^2$ WEs. In the figure, a CV scan is made every 5 min with the cell potential held at 0 V between scans. An increase in the area of the redox target signal is evident over time. Although equilibrium has not been reached after 35 min in the figure, it is expected that the maximum extent of hybridization will be about $7.0 \times 10^{12}\ cm^{-2}$.

Real-time detection permits the study of DNA hybridization kinetics, which is not possible using traditional fluorescence-based DNA microarrays. In addition, it may

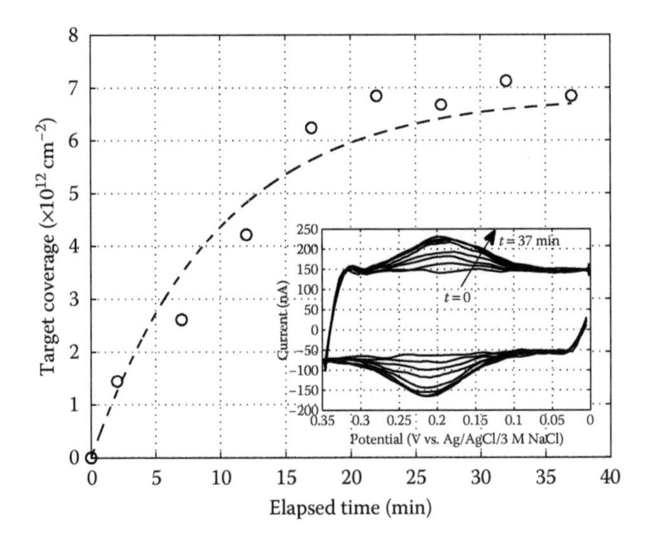

FIGURE 2.40 Real-time sensing of DNA hybridization at one 100 × 100 μm² WE in which a CV scan is run every 5 min. The measured data are fit to a first-order rate equation (*dashed line*) following Langmuir kinetics. Inset shows the results from each CV scan over time.

be possible to predict the extent of hybridization at equilibrium *before* this point is actually reached experimentally. This could significantly reduce the waiting period before readout, which is of great benefit to time-critical POC diagnostic applications.

The hybridization kinetics based on the data in Figure 2.40 are determined by ignoring mass-transport limitations, the effects of finite reaction volume, and interactions among surface sites, and by assuming that no DNA probes have hybridized at time $t = 0$. Therefore, the coverage of DNA duplexes on the WE surface as a function of time $S_T(t)$ can be expressed as [48]

$$S_T(t) = \left(\frac{S_P C_T}{K_a^{-1} + C_T}\right)\left[1 - \exp\left(\frac{-t}{\tau}\right)\right]. \tag{2.44}$$

The time constant τ over which the system reaches equilibrium is given by

$$\tau = \frac{1}{k_f\left(C_T + K_a^{-1}\right)}, \tag{2.45}$$

where $K_a = k_f/k_r$.

Performing a nonlinear, least-squares fit of the real-time curve using Equations 2.44 and 2.45 (also plotted in Figure 2.40) gives a τ of about 590 s. From this and the value of K_a determined previously, k_f and k_r are calculated to be 2.4×10^4 M^{-1} s^{-1} and 2.4×10^{-4} s^{-1}, respectively. These forward and reverse rate constants fall in the same order of magnitude as those observed by others using QCM [96] and surface plasmon fluorescence spectroscopy [97]. However, the K_a measured here is relatively smaller, most likely because avidin–biotin spacers are used for probe immobilization in the referenced works.

2.4.5.4.4 *Genetic Mutation Detection*

Sensor specificity in the presence of SNPs is important when screening for genetic mutations that could give rise to human disease. Figure 2.41 displays the results of two separate experiments in which the same density of P1 is immobilized on a $100 \times 100\ \mu m^2$ WE. In the first experiment, P1 is hybridized to 6 nM of the fully complementary T1, and in the second, P1 is hybridized to the same concentration of T1s. Since the latter target is one base shy of having full complementarity with P1, these ssDNA molecules have less affinity for one another compared with that of T1 for P1. As a result, fewer T1s targets hybridize to P1.

This can be easily seen in Figure 2.41, in which the peak current level for the fully complementary target is noticeably higher than that for the target with an SNP, when both are measured at equilibrium after 60 min following target addition. In the former case, the surface target coverage is approximately $3.0 \times 10^{12}\ cm^{-2}$, whereas it is about $1.6 \times 10^{12}\ cm^{-2}$ in the latter. This demonstrates the use of the active CMOS biochip for genetic mutation detection.

2.4.5.4.5 *Multiplexed and Specific DNA Detection*

Multiplexed and specific DNA sensing using the active CMOS biochip is carried out by functionalizing the chip with two distinct probes and then hybridizing each with its complementary target. Probes P1 and P2 are spotted on four different WEs, each using a fluid microinjection system (IM-300, Narishige, East Meadow, NY) capable of delivering nL volumes of probe solution to the electrode surface.

Initially, 6 nM of target T1 is introduced, and Figure 2.42 shows the response at one of the $100 \times 100\ \mu m^2$ WEs functionalized with P1 (denoted "site A") after 60 min. A distinct current peak can be seen, indicating that hybridization has occurred. The

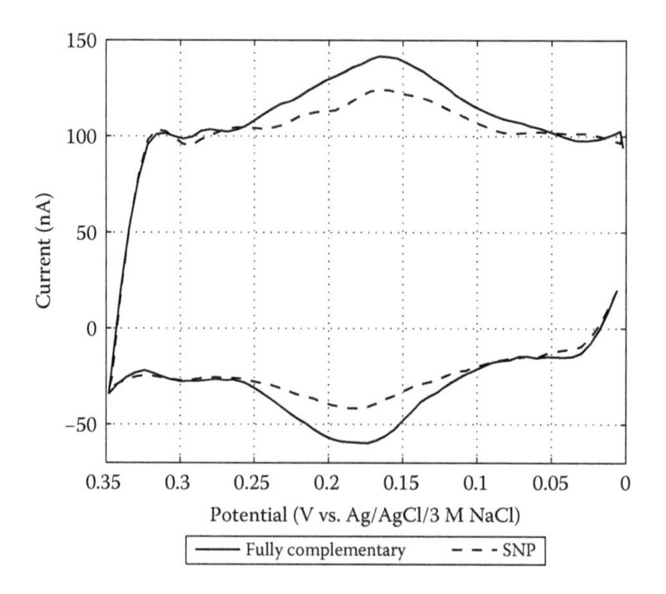

FIGURE 2.41 Detection of fully complementary target using the active CMOS biochip and that possessing a single-nucleotide polymorphism compared with the probe.

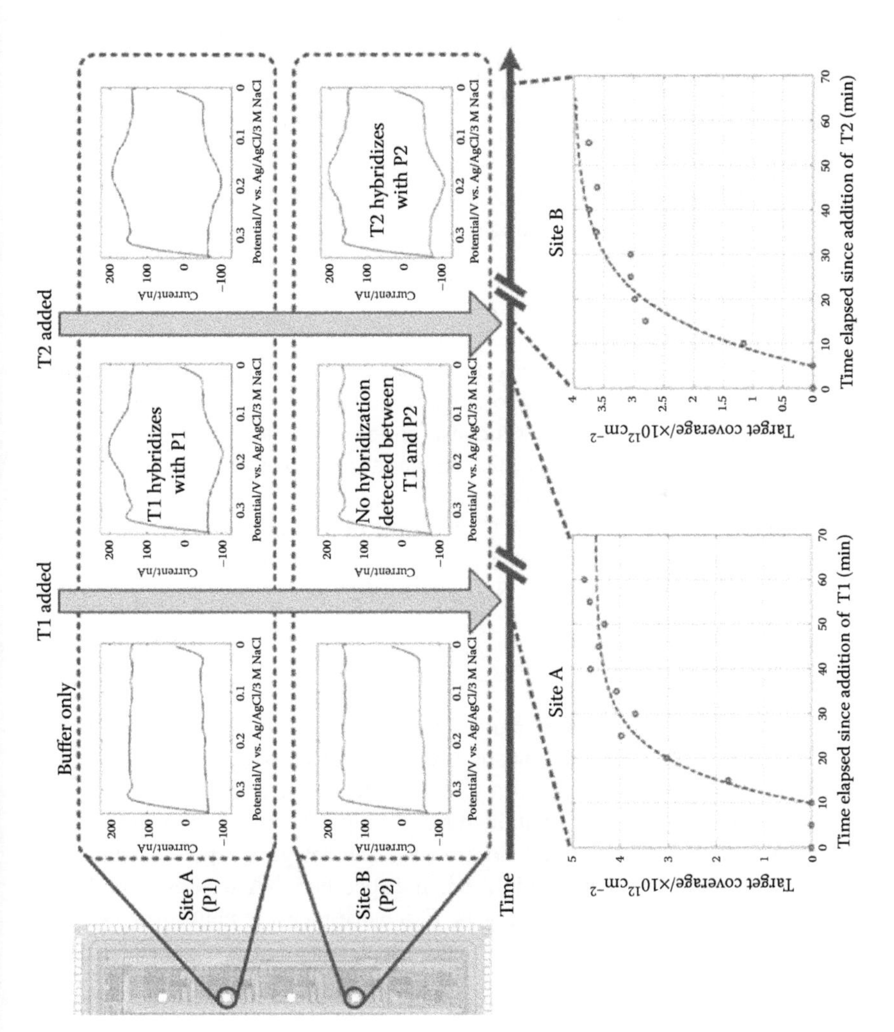

FIGURE 2.42 Demonstration of real-time, multiplexed, and specific DNA detection using the active CMOS biochip.

remaining WEs functionalized with P1 show similar behavior. Conversely, those WEs on which P2 is immobilized do not exhibit a hybridization signal, as P2 and T1 have little affinity for one another. The output from the sensor at one of the $100 \times 100\ \mu m^2$ WEs functionalized with P2 (denoted "site B") is also displayed in Figure 2.42. Next, 6 nM of target T2 is introduced. After 55 min, the hybridization signal at site B is evident, as shown in the figure. The signal at site A (measured at the same time) has not changed, however, since T1 is still present in solution. A separate experiment confirmed that sites functionalized with P1 do not exhibit any hybridization when T2 is added to the buffer first. Figure 2.42 also shows the measured target coverage at sites A and B in real time. The values of τ for the hybridization processes are approximately 540 and 740 s at site A and B, respectively. The slight shift of the data relative to the origin is attributed to mass-transport limitations of targets to the WE surface at the beginning of hybridization. These data demonstrate that the active CMOS biochip is capable of performing real-time monitoring of DNA hybridization in a multiplexed fashion.

2.4.5.4.6 Surface Heating Effects

Excessive heating of the WE surfaces on the active CMOS biochip could significantly reduce the extent of hybridization observed. The amount of heating is determined by such factors as the total power consumption of the biochip, chip packaging, and the volume of fluid in the reservoir. Temperature measurement of the analyte buffer, just above the chip surface, when all 16 sensor sites are operating simultaneously reveals an increase of about 3.5°C above the quiescent temperature (≈28°C). However, even with this increase, the absolute temperature is well below the melting temperature of the various DNA strands used (≈40°C).

2.5 SUMMARY

Active CMOS-integrated electrochemical biochips are an enabling technology in the development of genetic diagnostic platforms for clinical and POC medical applications. This chapter presented a CMOS biochip intended for use in these diagnostic systems. The biochip contained an on-chip 4×4 array of postprocessed Au WEs, each connected to a dual-slope ADC, and integrated potentiostat electronics to perform quantitative, multiplexed, and specific DNA sensing in real time using ferrocene-conjugated ssDNA targets. Extensive measurement results involving real-time electrochemical DNA detection demonstrated the proper functioning of this device.

REFERENCES

1. J. Shendure, R. D. Mitra, C. Varma, and G. M. Church, Advanced sequencing technologies: Methods and goals, *Nature Reviews. Genetics*, 5, 335–344, 2004.
2. International Human Genome Sequencing Consortium, Initial sequencing and analysis of the human genome, *Nature*, 409, 860–921, 2001.
3. J. C. Venter, M. D. Adams, E. W. Myers, P. W. Li, R. J. Mural, G. G. Sutton, and H. O. Smith, et al., The sequence of the human genome, *Science*, 291, 1304–1351, 2001.

4. R. H. Waterston, E. S. Lander, and J. E. Sulston, On the sequencing of the human genome, *Proceedings of the National Academy of Sciences USA*, 99(6), 3712–3716, 2002.
5. F. Sanger and A. R. Coulson, A rapid method for determining sequences in DNA by primed synthesis with DNA polymerase, *Journal of Molecular Biology*, 94(3), 441–448, 1975.
6. M. Ronaghi, S. Karamohamed, B. Pettersson, M. Uhlén, and P. Nyrén, Real-time DNA sequencing using detection of pyrophosphate release, *Analytical Biochemistry*, 242, 84–89, 1996.
7. M. Margulies, M. Egholm, W. E. Altman, S. Attiya, J. S. Bader, L. A. Bemben, J. Berka, et al., Genome sequencing in microfabricated high-density picolitre reactors, *Nature*, 437, 376–380, 2005.
8. J. M. Rothberg, W. Hinz, T. M. Rearick, J. Schultz, W. Mileski, M. Davey, J. H. Leamon, et al., An integrated semiconductor device enabling non-optical genome sequencing, *Nature*, 475(7356), 348–352, 2011.
9. F. S. Collins, M. Morgan, and A. Patrinos, The human genome project: Lessons from large-scale biology, *Science*, 300, 286–290, 2003.
10. J. M. Rothberg and J. H. Leamon, The development and impact of 454 sequencing, *Nature Biotechnology*, 26, 1117–1124, 2008.
11. M. Schena, D. Shalon, R. W. Davis, and P. O. Brown, Quantitative monitoring of gene expression patterns with a complementary DNA microarray, *Science*, 270, 467–470, 1995.
12. R. J. Lipshutz, D. Morris, M. Chee, E. Hubbell, M. J. Kozal, N. Shah, N. Shen, R. Yang, and S. P. A. Fodor, Using oligonucleotide probe arrays to access genetic diversity, *BioTechniques*, 19(3), 442–447, 1995.
13. M. Chee, R. Yang, E. Hubbell, A. Berno, X. C. Huang, D. Stern, J. Winkler, D. J. Lockhart, M. S. Morris, and S. P. A. Fodor, Accessing genetic information with high-density DNA arrays, *Science*, 274, 610–614, 1996.
14. X. Li, R. J. Quigg, J. Zhou, W. Gu, P. N. Rao, and E. F. Reed, Clinical utility of microarrays: Current status, existing challenges and future outlook, *Current Genomics*, 9, 466–474, 2008.
15. Microarrays find their way into the clinic, *Genetic Engineering & Biotechnology News*, 26(18), 2006.
16. J.-Y. Coppée, Do DNA microarrays have their future behind them? *Microbes and Infection*, 10, 1067–1071, 2008.
17. J. Kling, Moving diagnostics from the bench to the bedside, *Nature Biotechnology*, 24(8), 891–893, 2006.
18. U. A. Meyer, Pharmacogenetics and adverse drug reactions, *The Lancet*, 356, 1667–1671, 2000.
19. T. G. Drummond, M. G. Hill, and J. K. Barton, Electrochemical DNA sensors, *Nature Biotechnology*, 21(10), 1192–1199, 2003.
20. J. Wang, Electrochemical nucleic acid biosensors, *Analytica Chimica Acta*, 469, 63–71, 2002.
21. D. Johns and K. Martin, *Analog Integrated Circuit Design*. New York, NY: Wiley, 1997.
22. N. H. E. Weste and D. Harris, *CMOS VLSI Design: A Circuits and Systems Perspective*, 3rd edn. Boston, MA: Pearson, 2005.
23. J. Liu, B. A. Williams, R. M. Gwirtz, B. J. Wold, and S. Quake, Enhanced signals and fast nucleic acid hybridization by microfluidic chaotic mixing, *Angewandte Chemie International Edition*, 45, 3618–3623, 2006.
24. P. M. Levine, P. Gong, R. Levicky, and K. L. Shepard, Active CMOS sensor array for electrochemical biomolecular detection, *IEEE Journal of Solid-State Circuits*, 43(8), 1859–1871, 2008.

25. P. M. Levine, P. Gong, R. Levicky, and K. L. Shepard, Real-time, multiplexed electrochemical DNA detection using an active complementary metal-oxide-semiconductor biosensor array with integrated sensor electronics, *Biosensors and Bioelectronics*, 24(7), 1995–2001, 2009.
26. D. A. Skoog, F. J. Holler, and T. A. Nieman, *Principles of Instrumental Analysis*, 5th edn. Philadelphia, PA: Saunders, 1998.
27. J. S. Daniels and N. Pourmand, Label-free impedance biosensors: Opportunities and challenges, *Electroanalysis*, 19(12), 1239–1257, 2007.
28. M. Schena, R. A. Heller, T. P. Theriault, K. Konrad, E. Lachenmeier, and R. W. Davis, Microarrays: Biotechnology's discovery platform for functional genomics, *Trends in Biotechnology*, 16(7), 301–306, 1998.
29. C. A. Harrington, C. Rosenow, and J. Retief, Monitoring gene expression using DNA microarrays, *Current Opinion in Microbiology*, 3, 285–291, 2000.
30. C. Debouck and P. N. Goodfellow, DNA microarrays in drug discovery and development, *Nature Genetics*, 21, 48–50, 1999.
31. M. J. Cunningham, Genomics and proteomics: The new millennium of drug discovery and development, *Journal of Pharmacological and Toxicological Methods*, 44, 291–300, 2000.
32. J. G. Hacia, Resequencing and mutational analysis using oligonucleotide microarrays, *Nature Genetics*, 29, 42–47, 1999.
33. P. F. Macgregor and J. A. Squire, Application of microarrays to the analysis of gene expression in cancer, *Clinical Chemistry*, 48(8), 1170–1177, 2002.
34. G. Gibson, Microarrays in ecology and evolution: A preview, *Molecular Ecology*, 11, 17–24, 2002.
35. J. K. Peeters and P. J. Van der Spek, Growing applications and advancements in microarray technology and analysis tools, *Cell Biochemistry and Biophysics*, 43(1), 149–166, 2005.
36. M. Schena, *Microarray Analysis*. Hoboken, NJ: Wiley, 2003.
37. J. Liu, S. Tian, L. Tiefenauer, P. E. Nielsen, and W. Knoll, Simultaneously amplified electrochemical and surface plasmon optical detection of DNA hybridization based on ferrocene-streptavidin conjugates, *Analytical Chemistry*, 77(9), 2756–2761, 2005.
38. M. R. Linford and C. E. D. Chidsey, Alkyl monolayers covalently bonded to silicon surfaces, *Journal of the American Chemical Society*, 115(26), 12631–12632, 1993.
39. S. P. A. Fodor, R. P. Rava, X. C. Huang, A. C. Pease, C. P. Holmes, and C. L. Adams, Multiplexed biochemical assays with biological chips, *Nature*, 364, 555–556, 1993.
40. G. Patounakis, Active CMOS substrates for biomolecular sensing through time-resolved fluorescence detection, PhD. Dissertation, Columbia University, New York, NY, 2005.
41. R. Levicky and A. Horgan, Physicochemical perspectives on DNA microarray and biosensor technologies, *Trends in Biotechnology*, 23(3), 143–149, 2005.
42. R. P. Ekins and F. W. Chu, Multianalyte microspot immunoassay: Microanalytical "compact disk" of the future, *Clinical Chemistry*, 37(11), 1955–1967, 1991.
43. S. Draghici, P. Khatri, A. C. Eklund, and Z. Szallasi, Reliability and reproducibility issues in DNA microarray measurements, *Trends in Genetics*, 22(2), 101–109, 2006.
44. Affymetrix, Affymetrix GeneChip Scanner 3000 7G Data Sheet, 2005. scanner3000_datasheet.pdf. Available at: http://www.affymetrix.com/support/technical/datasheets/.
45. Molecular Devices. (2008) Axon GenePix 4300A and 4400A Scanners. gn_genepix_4300_4400.html. Available at: http://www.moleculardevices.com/Products/Instruments/Microarray-Scanners/GenePix-43004400.html.
46. G. Bhanot, Y. Louzoun, J. Zhu, and C. DeLisi, The importance of thermodynamic equilibrium for high throughput gene expression assays, *Biophysical Journal*, 84, 124–135, 2003.
47. A. W. Peterson, R. J. Heaton, and R. M. Georgiadis, The effect of surface probe density on DNA hybridization, *Nucleic Acids Research*, 29(24), 5163–5168, 2001.

48. H. Dai, M. Meyer, S. Stepaniants, M. Ziman, and R. Stoughton, Use of hybridization kinetics for differentiating specific from non-specific binding to oligonucleotide microarrays, *Nucleic Acids Research*, 30(16), e86, 2002.
49. H. Vikalo, B. Hassibi, and A. Hassibi, A statistical model for microarrays, optimal estimation algorithms, and limits of performance, *IEEE Transactions on Signal Processing*, 54(6), 2444–2455, 2006.
50. J. Wang, *Analytical Electrochemistry*, 3rd edn. Hoboken, NJ: Wiley, 2006.
51. A. J. Bard and L. R. Faulkner, *Electrochemical Methods: Fundamentals and Applications*, 2nd edn. New York, NY: Wiley, 2001.
52. E. Laviron, The use of linear potential sweep voltammetry and of a.c. voltammetry for the study of the surface electrochemical reaction of strongly adsorbed systems and of redox modified electrodes, *Journal of Electroanalytical Chemistry*, 100, 263–270, 1979.
53. K. Hashimoto, K. Ito, and Y. Ishimori, Sequence-specific gene detection with a gold electrode modified with DNA probes and an electrochemically active dye, *Analytical Chemistry*, 66(21), 3830–3833, 1994.
54. A. Sassolas, D. Leca-Bouvier, and L. J. Blum, DNA biosensors and microarrays, *Chemical Reviews*, 108, 109–139, 2008.
55. T. M. Herne and M. J. Tarlov, Characterization of DNA probes immobilized on gold surfaces, *Journal of the American Chemical Society*, 119(38), 8916–8920, 1997.
56. R. Levicky, T. M. Herne, M. J. Tarlov, and S. K. Satija, Using self-assembly to control the structure of DNA monolayers on gold: A neutron reflectivity study, *Journal of the American Chemical Society*, 120(38), 9787–9792, 1998.
57. J. J. Gooding, Electrochemical DNA hybridization biosensors, *Electroanalysis*, 14(7), 1149–1156, 2002.
58. T. de Lumley-Woodyear, C. N. Campbell, and A. Heller, Direct enzyme-amplified electrical recognition of a 30-base model oligonucleotide, *Journal of the American Chemical Society*, 118(23), 5504–5505, 1996.
59. L. Alfonta, A. K. Singh, and I. Willner, Liposomes labeled with biotin and horseradish peroxidase: A probe for the enhanced amplification of antigen-antibody or oligonucleotide-DNA sensing processes by the precipitation of an insoluble product on electrodes, *Analytical Chemistry*, 73(1), 91–102, 2001.
60. J. Wang, D. Xu, A.-N. Kawde, and R. Polsky, Metal nanoparticle-based electrochemical stripping potentiometric detection of DNA hybridization, *Analytical Chemistry*, 71(22), 5576–5581, 2001.
61. M. Ozsoz, A. Erdem, K. Kerman, D. Ozkan, B. Tugrul, N. Topcuoglu, H. Ekren, and M. Taylan, Electrochemical genosensor based on colloidal gold nanoparticles for the detection of Factor V Leiden mutation using disposable pencil graphite electrodes, *Analytical Chemistry*, 75(9), 2181–2187, 2003.
62. M. Chahma, J. S. Lee, and H.-B. Kraatz, Fc-ssDNA conjugate: Electrochemical properties in a borate buffer and adsorption on gold electrode surfaces, *Journal of Electroanalytical Chemistry*, 567, 283–287, 2004.
63. P. He, Y. Xu, and Y. Fang, A review: Electrochemical DNA biosensors for sequence recognition, *Analytical Letters*, 38, 2597–2623, 2005.
64. C. Xu, P. He, and Y. Fang, Electrochemical labeled DNA probe for the detection of sequence-specific DNA, *Analytica Chimica Acta*, 411, 31–36, 2000.
65. C. J. Yu, Y. Wan, H. Yowato, J. Li, C. Tao, M. D. James, C. L. Tan, G. F. Blackburn, and T. J. Meade, Electronic detection of single-base mismatches in DNA with ferrocene-modified probes, *Journal of the American Chemical Society*, 123(45), 11155–11161, 2001.
66. R. M. Umek, S. W. Lin, J. Vielmetter, R. H. Terbrueggen, B. Irvine, C. J. Yu, J. F. Kayyem, et al., Electronic detection of nucleic acids: A versatile platform for molecular diagnostics, *Journal of Molecular Diagnostics*, 3(2), 73–84, 2001.

67. S. D. Vernon, D. H. Farkas, E. R. Unger, V. Chan, D. L. Miller, Y.-P. Chen, G. F. Blackburn, and W. C. Reeves, Bioelectronic DNA detection of human papillomaviruses using eSensor: A model system for detection of multiple pathogens, *BMC Infectious Diseases*, 3(6), 2003.
68 GenMark (2014) The eSensor® XT-8 System. Available at http://www.genmarkdx.com/products/systems/index.php.
69. J. Li, M. Xue, Z. Zhang, C. Feng, and M. Chan, A high-density conduction-based micro-DNA identification array fabricated with a CMOS compatible process, *IEEE Transactions on Electron Devices*, 50(10), 2165–2170, 2003.
70. Y.-T. Cheng, C.-Y. Tsai, and P.-H. Chen, Development of an integrated CMOS DNA detection biochip, *Sensors and Actuators B: Chemical*, 120, 758–765, 2007.
71. M. Barbaro, A. Bonfiglio, L. Raffo, A. Alessandrini, P. Facci, and I. Baràk, A CMOS fully integrated sensor for electronic detection of DNA hybridization, *IEEE Electron Device Letters*, 27(7), 595–597, 2006.
72. J.-K. Shin, D.-S. Kim, H.-J. Park, and G. Lim, Detection of DNA and protein molecules using an FET-type biosensor with gold as a gate metal, *Electroanalysis*, 16(22), 1912–1918, 2004.
73. A. L. Ghindilis, M. W. Smith, K. R. Schwarzkopf, K. M. Roth, K. Peyvan, S. B. Munro, M. J. Lodes, et al., CombiMatrix oligonucleotide arrays: Genotyping and gene expression assays employing electrochemical detection, *Biosensors and Bioelectronics*, 22, 1853–1860, 2007.
74. M. Schienle, C. Paulus, A. Frey, F. Hofmann, B. Holzapfl, P. Schindler-Bauer, and R. Thewes, A fully-electronic DNA sensor with 128 positions and in-pixel A/D conversion, *IEEE Journal of Solid-State Circuits*, 39(12), 2438–2445, 2004.
75. F. Hofmann, A. Frey, B. Holzapfl, M. Schienle, C. Paulus, P. Schindler-Bauer, D. Kuhlmeier, et al., Fully electronic DNA detection on a CMOS chip: Device and process issues, in *Proceedings of the International Electron Devices Meeting*, pp. 488–491, 8–11 December 2002, IEEE, Piscataway, NJ.
76. N. Gemma, S. O'uchi, H. Funaki, J. Okada, and S. Hongo, CMOS integrated DNA chip for quantitative DNA analysis, in *Proceedings of the IEEE International Solid-State Circuits Conference Digest of Technical Papers*, pp. 560–561, May 2006, IEEE, Piscataway, NJ.
77. A. Hassibi and T. H. Lee, A programmable electrochemical biosensor array in 0.18 μm standard CMOS, in *Proceedings of the IEEE International Solid-State Circuits Conference Digest of Technical Papers*, pp. 564–565, 6–10 February 2005, IEEE, Piscataway, NJ.
78. A. Hassibi and T. H. Lee, A programmable 0.18-μm CMOS electrochemical sensor microarray for biomolecular detection, *IEEE Sensors Journal*, 6(6), 1380–1388, 2006.
79. C. Stagni, C. Guiducci, L. Benini, B. Riccò, S. Carrara, B. Samorí, C. Paulus, M. Schienle, M. Augustyniak, and R. Thewes, CMOS DNA sensor array with integrated A/D conversion based on label-free capacitance measurement, *IEEE Journal of Solid-State Circuits*, 41(12), 2956–2964, 2006.
80. F. Heer, M. Keller, G. Yu, J. Janata, M. Josowicz, and A. Hierlemann, CMOS electrochemical DNA-detection array with on-chip ADC, in *Proceedings of the IEEE International Solid-State Circuits Conference Digest of Technical Papers*, pp. 168–169, 3–7 February 2008, IEEE, Piscataway, NJ.
81. E. Anderson, J. Daniels, H. Yu, T. Lee, and N. Pourmand, A label-free CMOS DNA microarray based on charge sensing, in *Proceedings of the IEEE Instrumentation and Measurement Technology Conference Proceedings*, pp. 1631–1636, 12–15 May 2008, IEEE, Piscataway, NJ.
82. P. Gong and R. Levicky, DNA surface hybridization regimes, *Proceedings of the National Academy of Sciences U S A*, 105(14), 5301–5306, 2008.

83. R. J. Reay, S. P. Kounaves, and G. T. A. Kovacs, An integrated CMOS potentiostat for miniaturized electroanalytical instrumentation, in *Proceedings of the IEEE International Solid-State Circuits Conference Digest of Technical Papers*, pp. 162–163, 16–18 February 1994, IEEE, Piscataway, NJ.
84. R. G. Kakerow, H. Kappert, E. Spiegel, and Y. Manoli, Low-power single-chip CMOS potentiostat, in *Proceedings of the 8th International Conference on Solid-State Sensors and Actuators, and Eurosensors IX*, pp. 142–145, 25–29 June 1995, IEEE, Piscataway, NJ.
85. M. Breten, T. Lehmann, and E. Bruun, Integrating data converters for picoampere currents from electrochemical transducers, in *Proceedings of the IEEE International Symposium on Circuits and Systems*, pp. 709–712, 28–31 May 2000, IEEE, Piscataway, NJ.
86. H. S. Narula and J. G. Harris, A time-based VLSI potentiostat for ion current measurements, *IEEE Sensors Journal*, 6(2), 239–247, 2006.
87. R. Genov, M. Stanacevic, M. Naware, G. Cauwenberghs, and N. V. Thakor, 16-Channel integrated potentiostat for distributed neurochemical sensing, *IEEE Transactions on Circuits and Systems I, Fundamental Theory and Applications*, 53(11), 2371–2376, 2006.
88. B. Razavi, *Design of Analog CMOS Integrated Circuits*. New York, NY: McGraw-Hill, 2001.
89. Y. Tsividis, *Operation and Modeling of the MOS Transistor*, 2nd edn. New York, NY: Oxford University Press, 1999.
90. G. Shen, N. Tercero, M. A. Gaspar, B. Varughese, K. Shepard, and R. Levicky, Charging behavior of single-stranded DNA polyelectrolyte brushes, *Journal of the American Chemical Society*, 128(26), 8427–8433, 2006.
91. U. Bertocci and F. Huet, Noise analysis applied to electrochemical systems, *Corrosion Science*, 51(2), 131–144, 1995.
92. Y. Tu, G. Stolovitzky, and U. Klein, Quantitative noise analysis for gene expression microarray experiments, *Proceedings of the National Academy of Sciences USA*, 99(22), 14031–14036, 2002.
93. S. M. Martin, T. D. Strong, and R. B. Brown, Monolithic liquid chemical sensing systems, in *Materials Research Society Symposium Proceedings Volume 869: Materials, Integration and Technology for Monolithic Instruments*, pp. 109–118, March 2005, IEEE, Piscataway, NJ.
94. J. R. Vig, UV/ozone cleaning of surfaces, *Journal of Vacuum Science and Technology A*, 3(3), 1027–1034, 1985.
95. S. M. Martin, CMOS-integrated liquid chemical microdetection systems, PhD. Dissertation, University of Michigan, Ann Arbor, MI, 2005.
96. Y. Okahata, M. Kawase, K. Niikura, F. Ohtake, H. Furusawa, and Y. Ebara, Kinetic measurements of DNA hybridization on an oligonucleotide-immobilized 27-MHz quartz crystal microbalance, *Analytical Chemistry*, 70(7), 1288–1296, 1998.
97. K. Tawa, D. Yao, and W. Knoll, Matching base-pair number dependence of the kinetics of DNA-DNA hybridization studied by surface plasmon fluorescence spectroscopy, *Biosensors and Bioelectronics*, 21(2), 322–329, 2005.

3 Label-Free DNA Sensor Based on a Surface Long-Period Grating

Young-Geun Han

Biosensors have been attracting much attention in a variety of applications for biomarker detection in medical diagnostics and pathogen or toxin detection in food and water. Conventional biosensors based on the fluoroimmunoassay technique should exploit the fluorescence labeling of the antigen or target DNA and always require additional reagents [1]. Ordinary biosensors have many disadvantages, such as high cost, a complicated configuration, and the impossibility of real-time detection. To overcome these drawbacks, many methods have been proposed for developing label-free detection biosensors [2,3]. In particular, fiber-optic biosensors are promising devices for high-quality, label-free detection because they can overcome the limitations of bulk refractometers through their high sensitivity, fast response, convenience for *in situ* and remote sensing over long distances, and immunity to electromagnetic interference [4–8]. Various fiber-optic biosensors using the surface plasmon resonance (SPR) phenomenon have been proposed [1,9,10]. However, it is necessary to precisely design and fabricate the physical structure of SPR biosensors because SPR properties are highly sensitive to metal content, thickness, and biomolecules. Recently, biosensors based on fiber gratings have been widely investigated [11–15]. Most fiber grating–based biosensors have utilized long-period gratings (LPGs) with radiation mode coupling at resonance wavelengths that are very sensitive to variation in the external medium [16]. One simple technique for producing fiber grating–based index sensors is to use LPGs based on cladding mode coupling [4]. To improve the sensitivity of LPG-based biosensors, a variety of devices including colloidal gold modified LPGs [13], LPGs with etched cladding [14], and LPGs with an nanostructured overlayer [15] have been proposed. However, modifying LPGs in these ways is difficult because of the delicate and hazardous procedures and the requirement for additional materials. A more effective method to induce evanescent field coupling is based on using a D-shaped fiber. If part of the cladding region is removed, the external index change can directly interact with the evanescent field of the fundamental mode [17]. To enhance the performance of the D-shaped fiber-based biosensor, it was proposed that the surface LPG be inscribed in the D-shaped fiber to induce resonant coupling [18–22].

Two polishing processes were employed to fabricate a D-shaped fiber [19,23]. A single-mode fiber (SMF) is positioned on a quartz block. The quartz block has a

V-shaped groove with a curvature of 25 cm. The SMF is then fixed in the V-groove using an epoxy adhesive (NOA81). The properties of the SMF are as followed: relative index difference = 0.35%, core diameter = 8 μm, cladding diameter = 125 μm, and cutoff wavelength = 1180 nm. In the first polishing process, the SMF was placed on a brass plate and ground using a diamond suspension with grit sizes of 5, 16, and 27 μm [21,22]. The polished surface was then washed using deionized water and dried at 100°C for 10 min. During the first polishing process, we measured a transmission loss of 0.5 dB, which was caused by the roughness of the flat surface. Three D-shaped fibers (Samples 2, 3, and 4) were fabricated using the first polishing process. The remaining length of the cladding region is the distance between the core and the cladding after polishing the cladding region of the optical fibers. We indicated the remaining length of the cladding region in Figure 3.5. The second polishing process made the surface of the D-shaped fiber smoother. During the second polishing procedure, the first polished fiber was placed on a polyurethane plate and polished using a CeO_2 powder. The polished surface was washed using deionized water and dried at 100°C for 10 min. Sample 1 was fabricated by using the first and the second polishing processes simultaneously.

Figure 3.1 shows two-dimensional (2D) and 3D atomic force microscopy (AFM) micrographs of the polished surfaces of D-shaped fibers: (a) Sample 1 (root-mean-square [RMS] roughness [R_q] = 0.8 nm), (b) Sample 2 (R_q = 81.3 nm), (c) Sample 3 (R_q = 113.5 nm), and (d) Sample 4 (R_q = 172.0 nm). After ultrasonic cleaning with deionized water, we dried all samples at 100°C for 10 min. Next, each sample was mounted upright using a specially designed AFM holder. The surface roughness at the center of the polished surface for versatile D-shaped fibers was measured over an area of 4×4 mm^2. The 2D surface topography of Sample 1 was clearly the smoothest. The 2D surface topography of Sample 2 was also relatively smooth, and the valleys on the surface were much shallower than those of Samples 3 and 4. The surface of Sample 3 had valleys of moderate depth, and Sample 4 had the deepest valleys. These differences are also shown in the 3D images. The 3D surface topography of Sample 1 was plane, while the surface topographies of Samples 2, 3, and 4 were more rugged. The surfaces of Samples 2 and 3 had some gentle hills of moderate height, and the surface of Sample 4 had some cliffy, relatively high hills. The results of the experiments are shown in Table 3.1. The R_q is an average value of the surface roughness. The surfaces of the samples became rougher as the grit size was increased [23].

Figure 3.2 shows optical microscope photographs of the light scattered from the polished surfaces of D-shaped fibers with different surface roughness. Light from a He-Ne laser at a wavelength of 633 nm was directed into the core region of D-shaped fibers. In the image of Sample 1, the light scattered from the polished surface was invisible because the total internal reflection (TIR) condition was satisfied by removing the surface roughness. The light scattered from D-shaped fibers increased as the surface roughness increased because the intensity of the scattered evanescent field is proportional to the surface roughness of the D-shaped fibers. The image of Sample 2 shows that the scattered light was uniformly radiated from the relatively smooth surface, and the images of Samples 3 and 4 show that the scattered light had some circle fringes due to the high surface roughness of the D-shaped fibers.

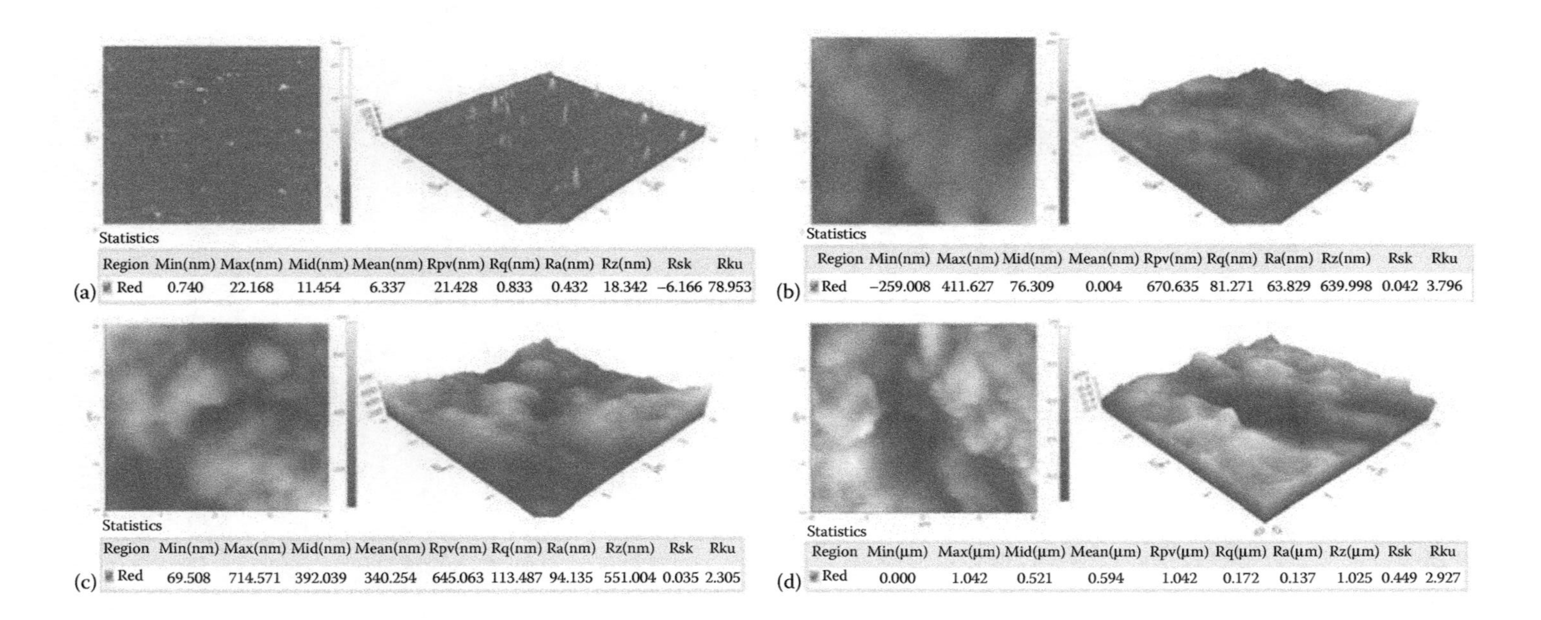

FIGURE 3.1 Two- and three-dimensional AFM micrographs of the flat surfaces of D-shaped fibers: (a) Sample 1 (R_q: 0.8 nm), (b) Sample 2 (R_q: 81.3 nm), (c) Sample 3 (R_q: 113.5 nm), and (d) Sample 4 (R_q: 172.0 nm). (From Kim, H.J., Kwon, O.J., and Han, Y.G., *Journal of the Korean Physical Society*, 56, 1355–1358, 2010.)

TABLE 3.1
Abrasive Diameters and Root-Mean-Square Roughness Values

Sample No.	Polishing Process	Diameter of Abrasives (μm)	RMS Roughness (R_q) (nm)
1-(a)	1st and 2nd polishing	<0.5	0.8
2-(b)	1st polishing	5.0	81.3
3-(c)	1st polishing	16.0	113.5
4-(d)	1st polishing	27.0	172.0

Figure 3.3a shows the transmission characteristics of various D-shaped fibers for changes in the ambient index for Samples 2, 3, and 4, which had different surface roughness values of 81.3, 113.5, and 172.0 nm, respectively. The sensitivities of the D-shaped fibers to variations in the ambient index were significantly decreased as the surface roughness was increased. Since the surface roughness induces additional loss, evanescent field coupling of the core mode to the modes in the ambient index overlay is severely reduced. Consequently, the sensitivity of the transmission power to changes in the ambient index is degraded by increasing the surface roughness of the D-shaped fibers. Figure 3.3b shows the transmission characteristics of two

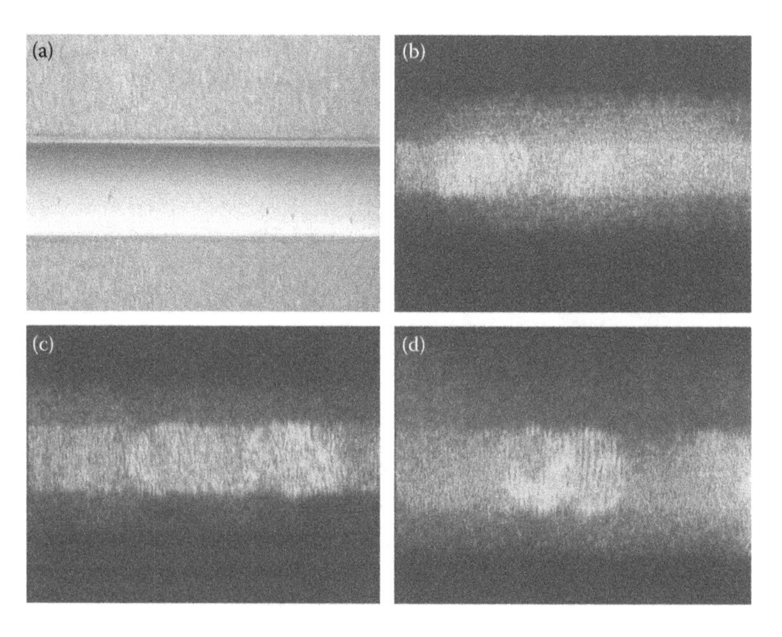

FIGURE 3.2 Optical microscope photographs of the light scattered from surfaces with different surface roughness values: (a) Sample 1 (R_q: 0.8 nm), (b) Sample 2 (R_q: 81.3 nm), (c) Sample 3 (R_q: 113.5 nm), and (d) Sample 4 (R_q: 172.0 nm). (From Kim, H.J., Kwon, O.J., and Han, Y.G., *Journal of the Korean Physical Society*, 56, 1355–1358, 2010.)

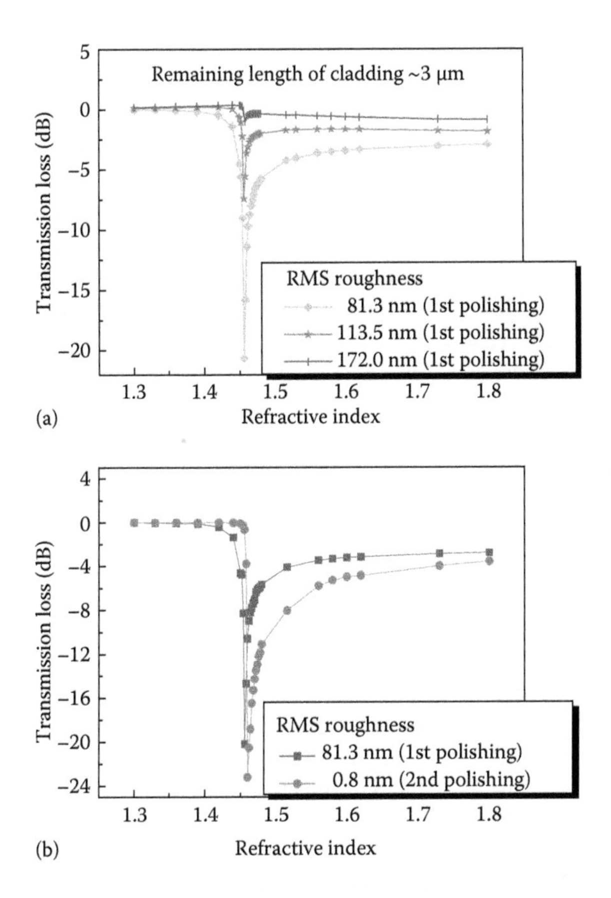

FIGURE 3.3 (a) Transmission loss of versatile D-shaped fibers as a function of the ambient index for various values of the surface roughness (Samples 2, 3, and 4) and (b) transmission loss as a function of the ambient index for various values of the surface roughness (Samples 1 and 2). (From Kim, H.J., Kwon, O.J., and Han, Y.G., *Journal of the Korean Physical Society*, 56, 1355–1358, 2010.)

D-shaped fibers (Samples 1 and 2) fabricated using two different polishing processes as the ambient index was changed. Sample 1 was fabricated using the first and second polishing processes simultaneously to remove the surface roughness. Sample 2 was fabricated using the first polishing process and had a relatively low surface roughness compared with the other samples. Since the surface roughness is effectively reduced by using the second polishing procedure, Sample 1 has a low surface roughness value. In the case of Sample 2, as the ambient index was increased from 1.3 to 1.45, the transmission loss was slightly increased due to the light scattered from the rough surface. For the case of Sample 1, however, as the ambient index was increased from 1.3 to 1.45, the transmission loss did not increase because the surface roughness had been removed, and the TIR condition had not been significantly modified. The sensitivity of transmission of the second polished fiber (Sample 1) was higher than that of the first polished fiber (Sample 2). It is evident that the performance of a chemical sensor based on a D-shaped fiber can be improved by reducing the surface roughness [23].

The coupling strength of the evanescent field is changed by using the remaining length of the cladding region in the D-shaped fiber, which is defined as the distance between the core and the cladding [19]. The coupling strength of the evanescent field between the core region and the flat surface of the D-shaped fiber can be estimated by measuring the transmission loss of the core mode. In the first polished fiber, the evanescent field of the core mode can be coupled to light scattered on the flat surface and transmission loss of the core mode occurs, which means that the coupling strength of the evanescent field from the core region to the flat surface of the D-shaped fiber is enhanced. The remaining length of the cladding region directly affects the coupling strength of the evanescent field. The coupling strength of the evanescent field increases as the remaining length of the cladding region is reduced. Therefore, the remaining length of the cladding region can be estimated by measuring the coupling strength of the evanescent field. As seen in Figure 3.4a, the coupling strength of the evanescent field is inversely proportional to the remaining length of the cladding region in the D-shaped fiber. As the coupling strength of the evanescent field of the SMF-based D-shaped fiber increases from 0.5 to 4 dB, the remaining length of the cladding region decreases from ~5 to ~−2 μm. The negative value of the remaining length means that the core region is partially removed.

Figure 3.4b and c show the transmission characteristics of the SMF-based D-shaped fiber for various values of the index after the first and the second polishing processes, respectively. In both cases, the coupling strength of the evanescent field mainly affects the transmission loss of the D-shaped fibers. When the ambient index of the overlay was increased from 1.3 to 1.4, the transmission loss was not degraded severely. Since the ambient index is lower than the effective index of the core mode in the D-shaped fiber, the TIR condition between the core region and the polished cladding region with a flat surface was not changed critically. The evanescent field coupling from the core to the cladding modes was not strong, and the transmission loss was not increased. However, the transmission loss became high once the ambient index neared the effective index of the core mode in the D-shaped fiber. Since the TIR condition between the core region and the polished cladding region is modified by the external index of the matching oil, most of the power of the core region is transferred to the overlay region. Therefore, the transmission loss is reduced when the ambient index is higher than the effective index of the core mode in the D-shaped fiber because the modified TIR condition becomes strong [19]. These phenomena were observed for all D-shaped fibers with different coupling strengths of the evanescent field. However, the sensitivity of the changes in the transmission loss to the changes in ambient index was improved as the coupling strength of the evanescent field increased, as shown in Figure 3.4. If the core region is polished and the remaining length of the cladding region is negative, the transmission loss becomes extremely high as the ambient index is similar to the effective index of the core mode, as seen in Figure 3.4d. Since the core region is directly affected by the ambient index change, the sensitivity to the ambient index changes increases.

Figure 3.5 shows the scheme for the surface LPG with periodic patterns of the photoresist on the surface of the D-shaped fiber. After optimizing the characteristics

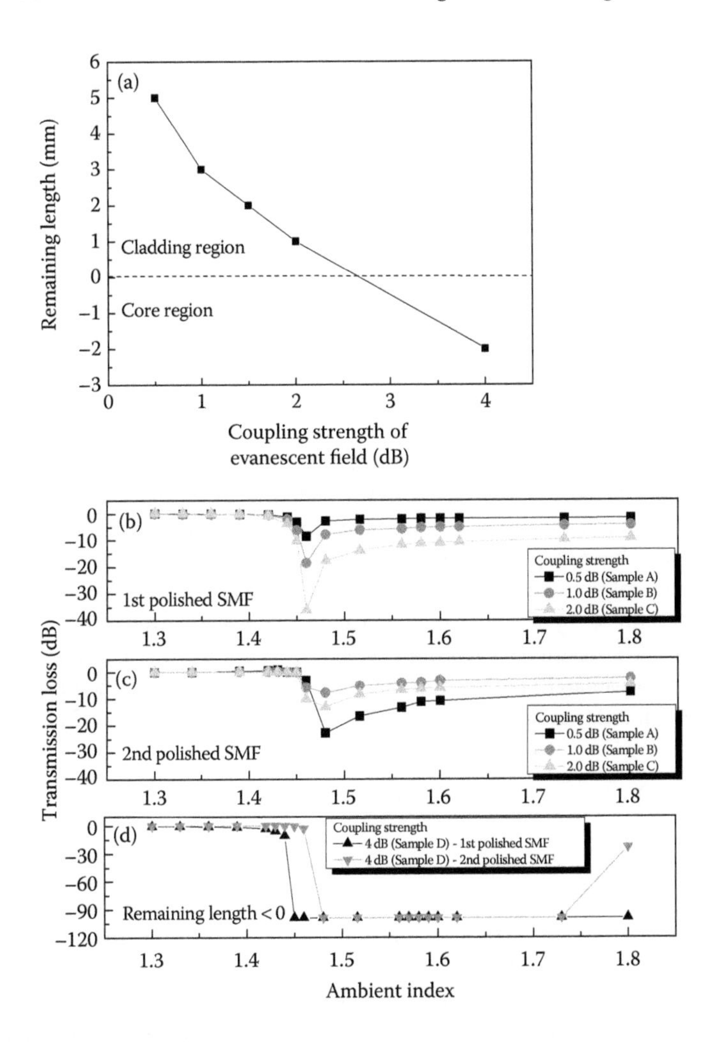

FIGURE 3.4 (a) Variation of the remaining length of the cladding region with the coupling strength of the evanescent field and transmission loss of the SMF-based D-shaped fiber as a function of the ambient index change: (b) the first polished fiber and (c) the second polished fiber. (d) Transmission loss of the SMF-based D-shaped fiber with a negative remaining length of the cladding region. (From Kim, H.J., Kwon, O.J., and Han, Y.G., *Journal of the Korean Physical Society*, 55, 1286–1289, 2009.)

of the D-shaped fiber by conducting the first and the second polishing procedures, we coated the photoresist on the surface of a D-shaped fiber using a spin coater. In order to induce periodic surface gratings, the photoresist on the surface of the D-shaped fiber covered by the shadow long-period mask was exposed to a UV lamp (295 nm) for 2 min. The sample was developed using a specific photoresist developer for 2 min, and the photoresist was periodically coated on the surface of the D-shaped fiber. After carrying out a postbaking process in a thermal oven for 30 min at 120°C, we achieved the surface LPG. The relevant physical parameters of the surface LPG

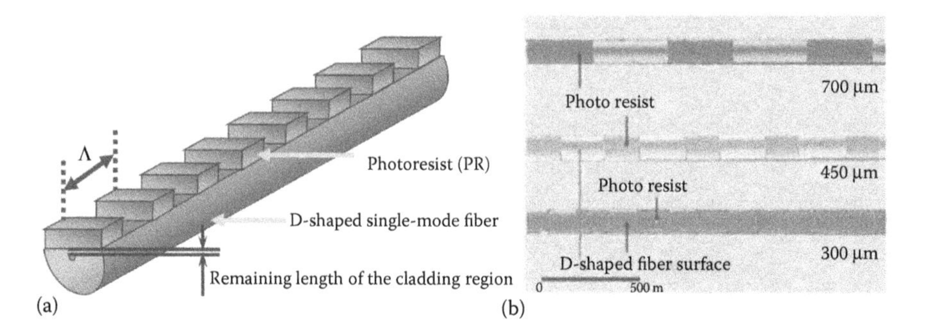

FIGURE 3.5 Structure (a) and photograph (b) of a surface LPG based on a D-shaped fiber.

are the refractive index and the thickness of the grating overlay based on the photoresist and the grating period. The resonant wavelengths in the surface LPG are induced at longer wavelengths as the three parameters increase. Since the refractive index of the grating overlay is higher than that of the core mode in the SMF, it couples the core mode to leaky modes excited by the surface LPG [18–22]. When the effective index of the *m*th leaky mode is matched with the effective index (n_{ef}) of the core mode in the SMF, resonant coupling between the core and the leaky mode occurs. The resonant wavelength (λ_m) of the surface LPG depending on the effective index (n_{eg}) of the *m*th order mode in the grating overlay, can be written as [21,22]

$$\lambda_m = \frac{2\pi d\sqrt{\left(n_{sgo}^2 - \left(n_{eg}^m\right)^2\right)}}{m\pi + \varphi}, \tag{3.1}$$

where d and n_{sgo} are the thickness and the refractive index of the surface grating overlay, respectively. Since the two polarization modes in the surface LPG have highly different dispersion relations, the phase-matching conditions are satisfied at different wavelengths. The phase shift (φ) associated with external index change can be expressed by

$$\varphi = \tan^{-1}\xi\frac{\sqrt{n_{eg}^2 - n_{ext}^2}}{\sqrt{n_{sgo}^2 - n_{eg}^2}}, \tag{3.2}$$

where ξ is a polarization-dependent constant and n_{ext} is an external index. In Equations 3.1 and 3.2, the external index (n_{ext}) decreases the amount of phase shift (φ), which contributes to the resonant wavelength shift to longer wavelengths. Based on a boundary condition of transverse electric (TE) and transverse magnetic (TM) modes, ξ can be determined to be 1 and $(n_{eg}/n_{ext})^2$ for TE and TM modes, respectively [21]. Since φ strongly depends on ξ, the ambient index sensitivities should be changed by the input polarization modes [21]. The ambient index sensitivity of TM mode is higher than that of TE mode because an electric field of TE mode is continuous at the boundary while an electric field of TM mode is discontinuous. That is, as the value of ξ for TM mode is larger than that of ξ for TE mode, the resonant wavelength shift of TM mode is more sensitively affected by the ambient refractive index change than that of TE mode [21]. The surface LPG can be also changed by temperature [22]. For temperature sensitivity, the resonant wavelength shift is dominantly caused by the thermo-optic effect of the surface grating overlay ($\Delta n_{sgo} = \kappa \Delta T$). The temperature sensitivity of the resonant wavelengths of the surface LPGs can be expressed by [22]

$$\frac{\partial \lambda_m}{\partial T} = \kappa \frac{2d}{m\left[1 - \left(\frac{n_{ef}^2}{n_{sgo}^2}\right)\right]^{1/2}}, \tag{3.3}$$

where κ is the thermo-optic coefficient of the surface grating overlay. Since the κ of the photoresist material has a negative thermo-optic coefficient, the resonant wavelength is shifted to a shorter wavelength as the applied temperature increases.

Figure 3.6a shows the transmission spectrum of the SMF-based surface LPG. The three different resonant peaks were generated in the transmission spectrum, which resulted from mode coupling between a fundamental core mode and three

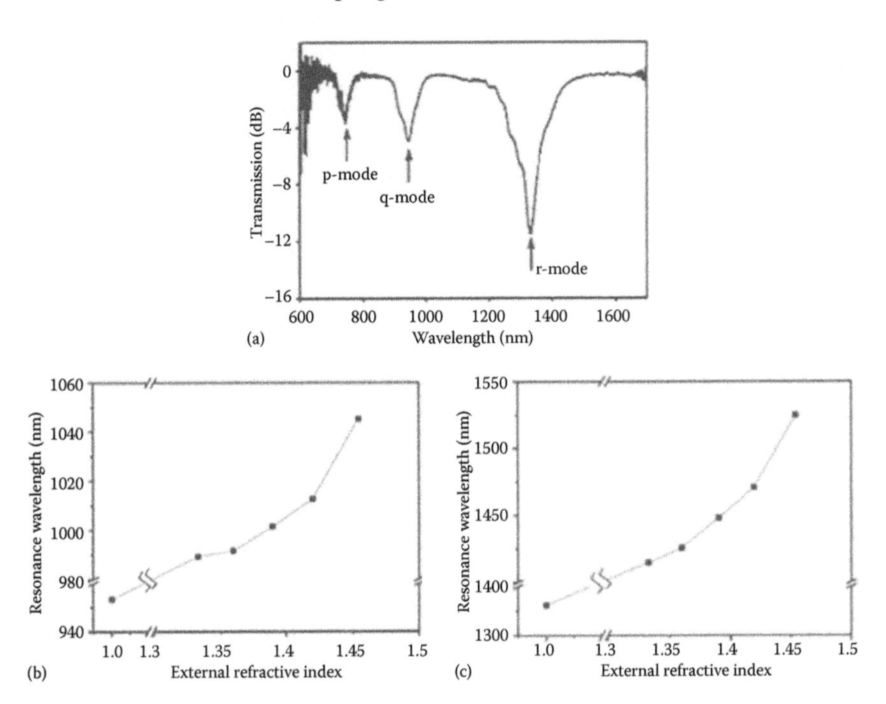

FIGURE 3.6 (a) Transmission spectrum of the SMF-based surface LPG, (b) resonance wavelengths of the q-mode, and (c) r-mode as functions of ambient indices. (From Jang, H.S., Park, K.N., Kim, J.P., Sim, S.J., Kwon, O.J., Han, Y.G., and Lee, K.S., *Optics Express*, 17, 3855–3860, 2009.)

leaky modes, such as p-, q-, and r-modes (named for convenience), in the surface LPG. Because the effective index of a high-order leaky mode is smaller than that of a low-order leaky mode, the mode number of the r-mode should be higher than that of the q-mode and the resonant peak of the r-mode should appear in a longer wavelength than the other two modes. The high extinction ratio for the r-leaky mode in Figure 3.6a indicates that strong coupling occurred in the surface LPG between the fundamental core mode and the higher-mode leaky order. Note that the resonance wavelengths, corresponding to the q- and the r-modes, are changed with variations in the external refractive indices as shown in Figure 3.6b. The increase in the ambient index causes the resonant wavelengths to shift to longer wavelengths. When the refractive index was changed in a range from 1.333 to 1.454, the resonance wavelengths for the q- and r-modes (as seen in Figure 3.6b and c, respectively) shifted toward a longer wavelength by 84.4 and 110 nm, respectively. The corresponding sensitivities were estimated to be 607.2 and 909.1 nm/RIU for the q- and r- modes, respectively. In this case, the resolutions of the biosensor were 1.6×10^{-5} and 1.1×10^{-5} RIU for the q- and r-modes, respectively. It is obvious that the r-mode has higher sensitivity than the q-mode because the higher-order modes have higher sensitivity than the low-order one [24].

Poly-L-lysine (PLL) has an extreme positive charge with NH_3^+ in the side chain, which is frequently utilized for adsorbing biomolecules with a negative charge like DNA (deoxynucleic acid) [25,26]. A PLL solution (0.1% w/v in water, molecular weight 150,000–300,000 g/mol; Sigma) was employed, which is commonly used in biology to treat glass slides. The surface LPG was initially cleaned using a phosphate-buffered saline solution (NaH_2PO_4/Na_2HPO_4 pH 7.4, 150 mM NaCl) (PBS) before being modified with poly-L-lysine. The PLL solution was applied to the surface LPG for 160 min at room temperature to make a PLL layer. The PLL layer functionalizes the surface LPG with an amino group at the free end, allowing negatively charged

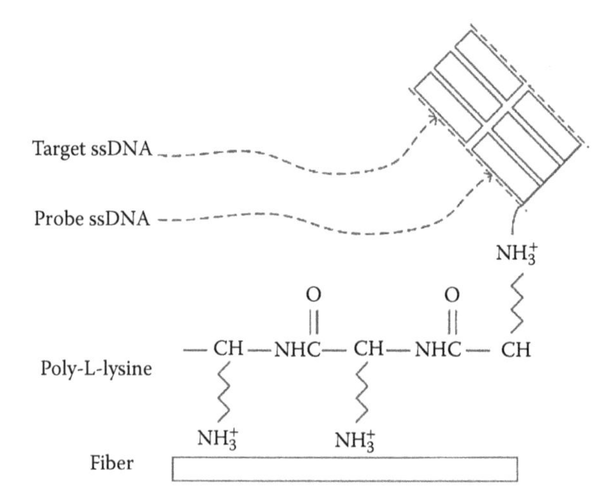

FIGURE 3.7 Molecular structure of poly-L-lysine, probe ssDNA, and target ssDNA immobilized on the surface LPG. (From Jang, H.S., Park, K.N., Kim, J.P., Sim, S.J., Kwon, O.J., Han, Y.G., and Lee, K.S., *Optics Express*, 17, 3855–3860, 2009.)

DNA to be immobilized on the surface LPG. Then, the surface LPG is washed again using PBS buffer to remove excess PLL layers that were not immobilized on the surface LPG. A 1 μM probe of single-stranded DNA in PBS buffer was then applied to the surface of the PLL layer for 130 min at room temperature. The cleaning process was then repeated. Finally, the 1 μM target single-stranded DNA in the PBS buffer was applied to the surface of a probe ssDNA layer and after 65 min the sensor was washed using PBS buffer. Approximately 100 μl of all biomolecules was added and removed using a micropipette. Figure 3.7 shows the molecular structure of the PLL, probe ssDNA, and target ssDNA. The DNA sequences were (5′-CAG CGA GGT GAA AAC GAC AAA AGG GG-3′) for the probe ssDNA and (5′-CCC CTT TTG TCG TTT TCA CCT CGC TG-3′) for the target ssDNA.

The wavelength shift of the r-mode in the surface LPG was measured, which was highly sensitive to the external refractive index in a range from 1.333 to 1.454.

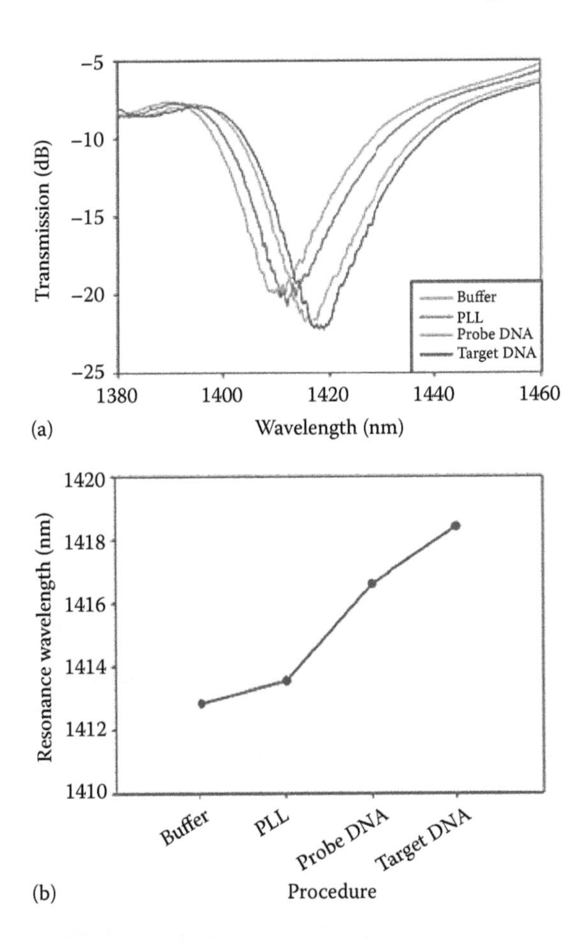

FIGURE 3.8 (a) Transmission spectra and (b) resonant wavelength shifts of the surface LPG after sequential procedures applying PBS buffer, PLL, probe ssDNA, and target ssDNA. (From Jang, H.S., Park, K.N., Kim, J.P., Sim, S.J., Kwon, O.J., Han, Y.G., and Lee, K.S., *Optics Express*, 17, 3855–3860, 2009.)

As shown in Figure 3.8a, the resonance wavelength shifted to a longer wavelength as the biomolecular layer was formed. When the PBS buffer was applied to the surface LPG, the measured resonance wavelength was 1411.69 nm. The immobilization of PLL on the surface LPG shifted the resonance wavelength to 1413.52 nm. Then, when the probe ssDNA was immobilized on the PLL layer on the surface LPG, the resonance wavelength shifted to 1416.61 nm. Finally, hybridization with the complementary target ssDNA induced a resonance wavelength shift of 1418.43 nm. The overall wavelength shift induced by the hybridization reaction was 1.82 nm, which is ~2.5 times higher than for the previously reported biosensor based on a dual-peak LPG [26] under the same 1 μM target DNA concentration. It is evident that the proposed surface LPG-based biosensor is highly sensitive to DNA hybridization in comparison with previously reported DNA fiber grating–based biosensors [13,26]. The resonance wavelengths of the fiber grating–based DNA biosensor after each procedure are shown in Figure 3.8b.

REFERENCES

1. P. T. Charles, G. J. Vora, J. D. Andreadis, A. J. Fortney, C. E. Meador, C. S. Dulcey, and D. A. Stenger, Fabrication and surface characterization of DNA microarrays using amine- and thiol-terminated oligonucleotide probes, *Langmuir*, 19, 1586–1591, 2003.
2. J. Homola, S. S. Yee, and G. Gauglitz, Surface plasmon resonance sensors: Review, *Sensors and Actuators B*, 54, 3–15, 1999.
3. T. M. Chinowskya, J. G. Quinn, D. U. Bartholomew, R. Kaiser, and J. L. Elkind, Performance of the Spreeta 2000 integrated surface plasmon resonance affinity sensor, *Sensors and Actuators B*, 91, 266–274, 2003.
4. H. J. Patrick, A. D. Kersey, and F. Bucholtz, Analysis of the response of long period fiber gratings to external index of refraction, *Journal of Lightwave Technology*, 16, 1606–1612, 1998.
5. H. Dobb, K. Kalli, and D. Webb, Measured sensitivity of arc-induced long-period grating sensors in photonic crystal fibre, *Optics Communications*, 260, 184–191, 2006.
6. Y. Zhu, Z. He, and H. Du, Detection of external refractive index change with high sensitivity using long-period gratings in photonic crystal fiber, *Sensors Actuators B Chemical*, 131, 265–269, 2008.
7. L. Rindorf and O. Bang, Highly sensitive refractometer with a photonic-crystal-fiber long-period grating, *Optics Letters*, 33, 563–565, 2008.
8. H. J. Kim, O. J. Kwon, S. B. Lee, and Y. G. Han, Measurement of temperature and refractive index based on surface long-period gratings deposited onto a D-shaped photonic crystal fiber, *Applied Physics B*, 102, 81–85, 2011.
9. M. Mehrvar, C. Bis, J. M. Scharer, M. Moo-Young, and J. H. Luong, Fiber optic biosensors: Trends and advances, *Analytical Sciences*, 16, 677–692, 2000.
10. J. F. Masson, M. Barnhart, T. M. Battaglia, G. E. Morris, R. A. Nieman, P. J. Young, C. L. Lorson, and K. S. Booksh, Monitoring of recombinant survival motor neuron protein using fiber-optic surface plasmon resonance, *Analyst*, 129, 855–859, 2004.
11. S. Maguis, G. Laffont, P. Ferdinand, B. Carbonnier, K. Kham, T. Mekhalif, and M. Millot, Biofunctionalized tilted fiber Bragg gratings for label-free immunosensing, *Optics Express*, 16(23), 19049–19062, 2008.
12. A. N. Chryssis, S. S. Saini, S. M. Lee, H. Yi, W. E. Bentley, and M. Dagenais, Detecting hybridization of DNA by highly sensitive evanescent field etched core fiber Bragg grating sensors, *IEEE Journal of Selected Topics in Quantum Electronics*, 11, 864–872, 2005.

13. J. L. Tang, S. F. Cheng, W. T. Hsu, T. Y. Chiang, and L. K. Chau, Fiber optic biochemical sensing with a colloidal gold modified long period fiber grating, *Sensors and Actuators B*, 119, 105–109, 2006.
14. J. Yang, P. Sandhu, W. Liang, C. Xu, and Y. Li, Label free fiber optic biosensors with enhanced sensitivity, *IEEE Journal of Selected Topics in Quantum Electronics*, 13(6), 1691–1696, 2007.
15. X. Wei, T. Wei, H. Xiao, and Y. S. Lin, Nano structured Pd-long period fiber gratings integrated optical sensor for hydrogen detection, *Sensors and Actuators B*, 134, 687–693, 2008.
16. K. S. Lee and T. Erdogan, Fiber mode coupling in transmissive and reflective tilted fiber gratings, *Applied Optics*, 39(9), 1394–1404, 2000.
17. T. Allsop, A. Gillooly, V. Mezentsev, T. E. Gould, R. Neal, D. J. Webb, and I. Bennion, Bending and orientational characteristics of long period gratings written in D-shaped optical fiber, *IEEE Transactions on Instrumentation and Measurement*, 53, 130–135, 2004.
18. H. S. Jang, K. N. Park, J. P. Kim, S. J. Sim, O. J. Kwon, Y. G. Han, and K. S. Lee, DNA biosensor based on a long-period grating formed on the side-polished fiber surface, *Optics Express*, 17(5), 3855–3860, 2009.
19. H. J. Kim, O. J. Kwon, and Y. G. Han, Effect of an ambient index change on transmission characteristics of versatile D-shaped fibers, *Journal of the Korean Physical Society*, 55(3), 1286–1289, 2009.
20. H. J. Kim, N. R. Jun, and Y. G. Han, Optical characteristics of a hybrid fiber grating based on surface long-period grating incorporating a fiber Bragg grating, *Journal of the Korean Physical Society*, 61(9), 1353–1357, 2012.
21. H. J. Kim, O. J. Kwon, S. B. Lee, and Y. G. Han, Polarization-dependent refractometer for discrimination of ambient refractive index and temperature, *Optics Letters*, 37(11), 1802–1804, 2012.
22. H. J. Kim, O. J. Kwon, S. B. Lee, and Y. G. Han, Measurement of temperature and refractive index based on surface long-period gratings deposited onto a D-shaped photonic crystal fiber, *Applied Physics B: Lasers and Optics*, 102, 81–85, 2011.
23. H. J. Kim, O. J. Kwon, and Y. G. Han, Effect of surface roughness variation on the transmission characteristics of D-shaped fibers with ambient index change, *Journal of the Korean Physical Society*, 56, 1355–1358, 2010.
24. Y. G. Han, S. B. Lee, C. S. Kim, J. U. Kang, U. C. Paek, and Y. Chung, Simultaneous measurement of temperature and strain using dual long-period fiber gratings with controlled temperature and strain sensitivity, *Optics Express*, 11(5), 476–481, 2003.
25. A. Iadicicco, A. Cusano, A. Cutolo, R. Bernini, and M. Giordano, Thinned fiber Bragg gratings as high sensitivity refractive index sensor, *IEEE Photonics Technology Letters*, 16(4), 1149–1151, 2004.
26. P. D. Sawant and D. V. Nicolau, Hierarchy of DNA immobilization and hybridization on poly-L-lysine: An atomic force microscopy study, *Smart Materials and Structures*, 15, S99–S103, 2006.

4 Measuring the Physical Properties of Cells

Shirin Mesbah Oskui and William H. Grover

CONTENTS

4.1 Introduction 102
4.1.1 The Importance of Water 103
4.1.2 The Importance of Cell Size 104
4.1.3 Sorting versus Measuring 104
4.2 Measuring Cell Volume 105
4.2.1 The Promise of Microfluidic Coulter Counters 106
4.2.2 Increasing Dynamic Range 107
4.2.3 Increasing Throughput 109
4.2.4 Low-Cost Coulter Counters 109
4.2.5 Sensitivity of Coulter Counter Volume Measurements to Aperture Geometry and Cell Shape/Orientation 112
4.3 Measuring Cell Mass 112
4.3.1 Suspended Microchannel Resonators 113
4.3.2 Measuring the Mass of Single Cells 115
4.3.3 Measuring Cell Growth 115
4.4 Measuring Cell Density 117
4.4.1 Measuring Average Cell Density 117
4.4.2 Measuring Single-Cell Density 118
4.4.3 Mass-Based Measurement of Cell Volume 120
4.5 Measuring Cell Deformability 120
4.5.1 Modeling Viscoelastic Behavior of Cells 120
4.5.2 Measuring Single-Cell Deformability 122
4.5.3 Decoupling Cell Deformability and Cell Size 123
4.5.4 Hydrodynamic Stretching of Cells 124
4.6 Future Opportunities 126
4.6.1 Measuring the Properties of Adherent Cells 126
4.6.2 Physical Sorters as Front Ends for Physical Measurements 126
4.6.3 Developing a "Physical FACS" 127
4.6.4 Decoupling Physical Properties 127
4.6.5 Toward "Physical Fingerprinting" of Cells 128
References 128

4.1 INTRODUCTION

Every living cell is a physical object. This rather obvious statement actually has some significant consequences. It means that every living cell—from a simple bacterium to any cell in your body—is fundamentally a *thing*, a little discrete piece of matter, like an apple or a rock. And as a *thing*, any cell can be described using the properties we use to describe all other pieces of matter, properties like *mass*, *volume*, *density*, *deformability*, and so on.

At first glance, this reductionist view of "cells as physical objects" seems to ignore the important *biological* differences between different cells: animal versus plant, prokaryote versus eukaryote, and even live versus dead. But these biological differences are *physical* differences as well: cell walls make plant cells *stiffer* than animal cells, more complex cellular machinery gives eukaryotic cells larger *masses* and *volumes* than prokaryotic cells, and osmosis following the loss of membrane integrity can give dead cells a different *density* than live cells.

Measurements of the physical properties of a cell can therefore provide useful information about the biological properties of the cell. Since all cells have these physical properties, this approach is useful for any cell. And since all cells *natively* have these physical properties, this approach is truly "label free" and requires no special preparation of the cells.

This approach is not new—indeed, some of the earliest biological measurements of cells were actually physical measurements. In many cases, observation of a physical difference between cells predated an understanding of the biological difference between the cells. For example, the different types of blood cells (erythrocytes, thrombocytes, and leukocytes) were distinguished by their different sizes and shapes long before their different biological functions were understood.

What *is* new—and the subject this chapter—is the recent emergence of tools for measuring the physical properties of single cells with unprecedented accuracy, precision, and speed. Though these tools measure different properties, they share many common characteristics:

- *Ability to measure single cells.* In a population of cells, the *average* value of a physical property can be readily obtained using techniques like filtration (for average cell size) and density gradient centrifugation (for average cell density). But these techniques are blind to cell-to-cell variation in these properties. Consequently, measurements of average cell properties cannot detect rare cells by their different physical properties. *Single-cell* measurements are necessary to detect these rare cell types. In this chapter, we will examine techniques that measure a wide range of different physical properties of single cells.
- *Extremely high sensitivity.* Cells are small objects, so tools for measuring the physical properties of cells must be very sensitive. The fields of *microelectromechanical systems* (MEMS) and *microfluidics* are extremely well suited for this task of high-sensitivity measurement of cell properties, and the instruments we discuss below arose from these fields.
- *Extremely high accuracy.* For example, the *diameter* or *area* of a cell (obtained via microscopy) is often a reasonable approximation of cell size.

But to measure subtle differences between cell sizes, a more accurate definition of cell size is needed, like *cell mass* or *cell volume*. We will examine several techniques for obtaining these accurate metrics of cell size.

- *Ability to deal with very small samples*. To measure a cell, we first need to manipulate a cell. Often this means isolating single micron-sized cells from a sample that could contain as many as 5 million cells per milliliter (as in whole blood). We will consider various techniques that exploit the unique physics of *laminar flow* in microfluidic devices to manipulate single cells.
- *High throughput*. For clinically meaningful statistical confidence and to identify rare cell types in a population, large numbers of cells must be measured. We will examine how microfluidic techniques can dramatically increase the throughput of instruments for measuring the physical properties of cells.

4.1.1 The Importance of Water

Before examining the various ways that the physical properties of cells can be measured, it is worthwhile to ask where the physical properties of a cell come from, and what ultimately determines the physical properties of a cell. At the most basic level, a cell's physical properties are a function of its chemical makeup. Figure 4.1 shows that mammalian cells are around 70% water by mass,[1] so it is reasonable to assume that the physical properties of water are the single most important determinant of a cell's properties. As a result, the *density* of a cell (usually around 1.05 g/mL) is very similar to the density of water (1.00 g/mL), the *mass* and *volume* of a cell are mostly the mass and volume of water in the cell, and the *deformability* of a cell can be approximately understood by modeling the cell as a tiny viscoelastic bag filled with water.

This sensitivity to water content can be both an advantage and a disadvantage for physically measuring cells. For a researcher interested in cellular water content (e.g., studying the role of erythrocyte dehydration in triggering a crisis in patients with sickle cell trait, or the osmotic permeability of different kinds of bacteria), then

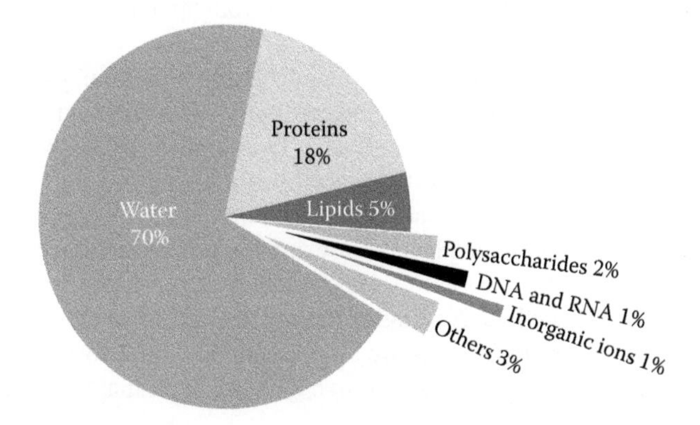

FIGURE 4.1 The contents of a mammalian cell, by mass. (Adapted from Alberts, B., *Molecular Biology of the Cell*, Garland Science, New York, 2008.)

physical measurements of cellular water content can be very useful. But for researchers interested in the other 30% of a cell (proteins, lipids, polysaccharides, nucleic acids, etc.), the variable water content of the cell can be a source of experimental uncertainty. In practice, we are constantly mindful of water flux when measuring the physical properties (mass, volume, and density) of a cell.

Some physical properties of cells are less sensitive to contributions by water content. For example, by weighing the *buoyant mass* of a cell in water, the water-weight of the cell is effectively canceled out and only the nonwater weight of the cell is measured.[2] These measurements are examined in detail later in this chapter.

4.1.2 The Importance of Cell Size

A cell's size has profound effects on its biology. For example, since all mass transfer between a cell and its environment occurs through the cell membrane or cell wall, a cell's *surface area* sets a fundamental limit on the flux of nutrients into and waste products out of a cell. Assuming a spherical cell shape, doubling the diameter of a cell increases the volume of the cell by a factor of 8 but increases the surface area of the cell by only a factor of 4. This mismatch between volume and surface area means that as a cell grows, it must either reduce its metabolic rate or work harder to transport mass into and out of it (or a combination of both). Maintaining this delicate balance between supply and demand is accomplished by the various cellular mechanisms that regulate cell growth, mechanisms that are profoundly important yet poorly understood.[3] The tools for ultraprecise measurement of cell size discussed in this chapter are providing novel insights into these mechanisms of growth regulation; they will prove especially valuable in cases where these growth mechanisms are disrupted, as in cancer.[4,5]

4.1.3 Sorting versus Measuring

As mentioned earlier, the field of microfluidics plays an important role in measuring the physical properties of single cells. Microfluidic devices provide the cell-sized sensitivity, single-cell manipulation, and high throughput necessary for making these measurements.

Of all the microfluidic devices that exploit differences in the physical properties of cells, the vast majority just *sort* cells according to their physical properties and do not actually *measure* cell properties in a quantitative sense. We will briefly summarize the major types of microfluidic cell sorters here.

Many on-chip sorters that separate cells based on their physical properties have off-chip (laboratory-scale) equivalents. For example, just as laboratory-scale filters can separate cells by their different sizes, on-chip channel constrictions can be designed that separate cells according to their sizes,[6] or separate cells from their surrounding fluid.[7] These techniques are also sensitive to the deformability of cells: a large but deformable cell may pass through a constriction that blocks a small but rigid cell.

Other on-chip sorters exploit the unique physical phenomena of fluid flow at low Reynolds numbers. For example, pinched-flow fractionation[8] separates cells by size but not stiffness. It accomplishes this by exploiting the fact that the center of mass of a cell in laminar flow tends to follow the streamlines of the surrounding fluid. If a fluid stream containing cells is "pinched" by joining a second fluid stream, small cells will remain in the pinched streamlines but large cells will extend into the second fluid's streamlines; these different trajectories can then be separated at a channel branch downstream. At high flow rates in micron-sized channels, inertial effects can alter a cell's tendency to follow flow streamlines.[9] These effects can be used to separate cells by size[10] and density.[11] Gossett et al. provide an excellent review of these and many other microfluidic cell sorters.[12]

Most of these cell sorters are sensitive to more than one physical property of a cell. This complicates the use of sorters as quantitative tools for measuring physical properties. Still, the high throughput of these sorters makes them useful for isolating rare cell types in a population, such as circulating tumor cells.[13] These cell sorters could also be useful as front ends for cell-measuring instruments, as discussed in Section 4.6.

4.2 MEASURING CELL VOLUME

For nearly 300 years after Antonie van Leeuwenhoek viewed cells with his microscope, *cell size* was synonymous with *cell diameter* (or *area*) obtained via microscopy. These estimates of cell size from microscopy made it possible to distinguish different cell types, but uncertainties in the estimates limited their usefulness for measuring the variation within a cell type. This finally changed in 1953 with the invention of the Coulter counter,[14] which made cell size measurement quantitative and made *cell volume* the first rigorous definition of cell size.

The function of a Coulter counter (or "resistive pulse particle sizer") is as simple as it is powerful. As shown in Figure 4.2, two fluid reservoirs are connected by a tiny fluid-filled hole called the *aperture*. If the fluid has a conductivity ρ and the aperture has length L and diameter D, then the electrical resistance R measured between the two fluid reservoirs will be[15]

$$R = \frac{4\rho L}{\pi D^2}. \tag{4.1}$$

If a nonconductive object like a cell enters the aperture from one of the reservoirs, it will partially block the flow of electric current and increase the resistance of the aperture temporarily. If the object is spherical with diameter d, the resistance of the aperture increases by ΔR:

$$\Delta R = \frac{4\rho d^3}{\pi D^4}. \tag{4.2}$$

The magnitude of this brief change in resistance as the particle passes through the aperture is therefore proportional to the cubed diameter of the particle (the particle's volume). In this manner, the Coulter counter can measure the volume of single cells.

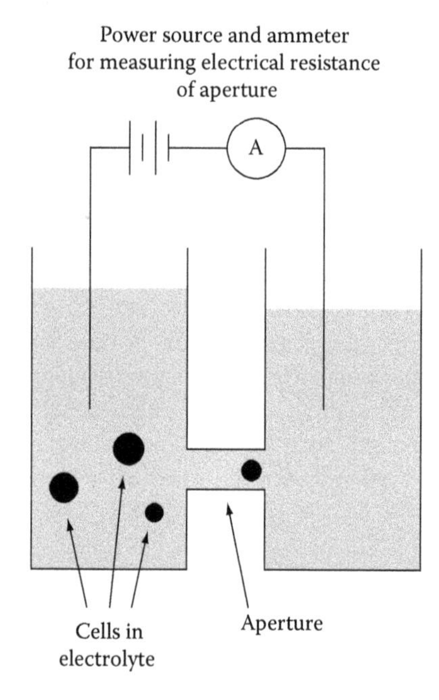

FIGURE 4.2 Design of a conventional Coulter counter. As a cell passes through the aperture, the electrical resistance of the aperture decreases by an amount proportional to the volume of the cell.

Now 60 years old, the modern Coulter counter has some deficiencies for certain applications. The modern Multisizer 4 from Beckman-Coulter is a 45 kg benchtop instrument that retails for tens of thousands of dollars. Since an aperture is only useful for a limited range of cell sizes, in practice a range of different-sized apertures is needed to analyze different sizes of cells. These apertures are too expensive to be disposable and must be rinsed clean by the instrument during operation, a process that consumes large volumes of reagents. The relatively large (several milliliter) sample volumes required by the Multisizer 4 complicate its use in analyzing small numbers of rare cells. Finally, while the Multisizer 4 can *measure* the volumes of single cells, it cannot *sort* measured cells according to their volumes (like a fluorescence-activated cell sorter [FACS] sorts cells according to their fluorescence).

4.2.1 The Promise of Microfluidic Coulter Counters

All of these deficiencies of conventional Coulter counters could be solved by a *microfluidic Coulter counter*. The reduced size and reduced reagent consumption of a microfluidic Coulter counter could open up applications for the technique in point-of-care or resource-limited settings. The decreased cost of a microfluidic Coulter counter chip could make apertures disposable, reducing the likelihood of clogging and eliminating the risk of cross contamination between samples. By including multiple aperture sizes on a single chip, a microfluidic Coulter counter could have a larger dynamic range than a conventional Coulter counter. The decreased sample volume in a microfluidic Coulter counter would facilitate its use in studying rare cell types. Finally, by including a sorting functionality, a microfluidic Coulter counter could select cells according to their volumes and save these cells for subsequent analysis by other techniques.

A handful of researchers have developed microfluidic Coulter counters. This research is reviewed elsewhere,[16–18] and most of it focuses on miniaturizing the conventional Coulter counter. But some researchers have leveraged the unique properties of microscale fluid flow to make microfluidic Coulter counters that successfully address some of the above problems. This section examines some of these microfluidic Coulter counters.

4.2.2 Increasing Dynamic Range

As mentioned above, one drawback of the conventional Coulter counter is the fixed size of the aperture tube. For maximum sensitivity, smaller cells (like bacteria) typically require a smaller aperture and larger cells (like mammalian cells) require a larger aperture. The dynamic range of any one aperture is thus low. The expense of obtaining multiple apertures, the manual labor required to swap them out when needed, and the effort required to unclog clogged apertures are all disadvantages of conventional Coulter counters.

All of these disadvantages can be addressed by making the aperture tube out of fluid, using *hydrodynamic focusing* in a microfluidic device. Hydrodynamic focusing exploits the viscous-dominated laminar flow often present in microfluidic channels. When two or more laminar flows merge in a microfluidic device, mixing between the two fluids is primarily by diffusion, a relatively slow process. The fluids therefore remain separated for a distance or time, even though they are in contact with each other. In hydrodynamic focusing, different channel arrangements and flow rates are used to control the geometry of the merged fluid streams.[19] The Reynolds number Re (the ratio of inertial forces to viscous forces in a flowing fluid) is used to predict whether flow in a microfluidic channel will be laminar (and therefore capable of supporting hydrodynamic focusing):

$$\mathrm{Re} = \frac{\rho v L}{\mu}, \tag{4.3}$$

where:

- ρ is the density of the fluid in the channel
- v is the mean linear velocity of the flow
- L is a characteristic dimension of the channel (e.g., its width)
- μ is the dynamic viscosity of the fluid

Flowing fluids with low Reynolds numbers ($\mathrm{Re} < 2300$) are laminar and will support hydrodynamic focusing; systems with high Reynolds numbers ($\mathrm{Re} > 4000$) are turbulent and will not support hydrodynamic focusing. Since most microfluidic devices contain dilute aqueous solutions with density and viscosity approximately equal to those of water ($\rho = 1000$ kg m^{-3} and $\mu = 0.001$ kg m^{-1} s^{-1}), a convenient approximation for the Reynolds number for a channel containing a dilute aqueous solution is

$$\mathrm{Re} \cong 10^6 vL, \tag{4.4}$$

where flow velocity v is in meters per second and channel width L is in meters.

Figure 4.3 shows how Rodriguez-Trujillo et al. used hydrodynamic focusing to create a microfluidic Coulter counter with a *liquid* aperture tube.[20] The top view shows the channel geometry used to surround a high-conductivity fluid on three sides with a low-conductivity fluid. On the fourth side, a pair of microfabricated electrodes makes electrical contact with the conductive fluid stream; when a cell in the conductive fluid passes the electrodes, it blocks the flow of electric current between the electrodes and causes a momentary increase in measured resistance that is proportional to the cell's volume (side view in Figure 4.3). The low-conductivity lateral and vertical focusing flows in the 3D view effectively serve as the walls of the aperture, and by simply adjusting the flow rates in the device, apertures of arbitrary dimensions can be created.

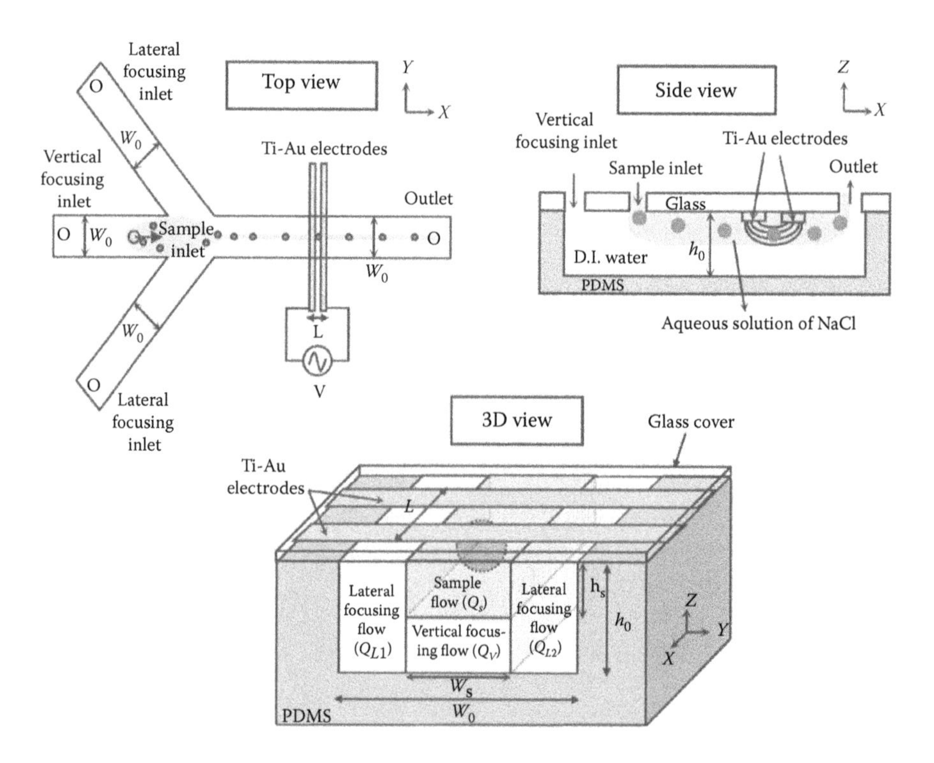

FIGURE 4.3 Design and operation of a flow-focusing Coulter counter. The effective size and shape of the aperture is controlled by adjusting the flow rates of the lateral and vertical focusing flows. D.I. water, deionized water; PDMS, polydimethylsiloxane. (From Rodriguez-Trujillo, R., et al., *Biosens. Bioelectron.*, 24, 290–296, 2008. With permission.)

Hydrodynamic focusing significantly expands the dynamic range of a microfluidic Coulter counter (the range of cell sizes that it can measure).[20–23] As a bonus, apertures made of fluid will pass, for example, dust particles that would clog the solid apertures used in conventional Coulter counters. A possible disadvantage of fluidic apertures is the sensitivity of the measured resistance to the dimensions of the conductive fluid stream: any variation in flow rates will change the effective diameter of the conductive fluid stream (*D* in the above equations) and subsequently affect the measured resistance.

Recently, Riordon et al. demonstrated an interesting alternative strategy for adjusting the aperture size (and therefore increasing the dynamic range) of a microfluidic Coulter counter.[24] By building their device out of a flexible elastomer and placing a microfluidic pinch valve above the current-sensing electrodes, an adjustable-sized aperture is formed. Pneumatic pressure applied to the pinch valve adjusts the height of the microfluidic channel above the electrodes, making it possible to measure a wider range of cell sizes with the same device.

4.2.3 Increasing Throughput

Conventional Coulter counters have a single aperture and a single pair of current-measuring electrodes, so they can only measure one cell at a time. In spite of this limitation, high flow rates and high cell concentrations still lead to impressive throughputs of a few hundred cells per second. But for dilute solutions of rare cells, or for fragile cells that cannot be exposed to high shear stress, it may be necessary to increase throughput while decreasing sample volume and flow rate, a combination that is incompatible with conventional Coulter counters.

To address this need, microfluidic Coulter counters have been developed that utilize multiple apertures in parallel. Zhe et al. demonstrated a four-aperture microfluidic Coulter counter that utilizes five microfabricated electrodes and a network resistance model to determine the resistance change (and hence cell volume) at each aperture.[25] More recently, Jagtiani et al. used the principle of frequency-division multiplexing to measure resistance changes at multiple apertures simultaneously: by applying an AC signal with a different frequency to each aperture, the resistance across the entire array of apertures can be measured and analyzed to extract the resistance changes at each aperture.[26]

4.2.4 Low-Cost Coulter Counters

The relative simplicity of the Coulter counter measurement—measuring a resistance across a tiny hole—should make it possible to create small and inexpensive Coulter counters, thereby opening applications for Coulter counters in resource-limited or point-of-care settings. But as mentioned earlier, modern Coulter counters remain large and expensive instruments. We offer three possible reasons for this:

1. Conventional instruments require large sample volumes.
2. Conventional apertures (typically micromachined sapphire disks) are expensive to make.

3. Conventional instruments require a large amount of fluid-handling hardware (pumps, valves, pressure sensors, etc.) to send samples and rinse fluids through the aperture.

Problem 1 is solved by all the microfluidic Coulter counters examined above: they require only a tiny fraction of the sample volume used by conventional Coulter counters.

Problem 2 should also be solved in microfluidics: micron-sized features are simple and inexpensive to make using microfabrication. But each of the previously mentioned microfluidic Coulter counters uses microfabricated on-chip metal electrodes to measure the current flow through the aperture. These integrated electrodes add additional steps to device fabrication processes and raise the cost of the final devices.

An interesting strategy for further reducing the cost of a Coulter counter chip involves using microfluidic channels as both the aperture *and the resistance-measuring electrodes.*[27–32] Figure 4.4 shows a very simple microfluidic Coulter counter developed by Song et al.[28] Instead of integrated metal electrodes, the chip adds two additional fluidic channels connected to reservoirs (labeled $\Delta V_{downstream}$ and $\Delta V_{upstream}$). Reusable wire electrodes make contact with the fluid in these reservoirs. By measuring the resistance (or more accurately the voltage drop) between these two electrodes, the volume of a cell in the aperture can be measured. In this instrument, the microfluidic chip consists of only a microchannel and four holes and could be manufactured cheaply enough for use as a single-use (disposable) component.

Finally, to solve Problem 3, off-chip fluid control pumps can be integrated on-chip. The device by Song et al. in Figure 4.4 accomplishes this using *electroosmotic flow*: the voltage applied at the sample and waste reservoirs (labeled V_+ and V_-) both powers the resistance measurement and drives the flow of fluid (and cells) through the aperture.[28] This dual use for the applied voltage eliminates a significant amount of off-chip pumping hardware and dramatically reduces the overall cost of the instrument.

McPherson and Walker used an even simpler pump in their microfluidic Coulter counter.[33] When fluid droplets of different radii r are placed on the fluid reservoirs of a microfluidic chip, the pressure inside the droplet increases by ΔP as described by the Young–Laplace equation:

$$\Delta P = \gamma \frac{2}{r}, \tag{4.5}$$

where γ is the surface tension of the fluid. Smaller droplets therefore have higher internal pressures than larger droplets and will pump their contents into the larger droplets. McPherson and Walker's chip uses this droplet-driven pumping to send cells through their on-chip aperture. By eliminating off-chip fluid pumps, this approach can be used to further reduce the cost of microfluidic Coulter counter instruments.

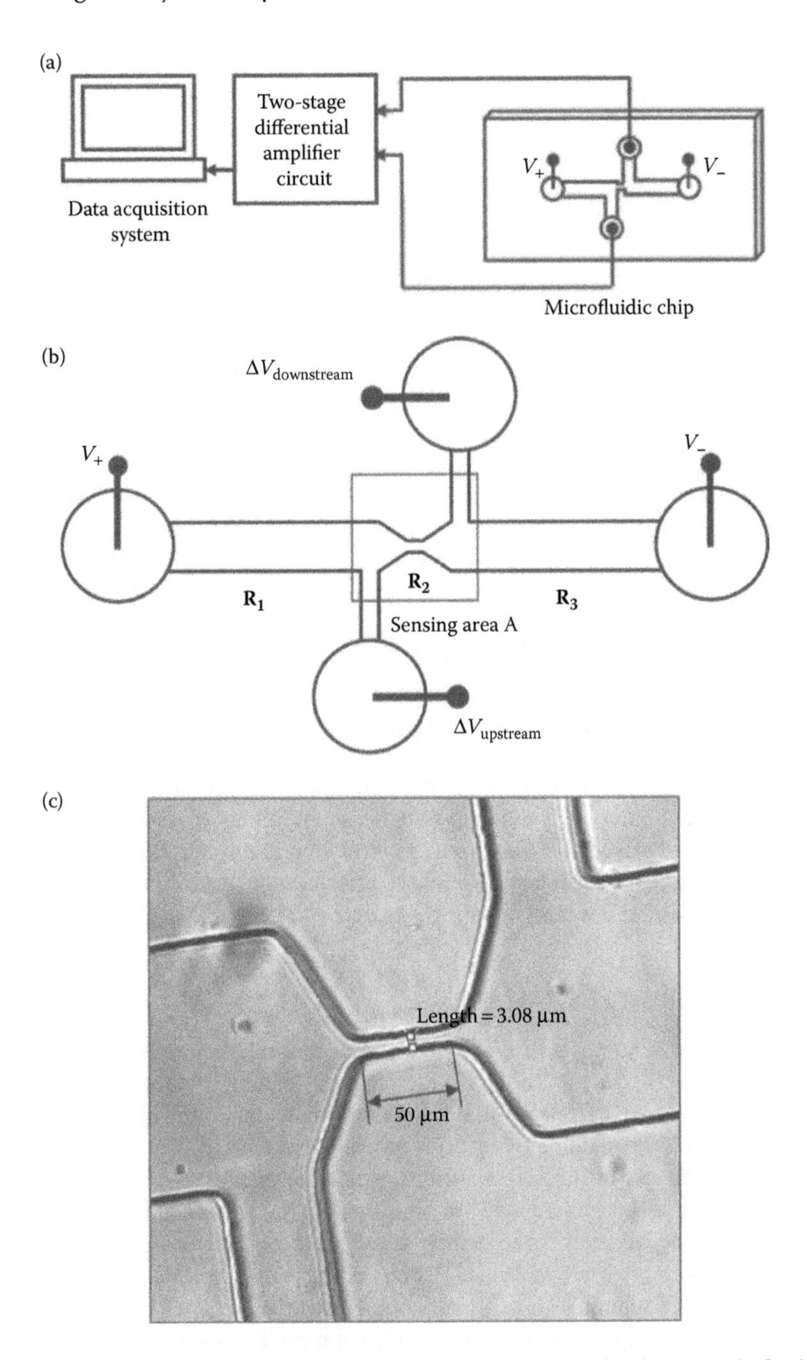

FIGURE 4.4 A low-cost Coulter counter chip that uses microfluidic channels for both introducing samples and measuring the resistance across the aperture. (From Song, Y., et al., *Anal. Chim. Acta*, 681, 82–86, 2010. With permission.)

4.2.5 Sensitivity of Coulter Counter Volume Measurements to Aperture Geometry and Cell Shape/Orientation

It is worth noting that *none* of the previously mentioned microfluidic Coulter counters has a cylindrical aperture (as was assumed in Equations 4.1 and 4.2). This reflects the practical limitations of microfabrication—it is difficult to make channels with round cross sections without orienting the channels vertically and adding additional layers to a device, and while this has been done in microfluidic Coulter counters,[34,35] it makes the fabrication of these devices significantly more difficult. Nonspherical apertures are easier to microfabricate, but their use complicates the analysis of data obtained using these devices: the relationship between resistance change and cell volume becomes much more complex than Equations 4.1 and 4.2, and the path followed by a cell through the aperture (e.g., down the center vs. along the edge) matters more.

Additionally, Equation 4.2 assumes that a cell is spherical. This assumption greatly simplifies the interpretation of experimental data because the orientation of a spherical cell has no effect on its electrical-current-blocking behavior in the aperture. But most cells are not spherical, and in extreme cases (e.g., rod-shaped bacteria or flagellate spermatozoa) the orientation of a cell can affect the cell volume measured by a Coulter counter: a rod-shaped cell that passes end-first through the aperture will block less current and appear to have a smaller volume than the same cell passing side-first through the aperture.

These complications suggest the need for a cell volume measurement technique that is *not* sensitive to cell shape or orientation. In the next section we examine a technique that measures the volume of a cell using measurements of cell mass.[36] Unlike Coulter counter measurements of cell volume, a mass-based measurement of cell volume is not affected by the shape or orientation of the cell. But before examining how cell mass measurements can be used to measure cell volume in Section 4.4, we will first consider how cell mass is measured.

4.3 MEASURING CELL MASS

Cell mass—the weight of a cell, usually on the order of picograms—is a relatively new metric of cell size. While the Coulter counter made it possible to measure the volumes of single living cells with a relatively simple apparatus in 1953, measurements of the masses of single living cells proved more elusive.

Perhaps one reason that cell mass historically received less attention than cell volume is that it is not immediately obvious that cell mass is meaningfully different from cell volume. Indeed, if the density of all cells were constant, then cell mass would be equivalent to cell volume and tools for measuring cell mass would provide no additional information over a Coulter counter. But different cell types *do* have different densities: compare the average densities of muscle tissue (1.06 g/mL) and adipose tissue (0.9 g/mL) and you will see why we weigh a person in water (1.0 g/mL) to measure their percentage of body fat. So since cell density is *not* constant, cell mass *can* provide information not obtainable from cell volume.

The other major barrier to measuring the mass of single cells is the extreme sensitivity required to measure picogram-scale masses. Conventional laboratory-scale mass sensors (analytical balances) can measure objects with microgram resolution, but this resolution remains six orders of magnitude too large to measure a cell with the required picogram resolution.

The advent of MEMS resonators finally made it possible to weigh single cells. These tiny sensors are made using the same photolithographic tools used to create integrated circuits. They typically consist of a small, suspended mass (e.g., a diving-board-shaped cantilever) made of silicon. Like a tuning fork, these resonators vibrate with a characteristic resonance frequency that is a function of their mass. So if a small object (like a cell) is attached to the cantilever, the resonator's resonance frequency will decrease by an amount proportional to the mass of the attached object.

The dependence of the resonance frequency f of a cantilever resonator on an added mass Δm is given by

$$f = \frac{1}{2\pi}\sqrt{\frac{k}{m+\alpha\Delta m}}, \tag{4.6}$$

where:

- m is the effective mass of the resonator
- k is the spring constant of the resonator
- α describes the location of the added mass (a mass added to the tip of the cantilever, where the vibrational amplitude is largest and $\alpha = 1$, will have a greater effect on the resonance frequency than the same mass added to the base, where the vibrational amplitude is smallest and $\alpha \cong 0$)[2]

From this relationship, the inverse dependence of resonance frequency on added mass is apparent.

By placing a cell at a known location on a tiny silicon cantilever and measuring the change in resonance frequency, the mass of the cell can be determined with femtogram resolution.[37] But this measurement must typically be performed in air or a vacuum, because immersing the cantilever in water dampens the oscillations and drastically reduces the sensitivity of the technique (imagine a tuning fork vibrating under water). Unfortunately, most living cells lose their water content and die when placed in air or a vacuum, so this approach is mostly limited to measuring the *dry mass* (the nonwater weight) of dead cells. Additionally, the time-consuming process of placing single cells on micron-sized resonators dramatically limits the throughput of this approach.

4.3.1 Suspended Microchannel Resonators

In 2007, Burg et al. removed these limitations by literally turning the resonating cantilever inside out: by placing the cell *inside* the cantilever, the cantilever could still be surrounded by vacuum for high-sensitivity mass measurements, but the cell could be surrounded by water and be kept alive inside the cantilever.[2] The resulting microfluidic device, the suspended microchannel resonator (SMR), is shown in Figure 4.5a. It consists of a microfabricated silicon beam with an embedded

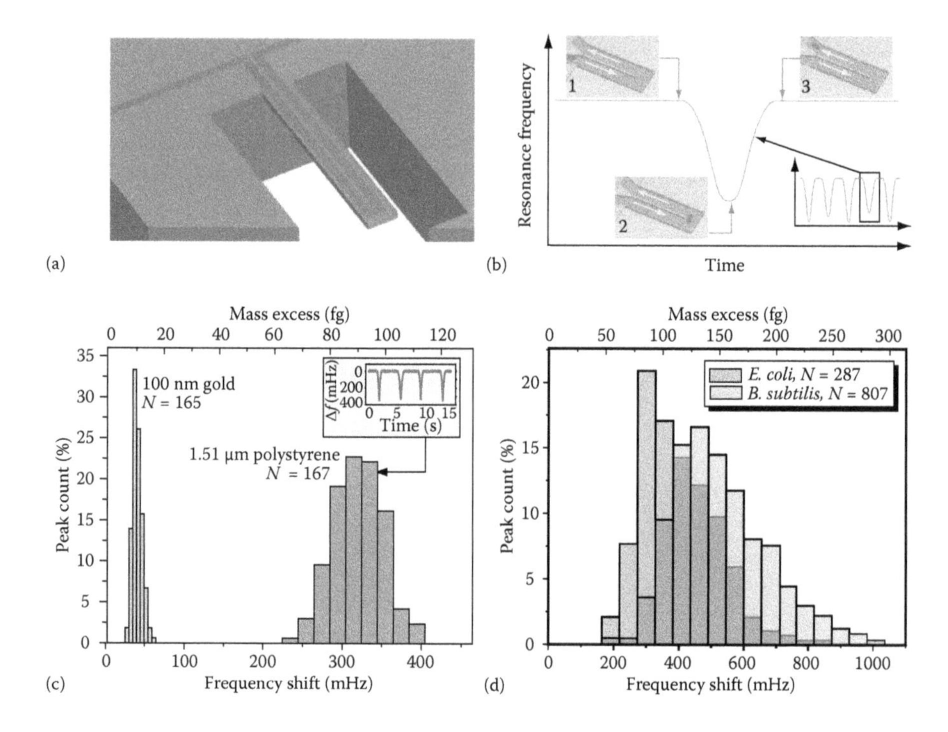

FIGURE 4.5 Illustration of a suspended microchannel resonator (SMR) mass sensor (a). The SMR consists of a micron-sized silicon cantilever that contains an embedded microfluidic channel. When a cell passes through the channel, the resonance frequency of the cantilever changes by an amount proportional to the buoyant mass of the cell (b). Using the SMR, single-particle or single-cell mass distributions can be obtained from particles of different sizes and materials (c) and bacteria of different types (d). (From Burg, T.P., et al., *Nature*, 446, 1066–1069, 2007. With permission.)

microfluidic channel. When filled with fluid, the resonance frequency of the cantilever is a function of the *density* of the fluid (as described by Equation 4.6, with $\alpha \cong 0.24$). But when a cell passes through the channel, the resonance frequency of the cantilever changes temporarily (Figure 4.5b). When the cell is at the tip of the cantilever ($\alpha \cong 1$), Equation 4.6 can be solved for Δm, the mass of the cell. Since the cell is still surrounded by fluid inside the cantilever, its mass can be measured by the SMR while it is still alive and growing. And since many cells can be weighed by flowing them through the cantilever, the throughput of the SMR is high enough to generate meaningful statistics about the population of cells being measured.

It is important to note that since the cell is weighed in fluid, the SMR does not measure the *in vacuo* mass m of the cell but rather measures the buoyant mass m_b of the cell:

$$m_b = m\left(1 - \frac{\rho_f}{\rho}\right), \tag{4.7}$$

where ρ_f is the density of the fluid and ρ is the density of the cell. Intuitively, the buoyant mass of a cell in water can be thought of as the *in vacuo* mass of the cell minus the mass of an equal volume of water. For a typical muscle cell ($\rho = 1.06$ g/mL) in water ($\rho_f = 1.00$ g/mL), the cell's buoyant mass is only about 6% of its true mass. This means that to measure the buoyant masses of cells, the SMR needs a resolution not of picograms (the *in vacuo* mass of a cell) but of femtograms (the buoyant mass of a cell). Recent SMRs actually have *attogram* resolution for buoyant mass measurements,[38] so in practice the buoyant mass of a cell in water is easily detectible by a suitably sized SMR.

Since their introduction in 2007, SMRs have been used in a variety of different mass measurements. In this section, we will examine the two main types of mass measurement performed using the SMR: obtaining the distribution of single-cell masses in a population, and measuring the change in mass of single cells as they grow or react to stimulus.

4.3.2 Measuring the Mass of Single Cells

The most basic mode of operation for the SMR is measuring the buoyant masses of single cells in a population. As cells pass through the resonator in Figure 4.5b, their buoyant masses are recorded from the changes in the resonance frequency of the resonator (the heights of the peaks in the plot of resonance frequency vs. time). Figure 4.5c and d shows distributions of single-particle and single-cell buoyant masses from four samples: 100 nm gold nanoparticles and 1.51 μm polystyrene particles (c), and *Escherichia coli* and *Bacillus subtilis* bacteria (d). Due to their dramatically different sizes and densities, the particles in (c) are easily distinguishable by their buoyant masses. But while the *E. coli* bacteria are on average lighter than the *B. subtilis* bacteria in (d), the two distributions overlap significantly and buoyant mass alone is not adequate for distinguishing the two types of bacteria.[2]

Figure 4.5 illustrates both the advantages and the disadvantages of mass measurements. For samples with significantly different sizes or densities (like the particles in Figure 4.5c), mass is an excellent discriminator of different types of particles. But for samples with similar sizes and densities (like the bacterial sample in Figure 4.5d), the intrinsic variability in cell or particle masses in the population can drown out any mass differences between subpopulations. In other words, for a given cell type, there is typically a wide range of cell sizes (masses and volumes). Mass alone (and volume alone) may not be enough to positively identify a cell type in a mixture of cells. But by using the SMR to measure cell mass, two other modes of measurement become possible: measuring the *change in mass of single cells over time* (discussed next), and measuring the *mass-to-volume ratio* (or *density*) of single cells (discussed in Section 4.4).

4.3.3 Measuring Cell Growth

By using the SMR to measure the buoyant mass of a cell not just once but repeatedly over time, the change in mass of a single cell can be readily obtained.[39,40] Figure 4.6

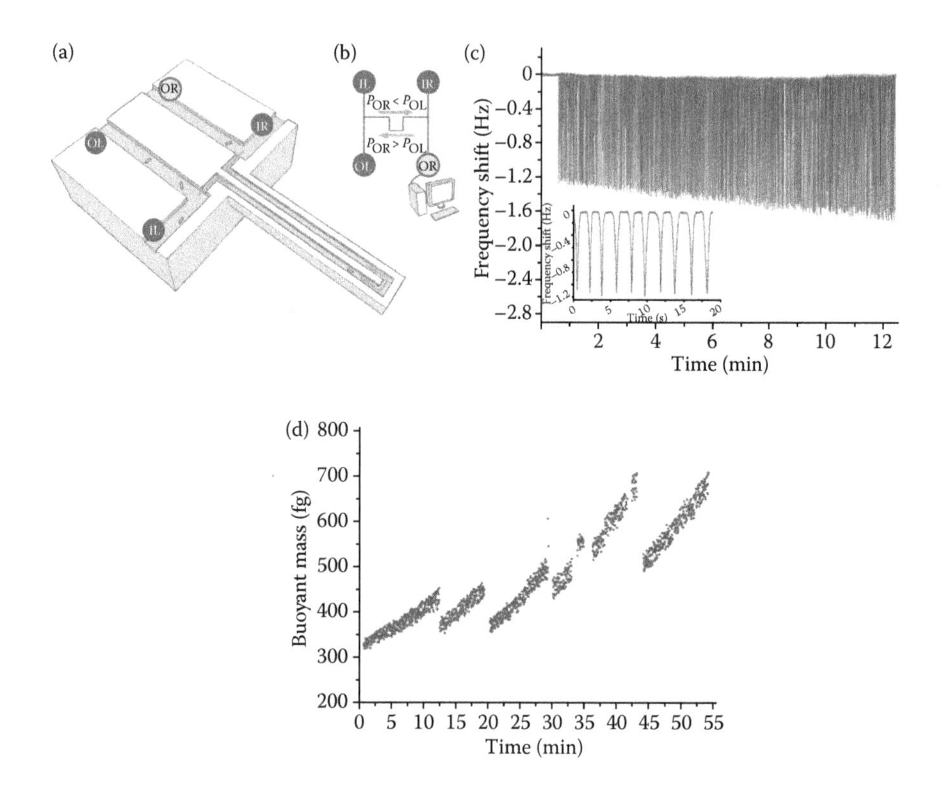

FIGURE 4.6 Measuring the growth of single cells using the SMR. By modulating the fluid pressure applied to points OL, IL, OR, and IR on the SMR, a single cell can be passed back and forth repeatedly through the cantilever (a and b). With each passage of the cell through the cantilever, the buoyant mass of the cell is recorded as a peak in the plot of SMR resonance frequency versus time (c). The growth rates of single cells can be measured using the technique (d). (From Godin, M., et al., *Nat. Methods*, 7, 387–390, 2010. With permission.)

shows how this is accomplished using the SMR. By carefully controlling the fluid pressure applied to four reservoirs on the SMR (labeled OL, IL, OR, and IR in Figure 4.6a and b), single cells can be passed back and forth through the cantilever. It is worth noting that this extremely precise control of the position of a single cell would be virtually unattainable outside of a microfluidic device. Each time the cell passes through the cantilever, its buoyant mass is recorded as a peak in the plot of resonance frequency versus time (Figure 4.6c). If the cell is growing or otherwise changing in mass over time, the height of these peaks will change proportionately over time. By plotting the buoyant mass of a cell versus time, the growth of single cells can be monitored with extraordinary accuracy and sensitivity (Figure 4.6d).

This technique has provided valuable insights into the fundamental mechanisms of cell growth, and how these growth mechanisms are disrupted by metastasis, drug exposure, nutrient deprivation, and so on. For example, using the SMR to monitor the weight of single cells as they grow, Godin et al. showed that for several different cell types, larger cells grow faster than smaller cells.[39] This observation supports an *exponential* model for cell growth: as cells increase in mass, they contain more cell growth "machinery" and can therefore grow even faster. Checkpoints in the cell cycle may serve to reduce this exponential growth, and Son et al. used the SMR to find evidence for a decrease in cell-to-cell growth rate variability at the G1–S transition in various types of cancer cells. These and other results indicate the usefulness of cell mass as a metric of cell growth.

4.4 MEASURING CELL DENSITY

As mentioned previously, the size (mass or volume) of individual cells in a population can be a rather "noisy" property of the cells. The natural variability in cell size is relatively large, so discriminating between two cell types by their masses or volumes alone can be difficult. Similarly, a small *change* in the sizes of cells in a population (say, a volume decrease accompanying apoptosis as cancer cells are treated with a drug) can be very hard to detect in the naturally wide distribution of cell sizes.

However, while cell mass and cell volume may be relatively noisy for cells in a population, their ratio (cell density) is remarkably noise-free. For example, there is a wide range of sizes of muscle cells in your body, but virtually all of your muscle cells have a density of about 1.06 g/mL. Likewise, fat cells come in a range of sizes, but all of them have a density of around 0.9 g/mL. The fact that cells of a certain type have similar densities is not surprising. As discussed previously, the density of a cell is a function of its chemical contents, and cells of the same type probably have similar chemical contents, so one would expect that cells of the same type would also have similar densities.

But this relative homogeneity of densities in cells of a given type is also powerful. Cell density can be used to distinguish different cell types much more easily than cell mass or cell volume. This is the basis for techniques like density gradient centrifugation, which separates mixtures of cells according to their different densities. And unlike changes in cell volume or cell mass, changes in cell density could be easily detected as a shift in the narrow distribution of cell densities.

Cell density is therefore a rather special physical property of cells. It is the product (well, ratio) of two other cellular properties, mass and volume. But to a greater extent than either mass or volume, cell density is a direct function of the chemical contents and biological state of a cell. So tools for measuring cell density could provide biologically and clinically meaningful information about the cells, information unavailable from other physical properties.

4.4.1 Measuring Average Cell Density

Bryan et al. made what is arguably the first precision measurement of the *average* density of a population of cells.[41] By using the SMR to measure the average mass of a population of yeast cells and then using a conventional Coulter counter to measure the average volume of the yeast cells, the average density can be calculated. While this technique cannot be used to measure single-cell density or obtain statistics on the distribution of densities in a cell population, it is sensitive enough to detect tiny (<1%) differences between the average densities of different species of yeasts and changes in density during the cell cycle of yeast.

4.4.2 Measuring Single-Cell Density

More recently, Grover et al. used the SMR to make the first precision measurements of single-cell density for statistically meaningful numbers of cells.[36] Using a technique first described by Archimedes around 250 B.C., measurements of a cell's buoyant mass in two fluids of different densities are used to calculate the density of the cell.

Weighing a cell in two different fluids presents an interesting challenge for a microfluidic device. It requires that a cell be weighed once, and then the fluid around the cell exchanged for a different fluid before the cell is weighed a second time. To keep throughput high, this process should occur as quickly as possible (no more than ~1 s per cell). In practice, rapid exchange of the fluid around a cell is surprisingly difficult to attain in the viscous-dominated flow regime of a microfluidic device: imagine using honey to rinse maple syrup off an apple.

Grover et al. accomplished this rapid exchange of the fluid around a cell using a version of pinched-flow fractionation (described previously in Section 4.1). Their technique is shown in Figure 4.7, where the two fluids of different densities are shown as Fluid 1 and Fluid 2. In Step 1, Fluid 1 fills the cantilever, and the density of the fluid is determined from the resonance frequency of the cantilever. When a cell passes through the cantilever in Step 2, the height of the resulting peak in the resonance frequency is proportional to the buoyant mass of the cell in Fluid 1. The cell now enters a channel containing Fluid 2. At the channel intersection, dissimilar fluid flow rates pinch Fluid 1 into a thin stream that is narrower than the cell's diameter. By the principle of pinched-flow fractionation, the cell finds itself laterally displaced out of the thin stream of Fluid 1 and into Fluid 2. By the time the direction of flow is reversed and the cell passes through the cantilever a second time in Step 4, the thin layer of Fluid 1 has diffused away into Fluid 2 and the cell is surrounded by ~99% Fluid 2. The cell's buoyant mass in Fluid 2 is recorded from the second peak height.

From the four measurements made in Figure 4.7 (two fluid density measurements and two cell buoyant mass measurements), the mass, volume, and density of the cell can be calculated. Specifically, by plotting the buoyant masses of the cell versus the corresponding solution densities and fitting the two points to a line, the Y intercept of the line is the absolute (*in vacuo*) mass of the cell, the X intercept is the volume of the cell, and the slope is the density of the cell.[36] Using this technique, the densities of about 500 cells can be measured per hour with a resolution of 0.001 g/mL.

Grover et al. use their technique to measure the density of several cell types.[36] They discriminate *Plasmodium falciparum* malaria-infected erythrocytes from healthy erythrocytes based on their different densities, and discriminate blood cells made by a patient's own body from cells transfused into the patient's body (from a blood donor) based on the cells' different masses and densities. They also observe that for a particular type of cancer cell undergoing drug-induced apoptosis, a clear increase in single-cell density is apparent mere minutes after treating the cells with the drug. Though this cell density increase is actually smaller (in terms of percent change) than the accompanying decreases in cell volume and mass, the naturally narrow distribution of cell densities makes the postdrug density

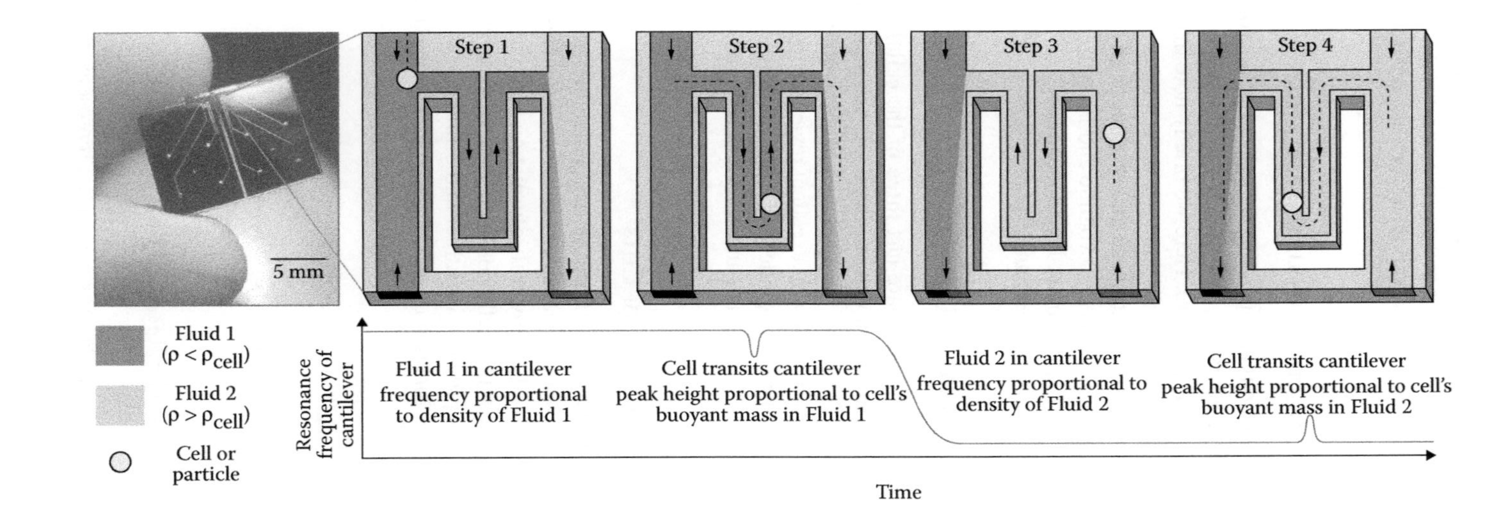

FIGURE 4.7 Measuring the density of single cells using the SMR. After determining the density of one fluid (Fluid 1) from the resonance frequency of the SMR (Step 1), the buoyant mass of the cell in Fluid 1 is measured (Step 2). The direction of fluid flow is reversed, and the density of the second fluid (Fluid 2) is measured (Step 3). Finally, the buoyant mass of the cell in Fluid 2 is measured (Step 4). From these measurements of fluid density and cell buoyant mass, the mass, volume, and density of the cell can be calculated. (From Grover, W.H., et al., *Proc. Natl Acad. Sci. USA*, 108, 10992–10996, 2011. With permission.)

increase clearly visible, even for single cells. In contrast, the naturally wide distribution of cell masses and volumes obscures the postdrug changes in mass and volume for single cells. These results suggest that single-cell density measurements could be well suited for screening libraries of chemical compounds for anticancer activity, for example.

4.4.3 Mass-Based Measurement of Cell Volume

As mentioned above, when using a Coulter counter (conventional or microfluidic) to measure cell volume, the measured volume can be affected by the shape and orientation of a cell as it passes through the aperture. Here we briefly note that since the cell volume measured by Grover et al. is obtained from two mass measurements, and those mass measurements are insensitive to the shape and orientation of the cell, the cell volume obtained by Grover et al. is a true cell volume and is unaffected by the shape and orientation of the cell. Using this mass-based technique, the volumes of even highly irregularly shaped cells or particles can be obtained.[42]

4.5 MEASURING CELL DEFORMABILITY

Cell size (mass and volume) and cell density are just three of the cellular physical properties that have been measured in microfluidic devices. Of the remaining properties, *cell deformability* (how a cell reacts to mechanical stretching or pinching) is perhaps the most interesting. Di Carlo provides an excellent review of the use of cell deformability as a biomarker.[43] Changes in cell deformability have been observed in a variety of diseases or cell states, including cancer,[44] malaria,[45] and pluripotent stem cells.[46]

4.5.1 Modeling Viscoelastic Behavior of Cells

Cell deformability is often explained by *viscoelastic models*, combinations of springs and shock absorber–like dashpots, whose mechanical behavior resembles that of a cell. To convey the remarkable amount of physical information contained within a cell-deformability measurement, we will briefly consider some of these viscoelastic models. A more rigorous analysis is provided by Ethier and Simmons.[47]

First, when we apply a force as a function of time $F_{\text{spring}}(t)$ to a spring, the spring will deform by an amount x_{spring}:

$$F_{\text{spring}}(t) = k\,x_{\text{spring}}(t), \tag{4.8}$$

where k is the spring constant. Applying a force $F_{\text{dashpot}}(t)$ to a dashpot causes a deformation x_{dashpot}:

$$F_{\text{dashpot}}(t) = \eta\,\dot{x}_{\text{dashpot}}(t), \tag{4.9}$$

where η is the damping coefficient of the dashpot and $\dot{x}_{\text{dashpot}}$ represents the differentiation of x_{dashpot} over time.

One of the simplest viscoelastic models for the cell being stretched (Figure 4.8a) is the *Maxwell body*, a spring in series with a dashpot (Figure 4.8b). The displacement $x(t)$ of a Maxwell body with applied force F is

$$\frac{x(t)}{F} = \frac{1}{k} + \frac{1}{\eta} t. \tag{4.10}$$

In words, this means that when a force is applied to a Maxwell body, there is an instantaneous deformation from stretching the spring, followed by a slow additional deformation from driving the dashpot.

In practice, the Maxwell body is usually too simple to accurately model the deformability of a cell—it does not adequately describe the tendency of a cell to *creep* or permanently deform under stress. The *Kelvin body* addresses this deficiency; it consists of a Maxwell body in parallel with a spring (Figure 4.8c) and has a more complex displacement with applied force:

$$\frac{x(t)}{F} = \frac{1}{k_1}\left(1 - \frac{k_0}{k_0 + k_1} e^{-t/\tau}\right) \quad \tau = \eta \frac{k_0 + k_1}{k_0 k_1}. \tag{4.11}$$

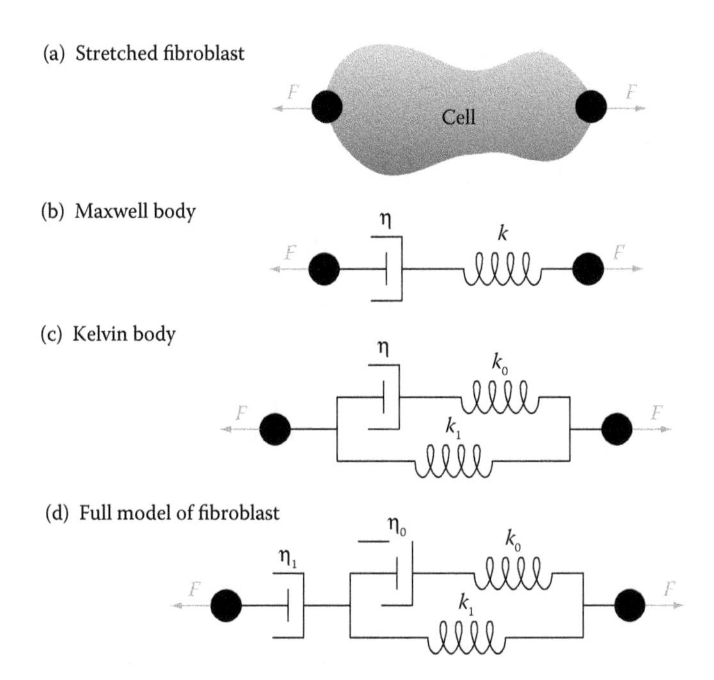

FIGURE 4.8 A stretched fibroblast (a) and three viscoelastic models for the stretched fibroblast (b–d) consisting of springs with spring constants k and dashpots with damping coefficients η. (Adapted from Ethier, C.R., *Introductory Biomechanics: From Cells to Organisms*, Cambridge University Press, Cambridge, 2007; Bausch, A.R., Ziemann, F., Boulbitch, A.A., Jacobson, K., and Sackmann, E., *Biophys. J.*, 75, 2038–2049, 1998.)

But experimental measurements show that even the Kelvin body is insufficient for modeling the deformation of a fibroblast. Only when an additional dashpot is added in series (Figure 4.8d) does the model successfully describe the deformation of a fibroblast[48]:

$$\frac{x(t)}{F} = \frac{1}{k_1}\left(1 - \frac{k_0}{k_0 + k_1} e^{-t/\tau}\right) + \frac{t}{\eta_1} \qquad \tau = \eta_0 \frac{k_0 + k_1}{k_0\, k_1}. \tag{4.12}$$

It is perhaps worthwhile to reflect on the information content of Equation 4.12 for a moment. Fitting experimental data to this model yields *four* physical constants (k_0, k_1, η_0, and η_1) for each measured cell. These four constants are the summation of a vast amount of biologically relevant information about the cell, its cytoplasm, its membrane, its cytoskeleton, and so on. Using measurements of deformability, different cells could be distinguished according to their locations on a *four-dimensional plot.* The richness of biological data provided from a single physical measurement (deformability) is thus truly impressive. And since all cells have a measurable deformability, this approach is potentially useful for all cell types.

4.5.2 Measuring Single-Cell Deformability

Four techniques have provided most of the data we have on cell deformability. First, the probe of an *atomic force microscope* (AFM) can be used to apply known forces to a cell and measure the resulting deflection.[49] Second, *optical traps* (or *optical tweezers*) use the momentum of photons to trap and stretch single cells or beads attached to cells.[50] Third, *magnetic bead microrheometry* is similar to optical trapping but uses a magnetic field to manipulate magnetic beads attached to cells.[48] Fourth, *micropipette aspiration* uses a carefully controlled pressure difference to stretch a cell's membrane and cytoplasm into the tip of a small pipette.[51]

These established techniques for measuring cell deformability have disadvantages that preclude their use in most clinical applications. In particular, the current throughput of these techniques is very low, and prospects for improving their throughput are poor. Low throughput renders these four techniques unsuitable for studies involving rare cell types present at low concentrations in a sample with a high background of other cells (as is often the case in cancer studies).

Motivated by these disadvantages, researchers have developed higher-throughput microfluidic tools for measuring the deformability of cells. Most of these tools infer deformability from the amount of time required for a cell to squeeze through tiny constrictions.[52–55] For example, the device developed by Bow et al.[55] contains an array of 3 μm wide constrictions (Figure 4.9a). When erythrocytes flow through the device, healthy cells squeeze through the constrictions quickly but erythrocytes infected with *P. falciparum* (malaria) are less deformable and take more time to squeeze through the constrictions (Figure 4.9b). The decreased deformability of infected cells is attributed to the presence of the (relatively stiff) parasite in the infected cells (Figure 4.9c). By measuring the time required for a cell to pass through the constriction array, the deformability of the cell can be measured (and the diagnosis of malaria confirmed).

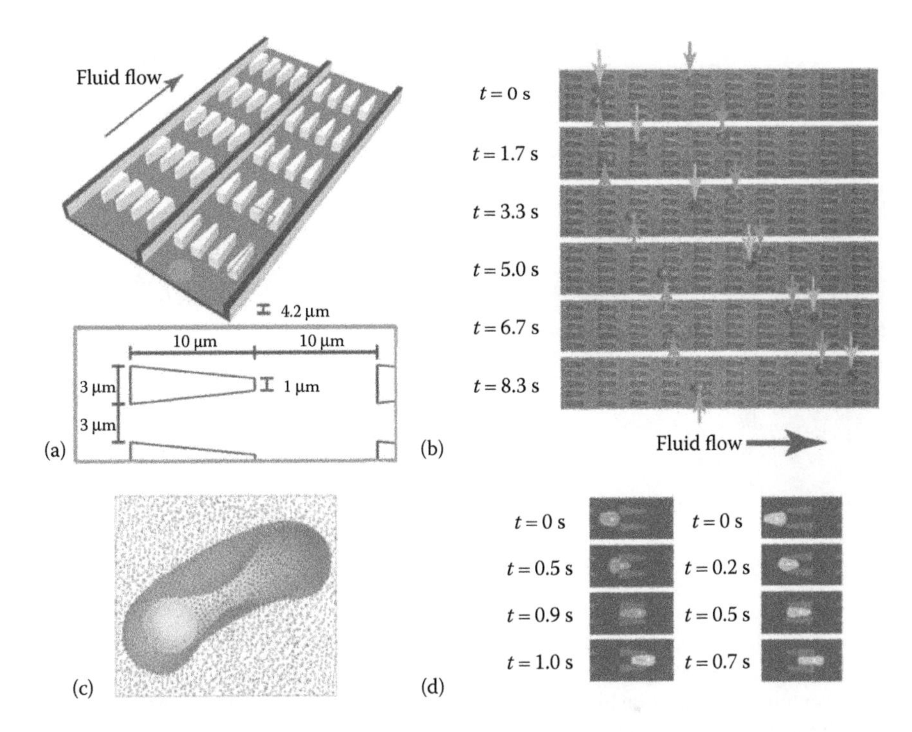

FIGURE 4.9 A constriction-based microfluidic chip for measuring the deformability of erythrocytes (a). Malaria-infected erythrocytes (dark gray arrows in b) are less deformable than healthy erythrocytes (light gray arrows) and consequently take more time to squeeze through the constrictions (d). The presence of the relatively stiff *Plasmodium falciparum* parasite in the infected erythrocytes (c) is the origin of the decreased stiffness of the infected cells. (From Bow, H., et al., *Lab Chip*, 11, 1065–1073, 2011. With permission.)

These constriction-based microfluidic tools for measuring cell deformability have much higher throughputs than earlier techniques. This enhanced throughput translates into greater clinical utility: the device by Bow et al. can measure ~100 cells per minute, making it potentially capable of diagnosing even low-parasitemia cases of malaria where only 1%–2% of erythrocytes are infected.[55]

4.5.3 Decoupling Cell Deformability and Cell Size

However, constriction-based devices for measuring cell stiffness[52–55] are sensitive to more than just cell deformability. The speed at which a cell squeezes through a constriction is also heavily influenced by *cell size*: a small cell will squeeze through a constriction faster than a large cell of equal deformability. Measurements of cell deformability can therefore be confounded by variation in cell size.

Byun et al. solved this problem by simultaneously measuring both *cell deformability* and *cell mass*.[56] Using a specialized SMR mass sensor containing a 6 μm wide constriction in the cantilever channel (Figure 4.10a), Byun et al. measured the mass and deformability of ~10^5 cells per hour. Because the SMR is sensitive to not

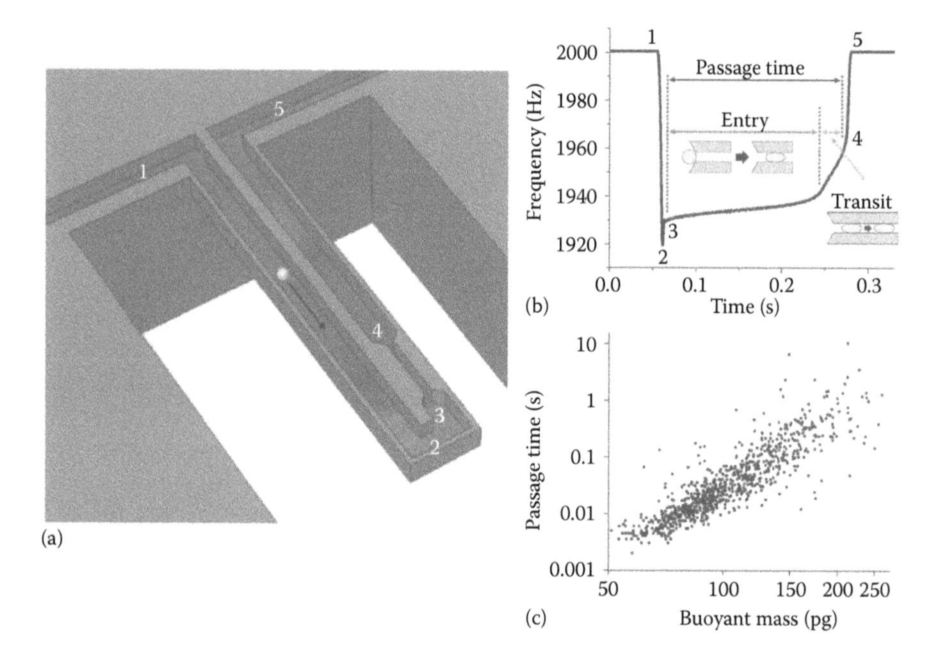

FIGURE 4.10 Measuring the mass and deformability of single cells. A specialized SMR sensor contains a constriction in the cantilever fluid channel (from 3 to 4 in a). As a cell passes through the SMR, the plot of SMR resonance frequency versus time includes not only the cell mass (the peak height 2 in b) but also the time required to squeeze the cell through the constriction (features 3, 4, and 5 in b). In the resulting plots of single-cell deformability versus mass (c), certain types of cancer cells can be distinguished from healthy cells. (From Byun, S., et al., *Proc. Natl Acad. Sci. USA*, 110, 7580–7585, 2013. With permission.)

only the *mass* of a cell (Δm in Equation 4.6) but also the *location* of a cell in the cantilever channel (α in Equation 4.6), the resonance frequency of the SMR can be used to reconstruct both how much the cell weighs and how long it took the cell to squeeze through the constriction (Figure 4.10b). By plotting passage time versus buoyant mass for single cells, different cell types can be distinguished that would be indistinguishable by passage time or buoyant mass alone (Figure 4.10c). This approach is able to detect a small number of cancer cells in a background of healthy blood cells, based solely on the cells' deformabilities and masses.[56]

4.5.4 Hydrodynamic Stretching of Cells

All of the previously mentioned constriction-based approaches for measuring cell deformability are also sensitive to *friction*: a cell that adheres to the channel walls will squeeze through the constriction more slowly than an identical cell that does not adhere to the walls. Measurements of cell deformability can therefore be confounded by variation in the interactions between the cell surface and the constriction walls.

Additionally, constriction-based devices for cell deformability measurements have the same limitation as conventional Coulter counter apertures: different-sized cells require different-sized constrictions. For example, erythrocytes squeeze through a 3 μm wide constriction[55] but lung cancer cells require a 6 μm wide constriction.[56] So constriction-based devices must be sized appropriately for a cell and each new cell size requires a new device geometry.

Just as *hydrodynamic focusing* solved the limited dynamic range and clogging susceptibility of conventional Coulter counter apertures, *hydrodynamic stretching* solves the limited dynamic range and friction sensitivity of constriction-based deformability devices. Gossett et al.[57] demonstrated a microfluidic device that uses hydrodynamic stretching to deform single cells. The device (Figure 4.11a and b) uses inertial focusers[9] to center cells in a fast-moving fluid stream. When two such flows enter a cross-shaped intersection from opposite directions, the flows split and leave the intersection through channels orthogonal to the entry channels (Figure 4.11c). As a cell passes through the middle of this cross-shaped intersection, the hydrodynamic

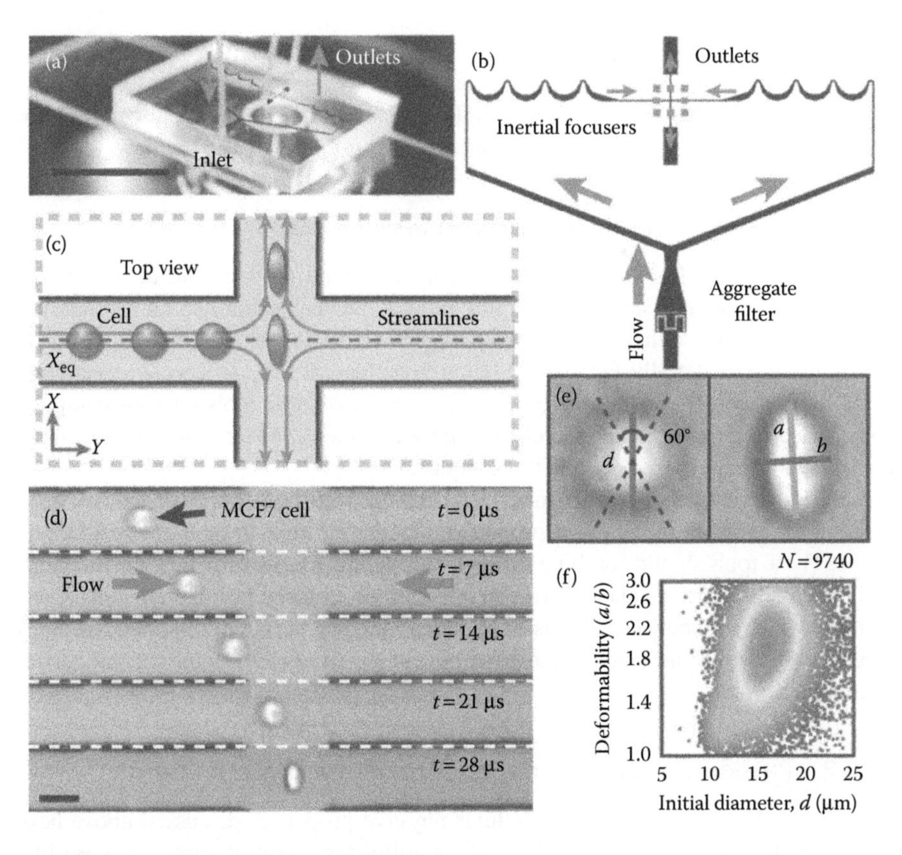

FIGURE 4.11 Measuring the deformability of cells by hydrodynamic stretching. The device (a and b) centers cells in two opposing fluid channels in a cross-shaped junction (c). Hydrodynamic forces in the junction stretch each cell. From microscope images of the cells before and after stretching (d and e), the deformability and size of single cells in a population can be measured (f). (From Gossett, D.R., et al., *Proc. Natl Acad. Sci. USA*, 109, 7630–7635, 2012. With permission.)

forces in the intersection briefly and reproducibly stretch the cell. A microscope and high-speed camera record an image of each stretched cell (Figure 4.11d). By analyzing these images (Figure 4.11e), different cell types can be discriminated based on their deformabilities and sizes (Figure 4.11f).

The hydrodynamic stretching technique has remarkably high throughput: the deformability and size of ~2000 cells per second can be measured using the microfluidic chip developed by Gossett et al.[57] Since the cells never touch the channel walls of the microfluidic device, the measured deformabilities are not confounded by friction, and the device is more resistant to clogging than devices that utilize cell-sized constrictions. And just as the effective aperture size could be adjusted in the hydrodynamic-focused Coulter counter chip,[20] the amount of cell stretching in the hydrodynamic-stretching chip can be adjusted by changing the fluid flow rates. The device is therefore capable of measuring the deformability of cells of virtually any size.

4.6 FUTURE OPPORTUNITIES

To conclude this chapter, we will briefly consider some future opportunities for microfluidic measurements of single-cell physical properties.

4.6.1 Measuring the Properties of Adherent Cells

The field of microfluidics can be justifiably criticized for focusing almost exclusively on nonadherent cells. All the microfluidic devices examined above measure nonadherent cells, either natively nonadherent cells (like blood cells) or adherent cells that are artificially forced into being nonadherent (e.g., by trypsin treatment) before measurement. Additionally, existing nonmicrofluidic tools for measuring the physical properties of adherent cells on surfaces (like AFM for deformability measurements[49]) have extremely low throughputs.

Since the vast majority of (nonbacteria) cells in your body are adherent, we need tools for measuring the physical properties of adherent cells in their native adherent state. These tools should combine the high-throughput of microfluidic measurements with the cell-to-surface and cell-to-cell interactions required by adherent cells. Cells attached to or embedded in beads or particles may play a role in this, but care must be taken to avoid the bead or particle influencing the cellular physical properties being measured.

4.6.2 Physical Sorters as Front Ends for Physical Measurements

Some of the tools for measuring cellular physical properties discussed above have relatively low throughput. This limits their utility for detecting rare cell types in a population.

The throughput of these tools could be increased significantly by using a tandem approach that combines devices for *sorting* cells according to their properties[6–13] with devices for *measuring* the physical properties of cells. By using a sorter as a front end to a measurement device, the fraction of cells of interest could first be isolated based on their physical properties, then the physical properties of only the cells of interest could be measured quantitatively using a measurement device. This tandem approach combines the high throughput of sorting devices with the quantitative results of measuring devices; it could enable novel instruments for detecting ultrarare cells (like circulating tumor cells in blood) based on their physical properties.

4.6.3 Developing a "Physical FACS"

The success of FACS instruments demonstrates the utility of *measurement followed by sorting*: by "gating" on a specific cell type, a FACS selects cells with a certain measured fluorescence, size, granularity, and so on, and saves this fraction for subsequent analysis by other techniques. A "physical FACS" could do the same, measuring the physical properties of a cell and then sorting the measured cells into fractions based on their properties. This is superficially similar to what flow-through cell sorters already do,[6–13] but by actually *measuring* the physical property of interest and then *sorting* in a binary fashion, the purity of the selected fraction of cells could be much higher (high-purity fractions of cells are often necessary for subsequent analysis by techniques such as mass spectrometry or DNA sequencing).

Recently, Sun et al. took a preliminary step toward this goal of a physical FACS by combining Coulter counter measurement with downstream dielectrophoretic cell separation.[58] Dielectrophoresis separates objects based in part on their electrical properties. Sun et al. use their device to both size a mixture of cells (via Coulter counting) and separate the sized cells by different types (via their different electrical properties using dielectrophoresis). It is important to note that their method is not a true physical FACS—the cells are *not* routed dielectrophoretically according to their measured volumes. But their work does show that two different electrical phenomena (Coulter counting and dielectrophoretic separation) can be closely integrated into a single microfluidic chip. Adjusting the dielectrophoretic force in real time to route single cells into different channels based on their measured volumes would seem to be a logical next step for their microfluidic device (possibly making it the first of many future physical FACS instruments).

4.6.4 Decoupling Physical Properties

As mentioned above, it can sometimes be difficult to measure a single physical property of a cell without that measurement being influenced by other physical properties. For example, constriction-based cell deformability measurements are also sensitive to cell size and friction, and Coulter counter measurements of cell volume are sensitive to the shape and orientation of a cell as it passes through the aperture. In this chapter, we have examined strategies for mitigating these effects, like using hydrodynamic stretching to deform cells without actually contacting the cells,[57] or measuring cell volume using mass-based measurements that are insensitive to cell shape or orientation.[36]

If we are to realize the vision of using the physical properties of cells as robust and quantitative biomarkers, we must continue to develop methods that measure single physical properties. Measured properties that conflate two or more physical aspects of a cell are inherently more difficult to interpret than fundamental physical properties like volume, mass, density, or deformability. As we have seen, the field of microfluidics is supremely well suited to provide techniques for decoupling physical properties. These current and future microfluidic tools will provide the quantitative rigor necessary for physical properties to become true universal biomarkers of cells.

4.6.5 Toward "Physical Fingerprinting" of Cells

While it is important to measure single physical properties of cells without interference from other properties, it is also valuable to measure *multiple* physical properties for each cell. For example, SMR techniques simultaneously measure the mass, volume, and density of each cell,[36] and hydrodynamic stretching simultaneously measures the deformability and size of each cell.[57] Each additional physical property measured provides an additional dimension in which different cell types may be distinguished. With enough physical properties, cells could be positively identified by their location in this multidimensional space, a sort of "physical fingerprint" for a given cell type. Since all cells have these properties, physical fingerprinting could be used for any type of cell, in an immense variety of different research and clinical applications. This vision of physical fingerprinting of cells is science fiction for now, but one could safely bet that microfluidic technologies will play a crucial role in making this vision a reality.

REFERENCES

1. Alberts, B. *Molecular Biology of the Cell.* Garland Science, New York, 2008.
2. Burg, T. P. et al. Weighing of biomolecules, single cells and single nanoparticles in fluid. *Nature* **446**, 1066–1069 (2007).
3. Marshall, W. F. et al. What determines cell size? *BMC Biol.* **10**, 101 (2012).
4. Cairns, R. A., Harris, I., McCracken, S. and Mak, T. W. Cancer cell metabolism. *Cold Spring Harb. Symp. Quant. Biol.* **76**, 299–311 (2011).
5. Cantor, J. R. and Sabatini, D. M. Cancer cell metabolism: One hallmark, many faces. *Cancer Discov.* **2**, 881–898 (2012).
6. Mohamed, H., Turner, J. N. and Caggana, M. Biochip for separating fetal cells from maternal circulation. *J. Chromatogr. A* **1162**, 187–192 (2007).
7. VanDelinder, V. and Groisman, A. Separation of plasma from whole human blood in a continuous cross-flow in a molded microfluidic device. *Anal. Chem.* **78**, 3765–3771 (2006).
8. Yamada, M., Nakashima, M. and Seki, M. Pinched flow fractionation: Continuous size separation of particles utilizing a laminar flow profile in a pinched microchannel. *Anal. Chem.* **76**, 5465–5471 (2004).
9. Di Carlo, D. Inertial microfluidics. *Lab Chip* **9**, 3038–3046 (2009).
10. Kuntaegowdanahalli, S. S., Bhagat, A. A. S., Kumar, G. and Papautsky, I. Inertial microfluidics for continuous particle separation in spiral microchannels. *Lab Chip* **9**, 2973–2980 (2009).
11. Huh, D. et al. Gravity-driven microfluidic particle sorting device with hydrodynamic separation amplification. *Anal. Chem.* **79**, 1369–1376 (2007).
12. Gossett, D. R. et al. Label-free cell separation and sorting in microfluidic systems. *Anal. Bioanal. Chem.* **397**, 3249–3267 (2010).

13. Ozkumur, E. et al. Inertial focusing for tumor antigen-dependent and -independent sorting of rare circulating tumor cells. *Sci. Transl. Med.* **5**(179), 179ra47 (2013).
14. Coulter, W. H. Means for counting particles suspended in a fluid. US Patent No. 2656508 (1953).
15. DeBlois, R. W. and Bean, C. P. Counting and sizing of submicron particles by the resistive pulse technique. *Rev. Sci. Instrum.* **41**, 909–916 (1970).
16. Zhang, H., Chon, C. H., Pan, X. and Li, D. Methods for counting particles in microfluidic applications. *Microfluid. Nanofluid.* **7**, 739–749 (2009).
17. Sun, T. and Morgan, H. Single-cell microfluidic impedance cytometry: A review. *Microfluid. Nanofluid.* **8**, 423–443 (2010).
18. Cheung, K. C. et al. Microfluidic impedance-based flow cytometry. *Cytom. A* **77**, 648–666 (2010).
19. Golden, J. P., Justin, G. A., Nasir, M. and Ligler, F. S. Hydrodynamic focusing: A versatile tool. *Anal. Bioanal. Chem.* **402**, 325–335 (2012).
20. Rodriguez-Trujillo, R. et al. High-speed particle detection in a micro-Coulter counter with two-dimensional adjustable aperture. *Biosens. Bioelectron.* **24**, 290–296 (2008).
21. Scott, R., Sethu, P. and Harnett, C. K. Three-dimensional hydrodynamic focusing in a microfluidic Coulter counter. *Rev. Sci. Instrum.* **79**, 046104-1–046104-3 (2008).
22. Bernabini, C., Holmes, D. and Morgan, H. Micro-impedance cytometry for detection and analysis of micron-sized particles and bacteria. *Lab Chip* **11**, 407–412 (2011).
23. Justin, G. A. et al. Hydrodynamic focusing for impedance-based detection of specifically bound microparticles and cells: Implications of fluid dynamics on tunable sensitivity. *Sens. Actuators B Chem.* **166–167**, 386–393 (2012).
24. Riordon, J., Mirzaei, M. and Godin, M. Microfluidic cell volume sensor with tunable sensitivity. *Lab Chip* **12**, 3016–3019 (2012).
25. Zhe, J., Jagtiani, A., Dutta, P., Hu, J. and Carletta, J. A micromachined high throughput Coulter counter for bioparticle detection and counting. *J. Micromech. Microeng.* **17**, 304 (2007).
26. Jagtiani, A. V., Carletta, J. and Zhe, J. A microfluidic multichannel resistive pulse sensor using frequency division multiplexing for high throughput counting of micro particles. *J. Micromech. Microeng.* **21**, 065004 (2011).
27. Sridhar, M. et al. Experimental characterization of a metal-oxide-semiconductor field-effect transistor-based Coulter counter. *J. Appl. Phys.* **103**, 104701 (2008).
28. Song, Y. et al. Counting bacteria on a microfluidic chip. *Anal. Chim. Acta* **681**, 82–86 (2010).
29. Sun, J., Stowers, C. C., Boczko, E. M. and Li, D. Measurement of the volume growth rate of single budding yeast with the MOSFET-based microfluidic Coulter counter. *Lab Chip* **10**, 2986–2993 (2010).
30. Song, Y., Zhang, H., Chon, C. H., Pan, X. and Li, D. Nanoparticle detection by microfluidic resistive pulse sensor with a submicron sensing gate and dual detecting channels-two stage differential amplifier. *Sens. Actuators B Chem.* **155**, 930–936 (2011).
31. Mei, Z. et al. Counting leukocytes from whole blood using a lab-on-a-chip Coulter counter. In *Proceedings of the 2012 Annual International Conference of the IEEE Engineering in Medicine and Biology Society* (*EMBC*), pp. 6277–6280. August 28–September 1, San Diego, CA. IEEE, Washington, DC, 2012.
32. Sun, J., Kang, Y., Boczko, E. M. and Jiang, X. A microfluidic cell size/density sensor by resistive pulse detection. *Electroanalysis* **25**, 1023–1028 (2013).
33. McPherson, A. L. and Walker, G. M. A microfluidic passive pumping Coulter counter. *Microfluid. Nanofluid.* **9**, 897–904 (2010).
34. Wei, J. et al. Design, fabrication and characterization of a femto-farad capacitive sensor for pico-liter liquid monitoring. *Sens. Actuators Phys.* **162**, 406–417 (2010).

35. Liu, J. et al. Selective individual primary cell capture using locally bio-functionalized micropores. *PLoS One* **8**, e57717 (2013).
36. Grover, W. H. et al. Measuring single-cell density. *Proc. Natl Acad. Sci. USA* **108**, 10992–10996 (2011).
37. Waggoner, P. S. and Craighead, H. G. Micro- and nanomechanical sensors for environmental, chemical, and biological detection. *Lab Chip* **7**, 1238–1255 (2007).
38. Lee, J. et al. Weighing nanoparticles and viruses using suspended nanochannel resonators. In *Proceedings of the 2011 IEEE 24th International Conference on Micro Electro Mechanical Systems (MEMS)*, pp. 992–994. January 23–27, Cancun. IEEE, Washington, DC, 2011.
39. Godin, M. et al. Using buoyant mass to measure the growth of single cells. *Nat. Methods* **7**, 387–390 (2010).
40. Son, S. et al. Direct observation of mammalian cell growth and size regulation. *Nat. Methods* **9**, 910–912 (2012).
41. Bryan, A. K., Goranov, A., Amon, A. and Manalis, S. R. Measurement of mass, density, and volume during the cell cycle of yeast. *Proc. Natl Acad. Sci. USA* **107**, 999–1004 (2010).
42. Godin, M. et al. Using buoyant mass to measure the growth of single cells. *Nat. Methods* **7**, 387–390 (2010).
43. Di Carlo, D. A mechanical biomarker of cell state in medicine. *J. Lab. Autom.* **17**, 32–42 (2012).
44. Suresh, S. Biomechanics and biophysics of cancer cells. *Acta Biomater.* **3**, 413–438 (2007).
45. Suresh, S. et al. Connections between single-cell biomechanics and human disease states: Gastrointestinal cancer and malaria. *Acta Biomater.* **1**, 15–30 (2005).
46. Pajerowski, J. D., Dahl, K. N., Zhong, F. L., Sammak, P. J. and Discher, D. E. Physical plasticity of the nucleus in stem cell differentiation. *Proc. Natl Acad. Sci. USA* **104**, 15619–15624 (2007).
47. Ethier, C. R. and Simmons, C. A. *Introductory Biomechanics: From Cells to Organisms*. Cambridge University Press, Cambridge, UK, 2007.
48. Bausch, A. R., Ziemann, F., Boulbitch, A. A., Jacobson, K. and Sackmann, E. Local measurements of viscoelastic parameters of adherent cell surfaces by magnetic bead microrheometry. *Biophys. J.* **75**, 2038–2049 (1998).
49. Ikai, A. et al. Nano-mechanical methods in biochemistry using atomic force microscopy. *Curr. Protein Pept. Sci.* **4**, 181–193 (2003).
50. Grier, D. G. A revolution in optical manipulation. *Nature* **424**, 810–816 (2003).
51. Shao, J. Y. and Hochmuth, R. M. Micropipette suction for measuring piconewton forces of adhesion and tether formation from neutrophil membranes. *Biophys. J.* **71**, 2892–2901 (1996).
52. Abkarian, M., Faivre, M. and Stone, H. A. High-speed microfluidic differential manometer for cellular-scale hydrodynamics. *Proc. Natl Acad. Sci. USA* **103**, 538–542 (2006).
53. Rosenbluth, M. J., Lam, W. A. and Fletcher, D. A. Analyzing cell mechanics in hematologic diseases with microfluidic biophysical flow cytometry. *Lab Chip* **8**, 1062–1070 (2008).
54. Chen, J. et al. Classification of cell types using a microfluidic device for mechanical and electrical measurement on single cells. *Lab Chip* **11**, 3174–3181 (2011).
55. Bow, H. et al. A microfabricated deformability-based flow cytometer with application to malaria. *Lab Chip* **11**, 1065–1073 (2011).
56. Byun, S. et al. Characterizing deformability and surface friction of cancer cells. *Proc. Natl Acad. Sci. USA* **110**(19), 7580–7585 (2013).
57. Gossett, D. R. et al. Hydrodynamic stretching of single cells for large population mechanical phenotyping. *Proc. Natl Acad. Sci. USA* **109**, 7630–7635 (2012).
58. Sun, J. et al. Simultaneous on-chip DC dielectrophoretic cell separation and quantitative separation performance characterization. *Anal. Chem.* **84**, 2017–2024 (2012).

5 Technologies for Low-Cost, Hall Effect–Based Magnetic Immunosensors

Simone Gambini, Karl Skucha, Jungkyu Kim, and Bernhard E. Boser

CONTENTS

5.1 Introduction 131
5.2 Assay Protocols and Microfluidic and Packaging Technologies for Magnetic Assays Using Micron-Sized Labels 134
5.2.1 Surface Chemistry for Immobilizing an Antibody 134
5.2.2 Microfluidic Integration and Assay Automation 135
5.2.3 Large Label Effects 136
5.3 Integrated Detection for Magnetic Immunoassays 138
5.3.1 State-of-the-Art Methods and Limitations 138
5.3.2 Design of Hall Effect–Based Detectors 142
5.4 Sensor Design and Optimization 142
5.5 Magnetization-Based Hall Effect Detection 143
5.6 Detectors Based on Magnetic Relaxation 144
5.7 Conclusions and Future Outlook 147
References 149

5.1 INTRODUCTION

Technologies for the high-sensitivity, specific detection of proteins are used routinely in a variety of laboratory tests, ranging from diagnostic tests for infectious disease detection to food integrity evaluation. In the current testing paradigm, samples are collected at the point of need and sent to a central laboratory for preprocessing and evaluation. Laboratories aggregate samples from a relatively large geographical area, and hence are able to amortize the cost of the instrumentation and personnel required to run each test by processing many samples in parallel. Today, the most common testing protocol is known as enzyme-linked immunosorbent assay (ELISA), and is summarized in Figure 5.1.

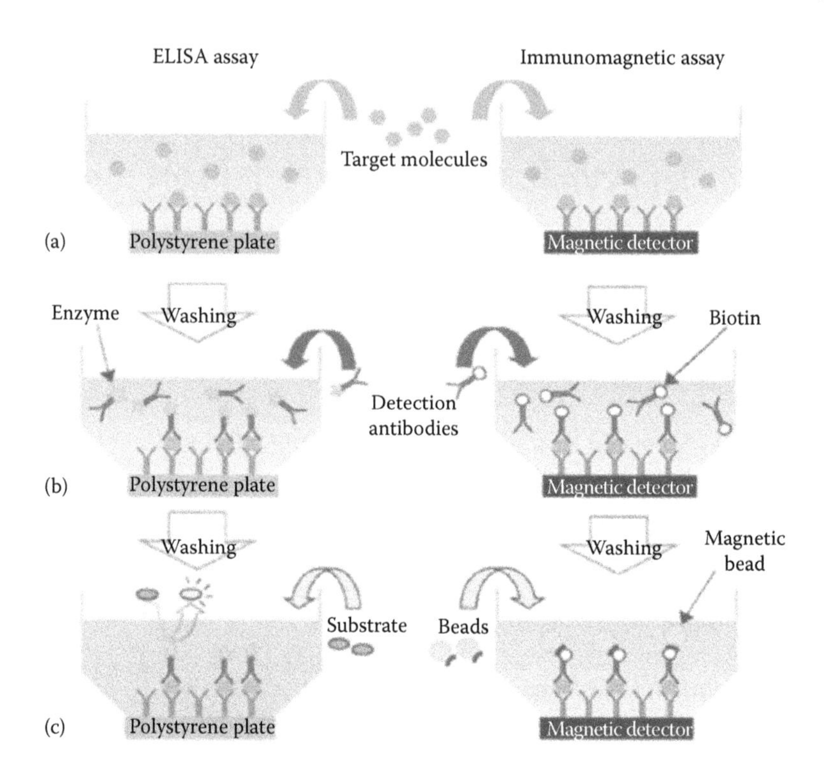

FIGURE 5.1 Comparison of 'standard' ELISA (*left*) to an immunomagnetic assay (*right*). The procedures are very similar, with major differences being the substrate and label.

The testing device has a substrate that has previously been functionalized with an antibody specific for the compound that needs to be detected (capture antibody). When the sample is introduced (a), molecules of the target bind to the capture antibodies, while other molecules do not. Some of the nontarget molecules might physically adsorb to the substrate, and are removed with a washing step. Finally, a detection antibody is introduced (b), which binds to a different epitope of the target compound. This antibody is usually labeled with an enzyme or a fluorescent marker, commonly known as the reporter label (c). The fluorescence intensity read from the device then provides a proxy for the amount of target compound. ELISA is an example of an affinity-based (because of the use of antibodies for chemical recognition), labeled assay protocol and state-of-art tests using this protocol achieve detection limits below 1 pg/ml.

The main shortcoming of the current test paradigm is the significant delay between sample collection and the processing and notification of the outcome to the end user. For example:

- In a disease detection application, patients typically are not notified of the test results before they leave the physician's facility and return home, a complication that can cause stress and generally decreases compliance. In addition, in applications such as the detection of early biomarkers of cardiac infarction (troponin) or stroke, the delay simply renders the test irrelevant.

- In food safety monitoring, this delay translates into shipping delays and ultimately increased cost, as trucks transporting bulk items need to wait for clearance before they can offload their content [1].
- In resource-limited or sparsely populated areas, the density of laboratories might be too low to ensure transport of the samples to a central facility in a cost-effective manner before they foul. This is the case in many developing countries, but also in parts of Australia, Russia, and Canada.

Moving the testing site from a central laboratory to the point of need would eliminate the delay and with it the limitations described above. Generally, a point-of-need testing device should not only have high sensitivity (the ability to detect small concentrations of sample) but should also be highly specific to minimize sample preprocessing and purification and hence the need for specialized staff. In addition, it should be inexpensive, small, and have a long shelf life.

Many point-of-need/point-of-care diagnostic devices have been proposed to address this potential market [2]. While these devices employ approaches that differ in both the biochemistry and the physical design, in this chapter we focus on labeled protocols that employ micron- or nanometer-sized super-paramagnetic particles as labels.

Figure 5.1 compares the main steps in a standard ELISA to a similar assay using magnetic beads as labels. With the exception of the different labels, both approaches proceed very similarly. The same antibodies are used in both cases, and the performance of both assays is comparable [3,4]. Magnetic particles can be functionalized with antibodies and have been used in the sciences for over 20 years [5] to purify or separate rare cells out of a population. When employed as labels, they have at least four potential advantages:

1. Blood has no low-frequency magnetic content, and does not attenuate low-frequency magnetic fields. In contrast, whole blood is optically opaque and presents significant autofluorescence. This fact suggests that magnetic labels have lower intrinsic background than optical labels, and could be used for direct testing in complex matrices such as whole blood. For the same reason, magnetic assays can potentially be performed on samples without prior centrifugation, possibly eliminating a step that is difficult to perform in a point-of-care setting.
2. Magnetic fields of up to tens of milliteslas can be generated by flowing current through a wire, creating spatially resolved magnetic forces [6]. These forces enable magnetic washing to be substituted for hydrodynamic washing in assays where microfluidic support for the washing step is not available, obviating the extra costs associated with microfluidic systems. Additionally, this magnetic manipulation can be used to speed up assay times by allowing beads to sediment to the surface via magnetic force, which can be much faster than the diffusion used in ELISA and assays with nanosized labels.
3. Magnetic fields can be measured using a variety of miniaturized magnetic sensors. In contrast, fluorescent labels are usually excited by an LED, and

lenses and optical filters are often required to isolate the fluorescent signal and focus it on a detector. This fact suggests that magnetic labels have far greater potential for system miniaturization. In fact, systems employing a single silicon chip for detection have been reported by our group [7–9], and systems reported by other groups and using two chips and an external magnet have achieved outstanding performance [10].
4. Unlike enzymatic labels, magnetic beads are inert and have a long shelf life [11].

The rest of this chapter is organized as follows: in Section 5.2, we consider assay protocols, packaging, and sample interface technologies proposed for use with magnetic immunoassays; Section 5.3 focuses on the detection of magnetic labels, with an emphasis on electronic detection through Hall effect devices; and conclusions are drawn in Section 5.4.

5.2 ASSAY PROTOCOLS AND MICROFLUIDIC AND PACKAGING TECHNOLOGIES FOR MAGNETIC ASSAYS USING MICRON-SIZED LABELS

As shown in Figure 5.1 and anticipated in the previous section, assays exploiting magnetic labels (sometimes referred to as magnetic immunoassays or MIA) employ the same basic steps and principles as optical immunoassays. However, there are a few key differences:

1. The surface on which the capture antibodies are attached (substrate) is typically not a polymer but the surface of the integrated detector, consisting of SiO_2 or SiN_3.
2. The electronic detector must be packaged using the sample-manipulation system, while fluid is prevented from reaching the electrical leads.
3. Magnetic labels often have a diameter of a few microns, as compared to the diameter of tens of nanometers of enzymatic labels or fluorophores.

The impact of these differences on the assay protocol and the surface chemistry is discussed below.

5.2.1 Surface Chemistry for Immobilizing an Antibody

The first step of an immunoassay is to immobilize the capture antibodies on a substrate. For the standard ELISA, a microplate made out of polystyrene is generally used to attach the capture antibodies by physisorption. However, physisorption does not provide a sufficiently stable binding force on silicon or silicon dioxide surfaces to immobilize the capture antibodies. In this case, the biosensor surface must be functionalized using a linker molecule. The most common functional groups for attaching a capture antibody are reactive amine, epoxy, aldehyde, and carboxylic acid. For the amine functional group, 3-aminopropyl triethoxysilane (APTES) is one of the common linkers that has reactive amine groups. The siloxane groups of APTES react with a hydroxyl group on the silicon-based surface by condensation. Then, a glutaraldehyde as bilinker molecule interconnects between APTES and antibodies [12,13]. When the APTES linker is used, the antibodies can be attached on the surface electrostatically. The APTES-modified surface has positive charges that enable interaction with streptavidin (SA). The biotinylated capture antibodies are then added onto the functionalized surface coated with SA at room temperature [4,14,15]. Another common approach is to use a carbodiimide coupling method that uses 1-ethyl-3-(3-dimethylaminopropyl)carbodiimide (EDC) and *N*-hydroxysuccinimide (NHS) chemicals to generate a carboxylic acid functional group on the capture antibodies.

Alternatively, the overall process to immobilize the capture antibody on the solid substrate can be simplified by using aldehyde silane. By condensation, the aldehyde silane can be functionalized on the silicon-based surface and then allows coupling with amine on the antibodies [16–18]. The additional linker protein, such as protein A (PrA), allows control of the direction of the capture antibody. Use of the process described in [19] significantly improved the overall capture efficiency of the target molecule by attaching the F_{ab} binding site of the capture antibody.

Once the surface is functionalized with the capture antibodies, a simple sandwich assay can be started by passivating the overall surface with blocking solution (1% BSA in phosphate buffer saline [PBS]). Various concentrations of target molecules are then loaded onto the activated surface. After a washing step, a detection antibody conjugated with labeling molecules is introduced and incubated to create a full immunocomplex. After the washing step, fluorescence and color intensity can be measured using a microplate reader or a fluorescence microscope.

5.2.2 Microfluidic Integration and Assay Automation

The overall assay process requires various sample mixing, transport, and reaction procedures including multiple sequential steps. To perform these assays effectively, a microfluidic system for reagent delivery, washing, and bead storage can be combined with the integrated sensor. The magnetic bead detector can be placed inside microfluidic systems for automated sample preparation, reagent handling, and washing, facilitating a seamless integration with components. Figure 5.2 shows a magnetic detector chip embedded in a microfluidic system [20]. Since the chip is only a few millimeters high, a printed circuit board is used for assembly.

The normally closed membrane valve that is used in Figure 5.2 provides a high closing force and enables precise manipulation of nano/picoliter sample volumes [21]. However, integration of the microvalve structures often requires the fabrication of via holes to connect the layers [22]. These via holes increase the dead volumes of the device and device complexity.

A lifting gate microfluidic control system enables facile integration with various substrates [23,24]. First, a Hall sensor was mounted on a chip carrier and all the wires were connected using a wire bonding process. Then, the chip carrier was filled with uncured polydimethylsiloxane (PDMS) for interfacing with the lifting gate microfluidic device. Figure 5.3 shows a chip carrier integrated with the microfluidic system containing three bus valves for sample selection and three valves for

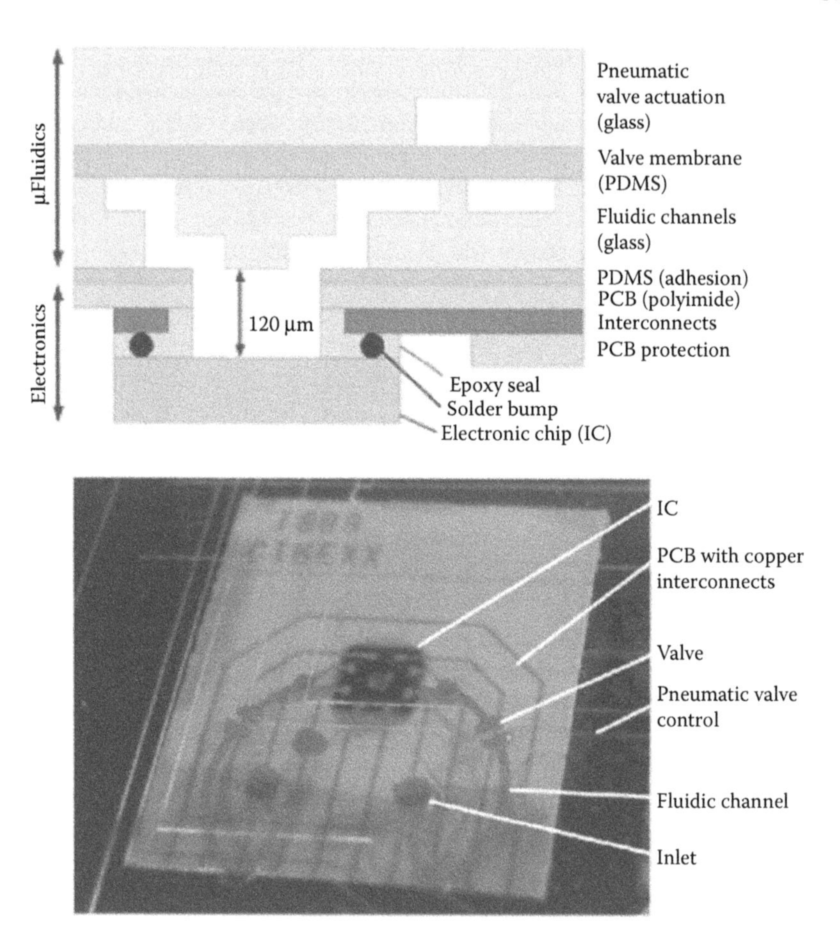

FIGURE 5.2 Immunomagnetic detector chip embedded in a microfluidic system. (*Top*) The electronic detector chip is flip-assembled onto a printed circuit board (PCB) bonded to the microfluidic device. (*Bottom*) Photo of the assembled microfluidic device with embedded chip. (From Wu, A., Wang, L., Jensen, E., Mathies, R., and Boser, B., *Lab on a Chip*, 10, 519–521, 2010.)

sample delivery without any via holes. To evaluate the overall functionality, various samples of colored water are pumped through the flow cell on the Hall sensor surface and then clear water is pumped through to demonstrate the washing capability. This technique provides rapid integration of the electronic sensor and automated microfluidic platform. The flex board method (Figure 5.2) [20] can be used for volume manufacturing.

5.2.3 Large Label Effects

As discussed earlier, magnetic labels tend to be several orders of magnitude larger than enzymatic labels in optical assays. Label size affects label diffusion behavior, binding behavior, and detectable signal strength; changing the size of the label

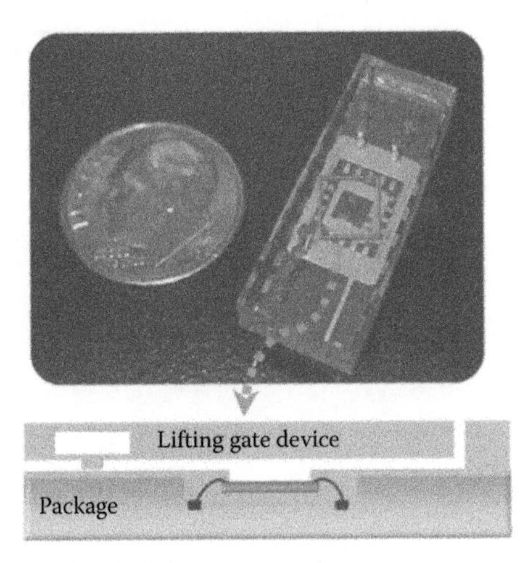

FIGURE 5.3 Lab-on-a-chip platform for biomarker screening that integrated a CMOS Hall sensor with the lifting gate microfluidic system. In the cross-sectional view, the lifting gate device was attached to the chip carrier containing the Hall sensor by manual alignment.

often requires changes to the entire assay protocol and is difficult to analyze beyond experimental means.

The larger size of labels can be exploited to speed up the dynamics of an assay. Large labels have the advantage of being easily manipulated via electromagnetic or hydrodynamic forces to enable mixing and to accelerate label sedimentation to the sensor surface, resulting in a faster assay protocol and enabling integrated washing [3,4,6]. Table 5.1 shows a comparison between the typical specifications of commercial ELISAs and two published immunoassays utilizing labels approximately a micrometer in diameter. The two large-label assays are on average 20 times faster than ELISA and have comparable detection limits, indicating that microlabels are especially suitable in situations where a fast assay protocol is important.

One particular concern with using large labels is increased noise: just 10–100 magnetic labels may constitute a positive signal, whereas in ELISA the label count is in the millions. Intuitively speaking, a result inferred from such a small number of labels may have significant errors caused by binding statistics and label

TABLE 5.1
Comparison of ELISA to Assays Utilizing Microbeads as Labels

	ELISA	Assay 1 [3]	Assay 2 [4]
Label size	5–15 nm	0.5 μm	2.8 μm
Detection limit	100 fM–10 pM	200 fM	1.7 pM
Assay time	2–4 h	4 min	13 min
Notes	Typical values	Magnetic mixing and washing	Hydrodynamic flow and washing

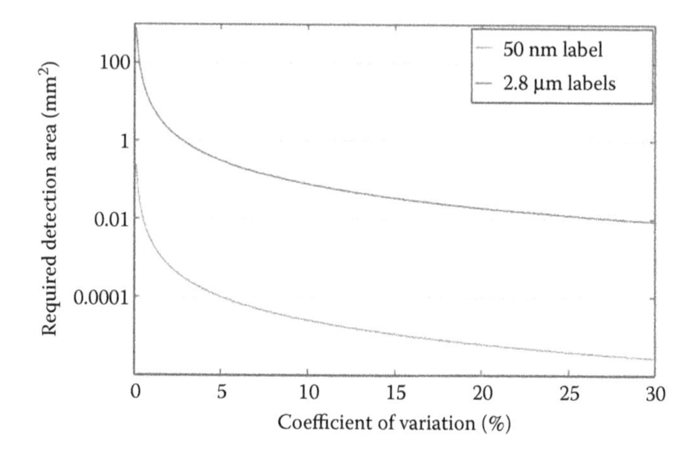

FIGURE 5.4 Minimum detection area for an assay with a dynamic range of 100 utilizing 50 nm and 2.8 μm labels versus the desired CV. When 2.8 μm labels are used, an area over 1000-fold larger is required to suppress biological shot noise.

variations. Labels bind to the surface with a Poisson distribution, as shown in [25]. Assay accuracy is typically measured using a coefficient of variation (CV), which is the standard deviation of a signal divided by the mean or the signal when the assay is performed many times under the same conditions. Typical CVs for commercial assay kits are in the range of 5% and better accuracies are rarely required for most immunodiagnostic applications. According to the Poisson distribution, to achieve a CV of 5% or less, at least 400 labels need to bind to the surface and be accounted for. In addition, typical assays require a dynamic range from approximately 100, so an assay area needs to accommodate as many as 40,000 labels on the surface, which translates to about 1 mm^2 for 2.8 μm diameter labels (see Figure 5.4). Such areas are over 1000 times larger than those required by optical labels with a 50 nm diameter, but are still practicably achievable on microchips and lab-on-chip systems.

5.3 INTEGRATED DETECTION FOR MAGNETIC IMMUNOASSAYS

5.3.1 State-of-the-Art Methods and Limitations

Since the original proposal in [26], several techniques have been proposed for the miniaturized quantification of the number of magnetic labels. In order to understand the challenges of this task, consider the problem of detecting a paramagnetic particle with volume susceptibility X and volume V, situated at a height h directly above a sensor (Figure 5.5).

A magnetic field generator creates a magnetizing field B_{mag} at the particle location, and a leakage field B_{leak} at the sensor location. The induced magnetic moment of the particle is $\vec{m} = \chi \cdot V \cdot \vec{H}_{\text{mag}}$. In order to simplify the mathematics, assume the particle is located directly above the sensor so that $\vec{h} = |h|\hat{z}$, and $B_{\text{mag}} = 1\ \text{mT} = B_{\text{leak}}$ is directed in the z direction. The magnetic field at the sensor evaluates to

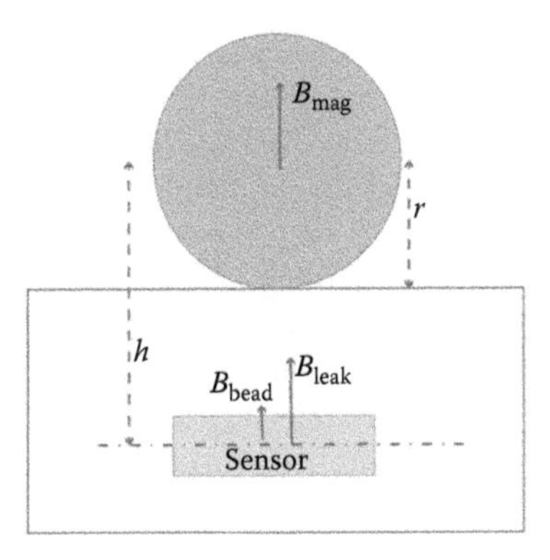

FIGURE 5.5 Generic magnetic bead detection setup.

$$B_{sens} = B_{leak} + B_{bead} = B_{leak} + \frac{\mu_0}{4\pi}\left(\frac{3\vec{h}\langle\vec{m}|\vec{r}\rangle}{h^5} - \frac{\vec{m}}{h^3}\right).$$

Last, assume the sensor only responds to magnetic fields directed along z. The magnetic field detected by the sensor is shown in Figure 5.6 as a function of the ratio of height h to particle radius r when the volume permeability of the particle $X = 1$.

As is apparent from Figure 5.6, the magnetic field from a bead is almost invariably smaller than the leakage field B_{leak} and as soon as the particle height over the sensor is significantly larger than its radius, this signal also becomes smaller than Earth's magnetic field.

Care must be taken to position the sensors as close as possible to the particle (reduce h/r), increase its magnetic sensitivity, and isolate the bead response from Earth's magnetic field through AC excitation. In addition, bead response must be separated from the magnetization field B_{mag}.

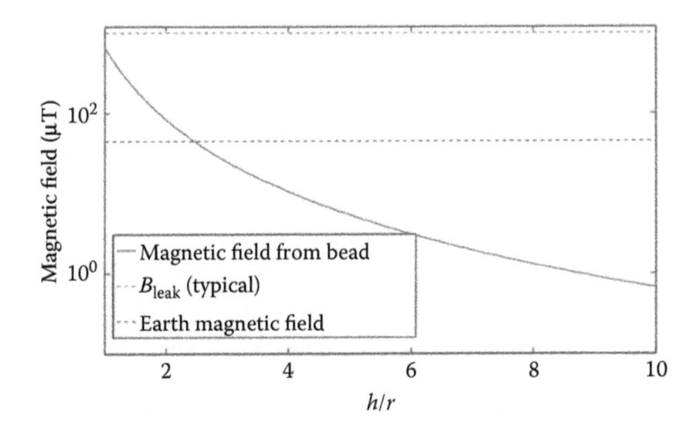

FIGURE 5.6 Typical values for bead signals.

Giant magnetoresistance (GMR) sensors were initially proposed for this purpose in [26,27] and have enjoyed popularity since [28]. They meet all of the guidelines described above. First, as the resistance of a GMR device changes by a few percentage points for each millitesla change in magnetic field, its sensitivity is excellent. In addition, GMR sensors are typically only separated from the sensing surface by a nanometer-thick passivation layer, making h/r close to 1 and effectively maximizing the signal. Last, GMR sensors do not respond to a magnetic field oriented along their hard magnetization axis. The system can then be designed to orient B_{mag} and B_{leak} along this hard axis, effectively rejecting the leakage component. In a typical configuration, a GMR device sensitive to the in-plane field and with an out-of-plane (z-directed) hard axis is employed. This has been shown to effectively reduce the leakage field, although it introduces a few issues.

First, the magnetization field must be highly uniform in the sensing area, as small components in the GMR sensitive plane can easily saturate the device. This often dictates the use of physically large, power-hungry, magnetizing coils or permanent magnets. Second, when this technique is used, the sensor response to particles located directly above it is zero, while the maximum occurs for particles off-axis (Figure 5.7) with the sign dependent on the particle location relative to the sensor. Particles located on opposite sides of the sensor induce opposite responses, and their effect is partially canceled.

In addition, fabrication cost and the difficulty of integrating high-performance GMR sensors directly on top of a CMOS are also often cited as disadvantages of this technology. GMR sensors are highly temperature sensitive, and either calibration [10] or analog compensation must be employed. Despite these drawbacks, the highest-sensitivity magnetic immunosensor reported to date utilizes GMR devices for magnetic transduction [29]. GMR sensors using permanent magnets to generate B_{mag} have also been successfully used to detect magnetic particles transiting over the sensor in a magnetic flow cytometer [30].

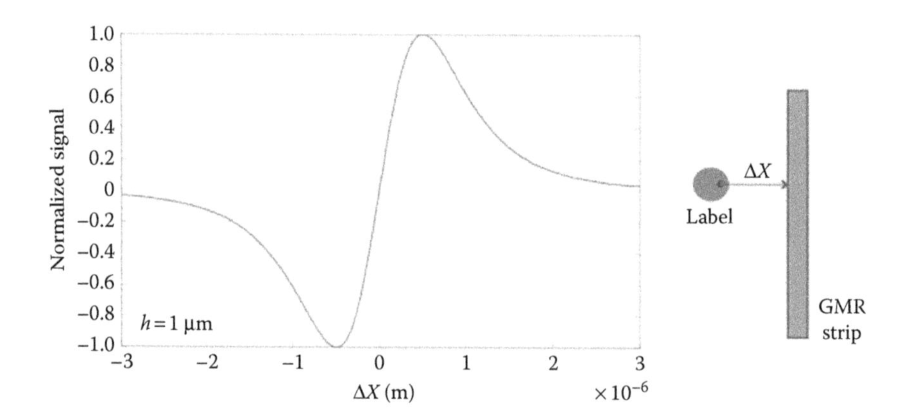

FIGURE 5.7 Normalized GMR sensor response to a 1 μm radius bead as a function of horizontal position. The bias field is in the x direction, while the sensitivity axis is z. Passivation thickness is assumed to be negligible.

Inductive sensors based on planar CMOS coils have also been proposed [31,32]. Current flowing in the coil generates a magnetic field. When a paramagnetic particle is introduced in the magnetic field, energy is stored in the particle by changing the orientation of the magnetic domains, increasing the effective inductance seen at the inductor input port. In [31,32], the inductance change is measured by building an integrated oscillator and measuring the change in its natural frequency with a technique reminiscent of first-generation metal-detector devices. The main advantages of this inductive sensor are its close proximity to the labels (the coil is realized in the top metal layer of a 45 nm CMOS process, so it is separated by roughly 300 nm from the sensing surface) and its low cost. However, this strategy has significant drawbacks. First, the sensing method responds not only to change in inductance but also to change in the loss associated with such inductance (this is indeed the principle of operation of early metal detectors). As a result, the sensor might perform poorly when the solution's ionic concentration fluctuates. Second, frequency-based detection through an integrated oscillator requires the use of an excitation field with frequency in the GHz range. Many super-paramagnetic labels have dynamics significantly slower than 1 GHz, resulting in reduced response. Third, the inductance change associated with the magnetic particles is very small (100 ppm/particle in [31]) and makes rejecting B_{leak} in an uncontrolled environment especially challenging.

Last, optical detection of magnetic labels has been proposed in [3,33]. This method has the advantage of a high signal-to-noise ratio when used in combination with large (4.5 μm) labels. The magnetic nature of the labels is in this case primarily used for manipulation, and the potential advantage of detection in whole blood is lost.

Our group [9,34] has primarily focused on magnetic detection using Hall effect sensors, which are also readily available in any CMOS process. The use of Hall effect sensors has similar cost and integration advantages to the use of inductive sensing, but without the sensitivity to loss or the mismatch between magnetization field and particle dynamics that affect the former. There are two crucial challenges when using Hall sensors:

1. Hall devices are separated from the sensing surface by the entire CMOS metal stack, which can be thicker than 12 μm for a 0.18 μm CMOS process. This results in a h/r ratio of approximately 7 for 4.5 μm particles, giving an extremely small signal of 2 μT.
2. Since the Hall effect is a manifestation of Lorenz's force, the sensitivity of Hall sensors is proportional to carrier velocity, and ultimately mobility. In fact, the output of a Hall effect sensor is given by [35]:

$$S = \frac{V_{\text{Hall}}}{B} = \mu G_H V_{\text{bias}}.$$

The factor G_H depends on the sensor geometry and has a maximum value around unity for an approximately square sensor. For $\mu = 1500\ \text{cm}^2/(\text{V s})$ and $V_{\text{bias}} = 1$ V, $S = 150$ mV/T, between 100 and 1000 times lower than what is achievable by GMR devices.

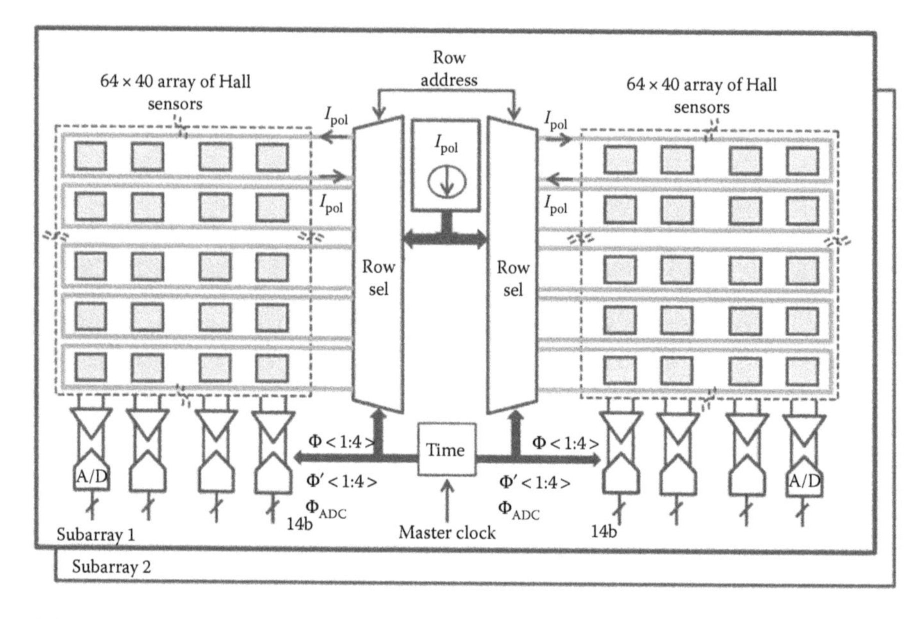

FIGURE 5.8 Block diagram of Hall effect detection platform.

As shown in the next section, system optimization and circuit design can be used to overcome these challenges and design a magnetic detection platform that, using these sensors, achieves input-referred noise within a factor of 10 of GMR-based systems.

5.3.2 Design of Hall Effect–Based Detectors

Our Hall effect detection platform is comprised of a Hall sensor array, a set of wires generating the magnetizing field B_{mag}, and readout electronics (Figure 5.8).

The Hall sensor array is 0.6 mm^2 and is chosen so that Poisson noise [36] causes the resulting assay results to have a CV of <5% at the detection limit [25], resulting in a minimum number of labels on the array of $N_{beads} = 400$. To sense the highly localized bead field, the sensing area is subdivided into four subarrays of 64×40 Hall sensor pixels, for a total of 10,240 sensors.

Column parallel electronics using 160 amplification and digitization chains ensures fast readout.

5.4 SENSOR DESIGN AND OPTIMIZATION

The first step toward the construction of a sensitive detection platform is sensor device optimization. The Hall sensor is realized using the N-well layer of the CMOS process, which offers the highest mobility. The most significant performance improvement is obtained by minimizing the r/h ratio, which can be done by etching through part of the metal stack. The devices described in this chapter use a DRIE step to reduce the distance between the Hall transducer and the sensing surface to 3.2 μm (cross section shown in Figure 5.9), reducing h/r to <2 for 4.5 μm labels,

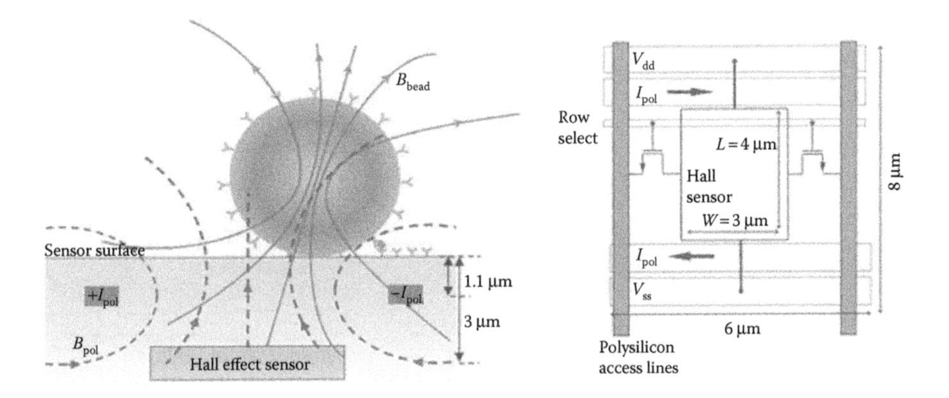

FIGURE 5.9 Cross section and top view of realized sensor pixel.

TABLE 5.2
Values of Baseline (B_{leak}) and Signal B_{bead} for Commercial Beads in Different Detection Modes

	Amplitude	Relaxation
Baseline	2.5 mT	5 μT
4.5 μm label (M450)	80 μT/bead	11 μT/bead
2.8 μm label (M280)	11.2 μT/bead	1.4 μT/bead
1 μm label (My1)	3.75 μT/bead	0.5 μT/bead

and to 4 for 1 μm labels, and improving the magnetic field B_{bead} to 80 and 11.2 μT, respectively (Table 5.2).

Since the same integrated circuit generates B_{mag} and detects B_{bead}, we further optimize system performance across the generator–detector interface. For a given sensor, when the system is limited by thermal noise, the energy consumed by the detector circuits is proportional to B_{bead}^{-2}. On the other hand, power consumed by the magnetic field generator is proportional to B_{bead}. As a result, the magnetizing field should be maximized. In practice, the use of a high magnetic field is limited by saturation effects in the magnetic particles (occurring around B_{mag} = 20 mT) and by increased thermal effects and layout complexities arising from high current levels [25].

5.5 MAGNETIZATION-BASED HALL EFFECT DETECTION

The most straightforward approach for bead detection is to evaluate the magnetic field at each sensor while the magnetizing field B_{mag} (and hence the leakage field B_{leak}) is present (Figure 5.10, *left*). In the following, we refer to this approach as magnetization detection. For our design, the values of B_{mag}, B_{leak}, and B_{bead} for different labels are reported in the first column of Table 5.2 for magnetization detection for several types of Dynal beads.

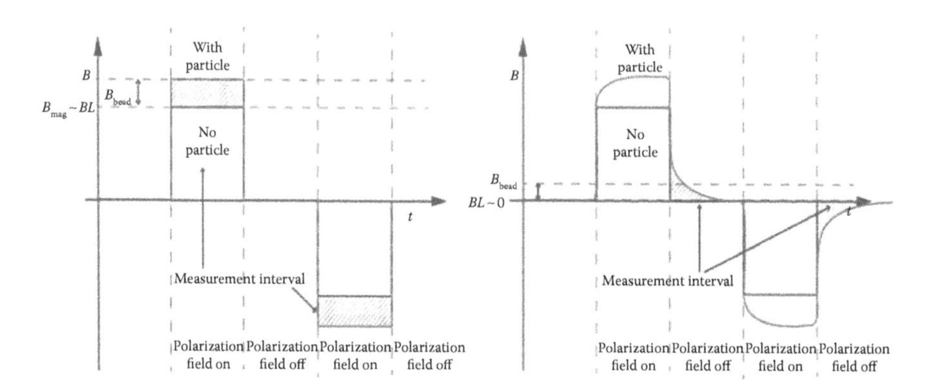

FIGURE 5.10 Relaxation measurement (*right*) compared to magnetization measurement (*left*) BL indicates the measurement baseline.

It is apparent from the values in Table 5.2 that a high degree of stability is required in the interface gain and offset. Taking as an example the case of 4.5 μm beads, the ratio of B_{leak} to B_{bead} is 30. As a result, a 3% change in the cascaded gain or offset of the transducer, electronic interface, and magnetic field generator gives an output as large as the signal from a bead. For the case of 1 μm beads, the ratio of B_{leak} to B_{bead} is 800 and the relative stability required is 0.12%. This scenario becomes even worse when the signal from the multiple sensors is averaged to obtain a quantification of the number of beads across the entire sensing array [25,37]. Since at the limit of detection the number of labels is much smaller than the number of Hall sensors being read out, in this case the required stability is enhanced by a factor at least equal to $\sqrt{N_{\text{sensor}}/N_{\text{beads}}} = 5$. The above issue can be mitigated by the use of reference sensors that have a nonfunctionalized surface. Due to the third-order dependence of the magnetic field on distance and the sensitivity of Hall effect sensors to stress, matching the reference sensor to the active sensor is only possible to within a few percentage points, making this technique inadequate for detection of particles smaller than 4.5 μm unless calibration is used [34].

5.6 DETECTORS BASED ON MAGNETIC RELAXATION

To overcome the stability challenges of magnetization detection, we proposed in [8] to exploit the intrinsic dynamics of magnetization in super-paramagnetic materials to obtain a baseline-free measurement. Figure 5.10 highlights this measurement. The time-dependent magnetization M_{dec} of a magnetic domain of volume V when an external field B_0 is applied for a long time and then suddenly switched off at time $t=0$ is given by [38]:

$$M_{\text{dec}}(t) = M_0 e^{-(t/\tau)},$$

$$\tau = \tau_0 e^{-K_m V/KT},$$

$$M_0 = \frac{B_0}{\mu}\chi V,$$

where K_m is the material magnetic anisotropy, χ its volume susceptibility, and $\tau_0 \sim 1$ ns. The particles used in this system and in [8] have $\chi \sim 1$ and $\tau \sim 100$–300 ns at room temperature. Since it is possible to switch off the on-chip magnetic field generator with a much faster timescale than this τ, we can measure the magnetic field generated by the labels free of any baseline by removing B_{mag} (and hence B_{leak}) immediately before connecting the readout electronics to the sensor. As shown in Table 5.2, we obtain an approximately 300-fold reduction in baseline. Because the signal B_{bead} is now decaying, this comes at the cost of an eightfold reduction in signal strength for commercial particles [37]. The electronic circuit used to perform this measurement is shown in Figure 5.11 together with a timing diagram.

The electronic circuit uses a combination of autozeroing and chopping to suppress the Hall sensor electrical offset, which can be as large as 400 mT, as well as 1/f noise from the electronics.

Only the signal from bead relaxation is converted into a DC current at the output of the V/I converter; this current is then digitized by a first-order incremental A/D converter. The A/D converter samples the signal current inside the loop, resulting in boxcar sampling [39] and further noise reduction. A nested chopper loop with

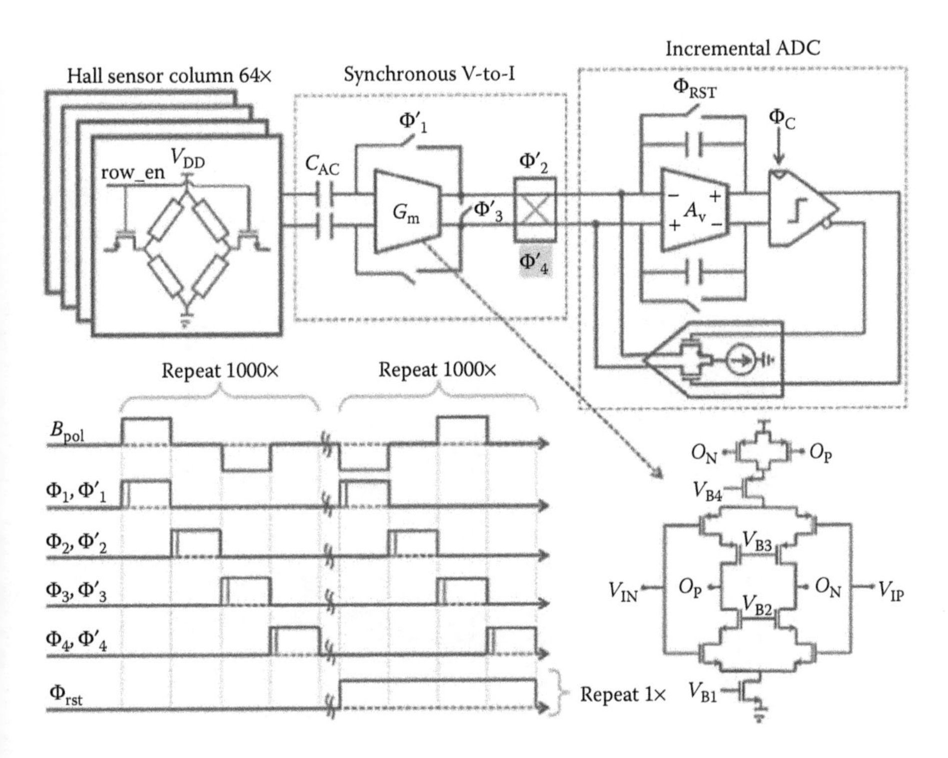

FIGURE 5.11 Electronics interface schematic and timing diagram.

digital demodulation eliminates the input offset due to the A/D converter (up to 500 μT). The up-conversion mechanism of this chopper is implemented, modulating the phase of the magnetization field (Figure 5.11). Each readout chain spans the width of two columns of the Hall sensor array. Electronics are therefore placed at the top and bottom periphery of the array to ensure that one readout serves each column without complex routing. Each chain consumes 330 μW from a 2 V supply to achieve an input-referred noise of 10.8 nV/rt(Hz), or (once referred to the Hall sensor input) 216 nV/rt(Hz). By comparison, the GMR-based detection system in [40] achieves an input-referred noise of 50 nT/rt(Hz) while using sensors that are >50-fold more sensitive. The total power consumption of the readout is 66 mW with all 160 channels active. An additional 128 mW is consumed by the magnetization field generator. Finally, the Hall sensor array consumes 352 mW, bringing the total power consumption to 546 mW. Since the system needs approximately 8 s to measure the number of 2.8 μm beads present on its surface, the energy cost of the detection operation is only 4 J.

The whole system is integrated in a 3.5 mm × 5.1 mm × 0.18 μm CMOS chip (Figure 5.12) and has on average a baseline of only 7 μT, a reduction of >300-fold compared to the magnetizing field of 2.5 mT.

Figure 5.13 shows the measured response of this interface as a function of the number of beads present on the sensor surface (optical count). Extrapolated noise floors range from 8 beads (for M450) to 200 beads (for 1 μm My1 beads) and are comparable with the noise floor from biological shot noise [37].

The ultimate ability of the system to detect field from small particles is evaluated through Allan deviation, which is shown in Figure 5.14 as a function of time for both magnetization and relaxation detection methods. The minimum Allan deviation for relaxation measurements is 9 nT and occurs at a measurement time of approximately 30 min, while the equivalent figures for magnetization are 120 nT and 30 s.

This limit is difficult to reach in practice as it only applies to temperature-stable laboratory conditions. As a result, the reduced baseline of the relaxation measurement

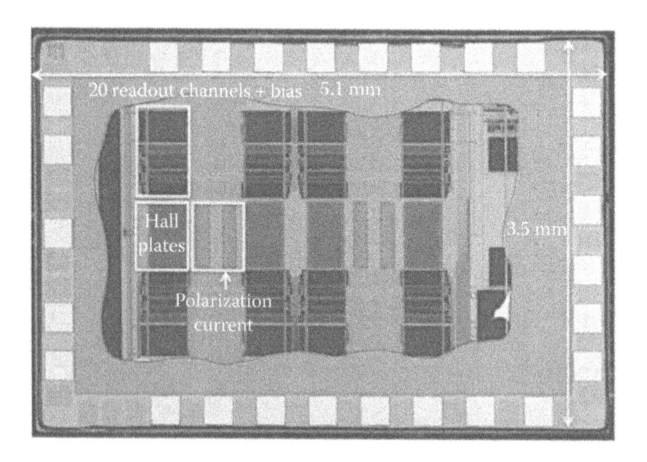

FIGURE 5.12 Detector chip photograph.

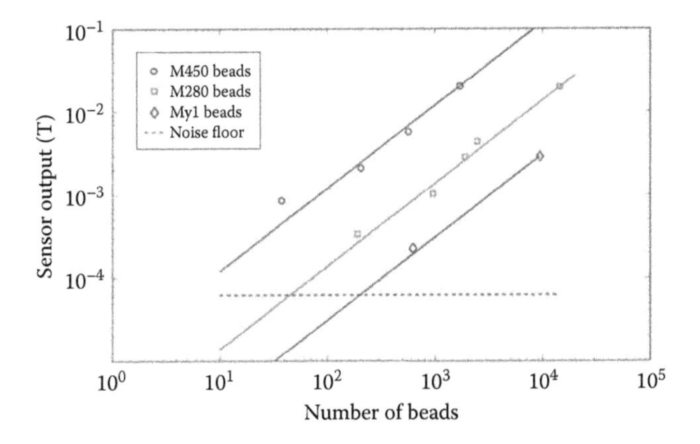

FIGURE 5.13 Digitized sensor output as a function of number of labels.

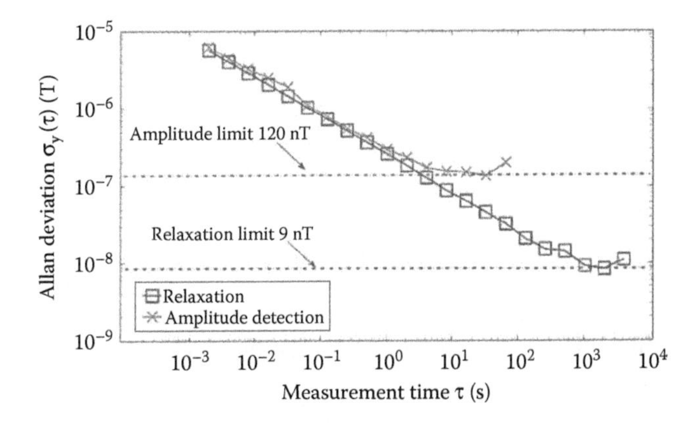

FIGURE 5.14 Allan deviation in relaxation and magnetization measurements.

is even more attractive as it also translates to simpler calibration and lower temperature sensitivity. In Figure 5.15 and in [8], we showed that the temperature sensitivity of this platform is approximately one 2.8 μm bead for a 1°C temperature increase in magnetization. For relaxation measurements, this figure is decreased approximately 40-fold.

5.7 CONCLUSIONS AND FUTURE OUTLOOK

Magnetic immunoassays could have substantial advantages compared to optical methods, giving the possibility to drastically reduce sample processing, combine separation and detection, and enable system miniaturization through nonoptical detection. In this chapter, we reviewed several research-grade detector technologies that are currently available, focusing on the low-cost Hall effect–based technology developed in our group.

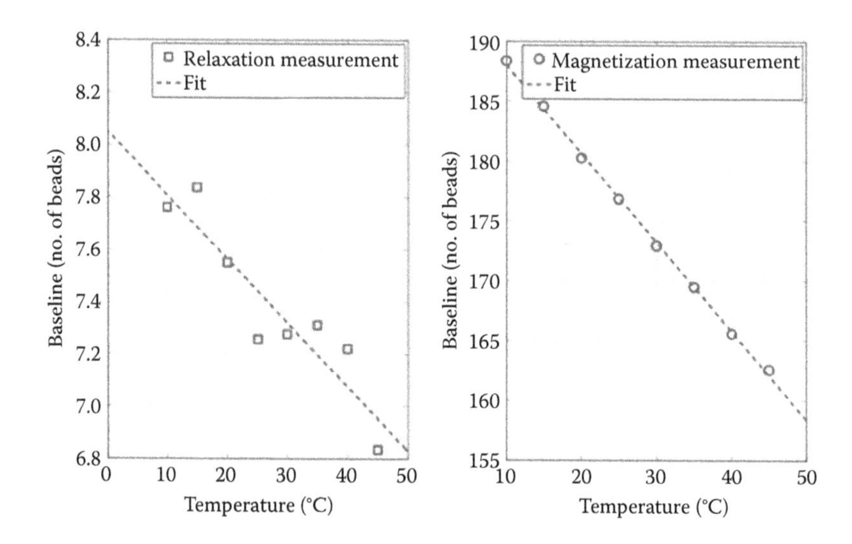

FIGURE 5.15 Temperature dependence of baseline for relaxation (*left*) and magnetization (*right*) measurements.

Further work is however required in a number of areas in order to make magnetic immunoassays practical. On the biochemistry side, there is still significant resistance toward the idea that micron-sized magnetic beads, when bound to nanometer-sized antibodies, do not alter their binding kinetics, degrading assay sensitivity or selectivity. Further research is required to either disprove or confirm this hypothesis. Demonstration of integrated detection with separation in a complex matrix is also still lacking, although progress in this direction has been achieved using electrochemical detection methods.

In the meantime, excellent results have been obtained by [28] using 50 nm nanoparticles as tags. Due to their small size, these labels are expected to have much less significant interaction with the antibody binding kinetics and are hence more likely to be accepted by the assay community as labels. The task of detecting such small particles in a completely integrated fashion is however far from accomplished. Fortunately, single-nanoparticle detection is not necessary. For example, in [10] a GMR-based integrated interface is reported that is capable of detecting 5000 nanoparticles while achieving a biological limit of detection of 10 fM. An external magnet is still required to generate a uniform magnetizing field.

Hall effect detection of nanometer-sized particles has not been reported. However, a calculation shows that the magnetization field from an aggregate of 5000 nanoparticles, in the system described in [37], is 7.2 μT. If the relaxation behavior of these particles were the same as that of Dynal beads, the average relaxation field would be 0.9 μT, well within the detection range of the system. Experiments with nanoparticles failed however to confirm this hypothesis, likely due to a different relaxation behavior. As a result, while the development of a Hall effect–based detector capable of sensing minute concentrations of nanoparticles is an open research problem, it is likely achievable with current technology.

The Hall effect sensing could also be potentially improved through technology scaling. For example, a metal-stack height of 90 nm is approximately half of that of 180 nm, giving an eightfold improvement in field strength for small particles. In practice, the improvement is less than eightfold because the thinner metals and lower supply voltage also require larger on-chip current-carrying wires and hence a larger pixel size, ultimately reducing the magnetizing field [25]. To date, no Hall effect bead detector has been realized in technologies with feature size smaller than 180 nm, making this a natural next goal for research.

REFERENCES

1. Palchetti, I. and M. Mascini. Electroanalytical biosensors and their potential for food pathogen and toxin detection. *Analytical and Bioanalytical Chemistry* 391(2): 455–471 (2008).
2. Myers, F. B. and L. P. Lee. Innovations in optical microfluidic technologies for point-of-care diagnostics. *Lab on a Chip* 8(12): 2015–2031 (2008).
3. Bruls, D. M., T. H. Evers, J. A. H. Kahlman, P. J. W. van Lankvelt, M. Ovsyanko, E. G. M. Pelssers, J. J. H. B. Schleipen, et al. Rapid integrated biosensor for multiplexed immunoassays based on actuated magnetic nanoparticles. *Lab on a Chip* 9(24): 3504–3510 (2009).
4. Kim, J., E. Jensen, M. Megens, B. E. Boser, and R. Mathies. Integrated microfluidic processor for solid phase capture immunoassays. *Lab on a Chip* 11(18): 3106–3112 (2011).
5. Miltenyi, S., W. Müller, W. Weichel, and A. Radbruch. High gradient magnetic cell separation with MACS. *Cytometry* 11(2): 231–238 (1990).
6. Florescu, O., K. Wang, P. Au, J. Tang, E. Harris, P. R. Beatty, and B. E. Boser. On-chip magnetic separation of superparamagnetic beads for integrated molecular analysis. *Journal of Applied Physics* 107(5): 054702 (2010).
7. Gambini, S., K. Skucha, P. Liu, J. Kim, R. Krigel, R. Mathies, and B. E. Boser. A CMOS 10 kPixel baseline-free magnetic bead detector with column-parallel readout for miniaturized immunoassays. In *Proceedings of the 2012 IEEE International Solid-State Circuits Conference Digest of Technical Papers (ISSCC)*, pp. 126–128. 19–23 February 2012, IEEE, San Francisco, CA.
8. Liu, P., K. Skucha, M. Megens, and B. Boser. A CMOS Hall-effect sensor for the characterization and detection of magnetic nanoparticles for biomedical applications. *IEEE Transactions on Magnetics* 47(10): 3449–3451 (2011).
9. Aytur, T., J. Foley, M. Anwar, B. Boser, E. Harris, and P. R. Beatty. A novel magnetic bead bioassay platform using a microchip-based sensor for infectious disease diagnosis. *Journal of Immunological Methods* 314(1): 21–29 (2006).
10. Gaster, R. S., D. A. Hall, C. H. Nielsen, S. J. Osterfeld, H. Yu, K. E. Mach, R. J. Wilson, et al. Matrix-insensitive protein assays push the limits of biosensors in medicine. *Nature Medicine* 15(11): 1327–1332 (2009).
11. Lee, H., D. Ham, and R. M. Westervelt. *CMOS Biotechnology*. Springer, New York (2007).
12. Diao, J., D. Ren, J. R. Engstrom, and K. H. Lee. A surface modification strategy on silicon nitride for developing biosensors. *Analytical Biochemistry* 343(2): 322–328 (2005).
13. Ouyang, H., L. A. DeLouise, B. L. Miller, and P. M. Fauchet. Label-free quantitative detection of protein using macroporous silicon photonic bandgap biosensors. *Analytical Chemistry* 79(4): 1502–1506 (2007).
14. Sivagnanam, V., B. Song, C. Vandevyver, and M. A. M. Gijs. On-chip immunoassay using electrostatic assembly of streptavidin-coated bead micropatterns. *Analytical Chemistry* 81(15): 6509–6515 (2009).

15. Sivagnanam, V., A. Sayah, C. Vandevyver, and M.A.M. Gijs. Micropatterning of protein-functionalized magnetic beads on glass using electrostatic self-assembly. *Sensors and Actuators B: Chemical* 132(2): 361–367 (2008).
16. Choi, S. H., J. W. Lee, and S. J. Sim. Enhanced performance of a surface plasmon resonance immunosensor for detecting Ab-GAD antibody based on the modified self-assembled monolayers. *Biosensors and Bioelectronics* 21(2): 378–383 (2005).
17. Kumeria, T., M. D. Kurkuri, K. R. Diener, L. Parkinson, and D. Losic. Label-free reflectometric interference microchip biosensor based on nanoporous alumina for detection of circulating tumour cells. *Biosensors and Bioelectronics* 35(1): 167–173 (2012).
18. Kim, J., J. Elsnab, C. Gehrke, J. Li, and B. K. Gale. Microfluidic integrated multi-walled carbon nanotube (MWCNT) sensor for electrochemical nucleic acid concentration measurement. *Sensors and Actuators B: Chemical* 185: 370–376 (2013).
19. Yuan, Y., H. He, and L. J. Lee. Protein A-based antibody immobilization onto polymeric microdevices for enhanced sensitivity of enzyme-linked immunosorbent assay. *Biotechnology and Bioengineering* 102(3): 891–901 (2009).
20. Wu, A., L. Wang, E. Jensen, R. Mathies, and B. Boser. Modular integration of electronics and microfluidic systems using flexible printed circuit boards. *Lab on a Chip* 10(4): 519–521 (2010).
21. Grover, W. H., A. M. Skelley, C. N. Liu, E. T. Lagally, and R. A. Mathies. Monolithic membrane valves and diaphragm pumps for practical large-scale integration into glass microfluidic devices. *Sensors and Actuators B: Chemical* 89(3): 315–323 (2003).
22. Skelley, A. M., J. R. Scherer, A. D. Aubrey, W. H. Grover, R. H. Ivester, P. Ehrenfreund, F. J. Grunthaner, J. L. Bada, and R. A. Mathies. Development and evaluation of a microdevice for amino acid biomarker detection and analysis on Mars. *Proceedings of the National Academy of Sciences of the United States of America* 102(4): 1041–1046 (2005).
23. Kim, J., M. Kang, E. C. Jensen, and R. A. Mathies. Lifting gate polydimethylsiloxane microvalves and pumps for microfluidic control. *Analytical Chemistry* 84(4): 2067–2071 (2012).
24. Kim, J., E. C. Jensen, A. M. Stockton, and R. A. Mathies. Universal microfluidic automaton for autonomous sample processing: Application to the Mars organic analyzer. *Analytical Chemistry* 85(16): 7682–7688 (2013).
25. Skucha, K., S. Gambini, P. Liu, M. Megens, J. Kim, and B. E. Boser. Design considerations for CMOS-integrated Hall-effect magnetic bead detector for biosensor applications. *IEEE Journal of Microelectromechanical Systems* 22(6): 1327–1338 (2013).
26. Edelstein, R. L., C. R. Tamanaha, P. E. Sheehan, M. M. Miller, D. R. Baselt, L. J. Whitman, and R. J. Colton. The BARC biosensor applied to the detection of biological warfare agents. *Biosensors and Bioelectronics* 14(10): 805–813 (2000).
27. Megens, M. and M. Prins. Magnetic biochips: A new option for sensitive diagnostics. *Journal of Magnetism and Magnetic Materials* 293(1): 702–708 (2005).
28. Osterfeld, S. J., H. Yu, R. S. Gaster, S. Caramuta, L. Xu, S.-J. Han, D. A. Hall, et al. Multiplex protein assays based on real-time magnetic nanotag sensing. *Proceedings of the National Academy of Sciences of the United States of America* 105(52): 20637–20640 (2008).
29. Gaster, R. S., D. A. Hall, and S. X. Wang. nanoLAB: An ultraportable, handheld diagnostic laboratory for global health. *Lab on a Chip* 11(5): 950–956 (2011).
30. Helou, M., M. Reisbeck, S. F. Tedde, L. Richter, L. Bär, J. J. Bosch, R. H. Stauber, E. Quandt, and O. Hayden. Time-of-flight magnetic flow cytometry in whole blood with integrated sample preparation. *Lab on a Chip* 13(6): 1035–1038 (2013).
31. Wang, H., Y. Chen, A. Hassibi, A. Scherer, and A. Hajimiri. A frequency-shift CMOS magnetic biosensor array with single-bead sensitivity and no external magnet. In *Proceedings of the IEEE International Solid-State Circuits Conference-Digest of Technical Papers, ISSCC*, pp. 438–439. 8–12 February 2009, IEEE, San Francisco, CA.

32. Wang, H., C. Sideris, and A. Hajimiri. A frequency-shift based CMOS magnetic biosensor with spatially uniform sensor transducer gain. In *Proceedings of the IEEE Custom Integrated Circuits Conference* (*CICC*), pp. 1–4. 19–22 September 2010, IEEE, San Jose, CA.
33. Lehmann, U., M. Sergio, S. Pietrocola, E. Dupont, C. Niclass, M. A. M. Gijs, and E. Charbon. Microparticle photometry in a CMOS microsystem combining magnetic actuation and in situ optical detection. *Sensors and Actuators B*: *Chemical* 132(2): 411–417 (2008).
34. Florescu, O., M. Mattmann, and B. E. Boser. Fully integrated detection of single magnetic beads in complementary metal-oxide-semiconductor. *Journal of Applied Physics* 103(4): 046101–046101-3 (2008).
35. Popović, R. S. *Hall Effect Devices*. Taylor & Francis, Abingdon, UK (2004).
36. Hassibi, A., S. Zahedi, R. Navid, R. W. Dutton, and T. H. Lee. Biological shot-noise and quantum-limited signal-to-noise ratio in affinity-based biosensors. *Journal of Applied Physics* 97(8): 084701 (2005).
37. Gambini, S., K. Skucha, P. P. Liu, J. Kim, and R. Krigel. A 10 kPixel CMOS Hall sensor array with baseline suppression and parallel readout for immunoassays. *IEEE Journal of Solid-State Circuits* 48(1): 302–317 (2013).
38. Fannin, P. C. and S. W. Charles. On the calculation of the Neel relaxation time in uniaxial single-domain ferromagnetic particles. *Journal of Physics D, Applied Physics* 27(2): 185–188 (1994).
39. Van de Plassche, R. J. *CMOS Integrated Analog-to-Digital and Digital-to-Analog Converters*, vol. 2. Kluwer Academic Publishers, Dordrecht (2003).
40. Hall, D. A., R. S. Gaster, K. A. A. Makinwa, S. X. Wang, and B. Murmann. A 256 pixel magnetoresistive biosensor microarray in 0.18 μm CMOS. *IEEE Journal of Solid-State Circuits* 48(5): 1290–1301 (2013).

Part II

Integrated Microfluidic and Nanofluidic Systems

6 Two-Dimensional Paper Networks for Automated Multistep Processes in Point-of-Care Diagnostics*

Elain Fu, Barry Lutz, and Paul Yager

CONTENTS

6.1 Introduction to 2DPNs 155
6.1.1 Need for Assays with Improved Performance that are Designed for Use in Low-Resource Settings 155
6.1.2 Paper-Based Diagnostic Devices: A Potential Solution 156
6.1.3 Paper Networks for Automated Multistep Sample Processing 157
6.1.4 Paper Microfluidics Toolbox: Analogs of Pump Controls and Valves in Paper Networks 158
6.2 Applications of 2DPNs 162
6.2.1 Sample Dilution and Mixing 162
6.2.2 Small Molecule Extraction 163
6.2.3 Signal Amplification 164
6.2.4 Selected Advances That Complement Paper Microfluidics Technology 166
6.3 Summary 167
References 167

6.1 INTRODUCTION TO 2DPNS

6.1.1 Need for Assays with Improved Performance That Are Designed for Use in Low-Resource Settings

Gold-standard diagnostic assays are often high-performance laboratory-based tests that require multistep protocols for complex sample processing. Trade-offs

* The material in this chapter was published previously in Yallup, K. and K. Iniewski (eds), *Technologies for Smart Sensors and Sensor Fusion*. Boca Raton, FL: CRC Press, 2014.

for the high performance include long sample processing times, long times for samples to be transported to the laboratory and for results to be transmitted back to the patient or caregiver, the need for trained personnel to run the test and interpret the results, and the need for specialized instrumentation for processing samples and detecting analytes. Also assumed is access to electricity to power the instrumentation, to maintain strict environmental conditions, and to refrigerate reagents until use in the assay. The requirements of laboratory-based tests are often incompatible with the constraints of resource-limited settings. Constraints in these settings include patients with limited access to clinics and a short amount of contact time while there, limited training of test providers, testing environments with uncontrolled temperatures and humidity levels, and limited local infrastructure, including the absence of supporting laboratory equipment and a lack of cold chain for refrigeration of reagents [1–3]. The World Health Organization has coined an acronym for the characteristics of point-of-care diagnostics that are appropriate for even the lowest-resource global health settings: affordable, sensitive, specific, user-friendly, rapid and robust, equipment-free, and deliverable to users (ASSURED) [4]. The overall challenge has been and continues to be to create high-performance assays that are *appropriate* for the various multiconstraint settings relevant for global health applications, including the lowest-resource settings.

6.1.2 Paper-Based Diagnostic Devices: A Potential Solution

An especially compelling need in the lowest-resource settings is for equipment-free diagnostics for which ongoing maintenance and repair are not required. Simple lateral flow tests have been used in low-resource settings for decades. Though lateral flow tests fulfill many of the ASSURED criteria, they have been criticized for (1) their limited ability to multiplex (i.e., perform an assay for multiple analytes from a single biosample) and (2) their lack of sensitivity for many analytes of clinical importance [5,6]. The contrasting characteristics of the high-resource laboratory-based tests and the low-resource lateral flow tests are summarized in Figure 6.1. Recently, there has been a resurgence in work in the area of paper* microfluidics diagnostics with the goal of bringing high-performance testing to low-resource settings. In 2008, the Whitesides group pioneered the use of microfluidic paper-based analytical devices (μPADS), 2D and 3D paper-based structures that enable colorimetric assays (e.g., for the detection of glucose and protein) with multiplexing capability [7,8]. The original μPAD structures were created by photolithography [9], but since then, numerous alternative fabrication methods have been demonstrated, including wax printing [10,11], cutting with a knife blade [12], use of inkjet printing solvents [13], and etching or through-cutting with a CO_2 laser [14]. Additional work in the area of paper-based assay development has focused on implementing multiplexed assays for the detection of additional biomarkers using one-step colorimetric reactions (e.g., nitrite, uric acid, and lactate) [15,16] or performing the simultaneous analysis of multiple

* Note that we use the term "paper" broadly and include related porous materials.

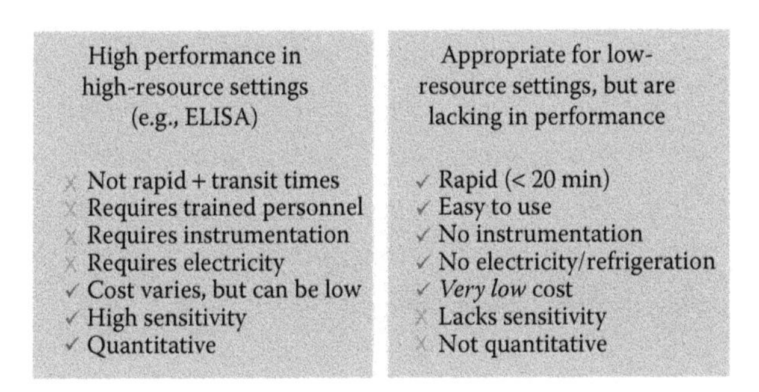

FIGURE 6.1 Two classes of diagnostic tests available today. There is a need for higher-performance tests that are appropriate for use in low-resource settings.

controls for on-device calibration [17]. In this brief review, we focus on a discussion of recent work from the collaboration of Fu, Lutz, and Yager, using two-dimensional paper networks (2DPNs) for automated multistep sample processing for high-sensitivity assays.

6.1.3 Paper Networks for Automated Multistep Sample Processing

As described above, a significant limitation of lateral flow devices is their low sensitivity with respect to the clinically relevant detection ranges for a number of analytes. This limitation effectively derives from an inability of these devices to perform multistep sample processing characteristic of high-performance gold-standard assays. For example, the poor sensitivity of rapid lateral flow tests for influenza has been highlighted recently. Those tests generally have an acceptable clinical specificity of >90%, but have poor clinical sensitivity of 11%–70% [18–21]. The Centers for Disease Control and Prevention even issued a statement during the influenza pandemic of 2009 that recommended discontinuing use of those tests [22].

The strength of the 2DPN is its ability to automatically perform multistep processes for increased performance, while still maintaining the advantages of conventional lateral flow tests, namely, a rapid time to result, ease of use, and low cost. A key feature of the 2DPN assay is the configuration of the network, composed of multiple inlets per outlet, which can function as a program for the controlled delivery of multiple reagent volumes within the network. The example of Figure 6.2 illustrates a multi-inlet paper network that can be used for the automated sequential delivery of reagents to a downstream detection region (left schematic). Upon the simultaneous application of reagents to the three inlets, the geometry of the network performs the automated delivery of the multiple reagent volumes to the detection region in order, first A, then B, then C (right schematic). Critical to the automated operation of multistep paper-based assays is a set of paper microfluidics tools, that is, analogs to the pump controls and valves of conventional microfluidics, to perform the desired manipulation of fluids within the network.

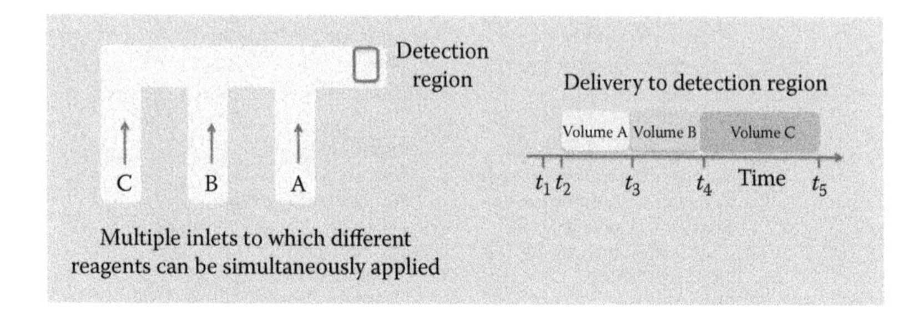

FIGURE 6.2 Two-dimensional paper networks with multiple inlets per detection region for automated multistep sample processing. The sequential delivery of multiple fluid volumes has been preprogrammed into the configuration of the network.

6.1.4 Paper Microfluidics Toolbox: Analogs of Pump Controls and Valves in Paper Networks

To enable automated multistep processing, there is a need for tools to precisely control the transport of multiple fluids within the paper networks [24]. As in conventional lateral flow assays, properties of the porous materials, including pore size, pore structure, and surface treatments, and properties of the fluid, can be used to tune the flow rate for the assay time and sensitivity requirements of a given application. Here we focus on some of our recently developed paper microfluidics tools. These tools serve to replace the costly and often complicated valves and pump controls in conventional microfluidics. This next section will discuss tools that have been developed for controlling flow rate, for switching on flow, and for switching off flow.* This last category is especially important to be able to independently meter discrete volumes of reagents within the paper networks.

One key paper microfluidics tool is the use of simple geometries to control the flow rates of fluids in paper networks [24,25]. One can investigate flow in the simplest 2D structures to create some basic design rules for transport in paper networks. For example, what happens to the fluid front in regions of expanding or contracting geometry? Figure 6.3 shows strips that contain a simple expansion at different locations downstream, and a constant width strip for comparison. Flow initially follows the Lucas–Washburn relation (i.e., the distance that the fluid front travels is proportional to the square root of time, with the proportionality factor dependent on the surface tension, the contact angle, the average pore size of the material, and the viscosity of the fluid) [26,27] in all the strips. Transition to a greater width results in a deviation from the Lucas–Washburn relation and a greater degree of slowing of the fluid front. The plot of Figure 6.3 shows the distance versus the square root of time for the two leftmost strips. Here one can see the initial Lucas–Washburn flow and then a further slowing that starts at the point of the expansion. The width of the expansion can also be used to control the speed of the fluid front, where for a greater width expansion, there is a greater degree of slowing of the fluid front. Thus, simple

* Recently, the authors discovered that Bunce et al. [23], in a patent issued in 1994, had disclosed many similar ideas on controlling flows in paper networks, but to our knowledge these methods were not carried forward.

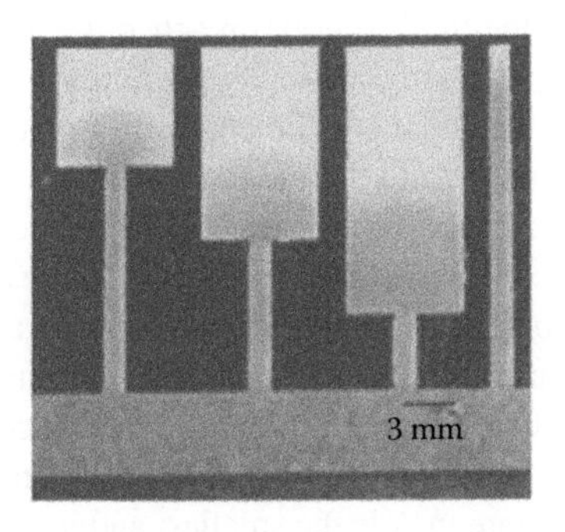

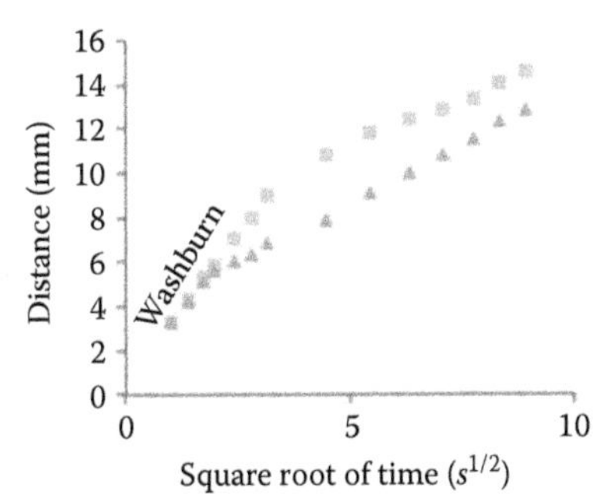

FIGURE 6.3 Transport of the fluid front in simple expansion geometries. The fluid front is slowed relative to a constant width strip and the degree of slowing depends on the location of the expansion. Squares represent data from the leftmost strip and triangles represent data from the strip second from the left. (Adapted from Fu, E., Ramsey, S.A., Kauffman, P., Lutz, B., and Yager, P., *Microfluidics and Nanofluidics*, 10, 29–35, 2011.)

control parameters for slowing down the transport time of the fluid front, the time that the fluid front takes to travel the length of the strip, are the downstream location of the expansion and the width of the expansion [25].

For the case of a contraction geometry (a transition to a smaller width), as shown in Figure 6.4, the flow starts out following the Lucas–Washburn prediction, increases transiently at the constriction, and then resumes Lucas–Washburn flow as the larger width section serves as a nonlimiting source for the smaller width section. The result is that the transport time of the fluid front is decreased relative to a constant-width strip. The downstream location of the constriction can be used to control the transport time of the fluid front, and this time is minimized when the lengths of the two sections of different widths are equal [25].

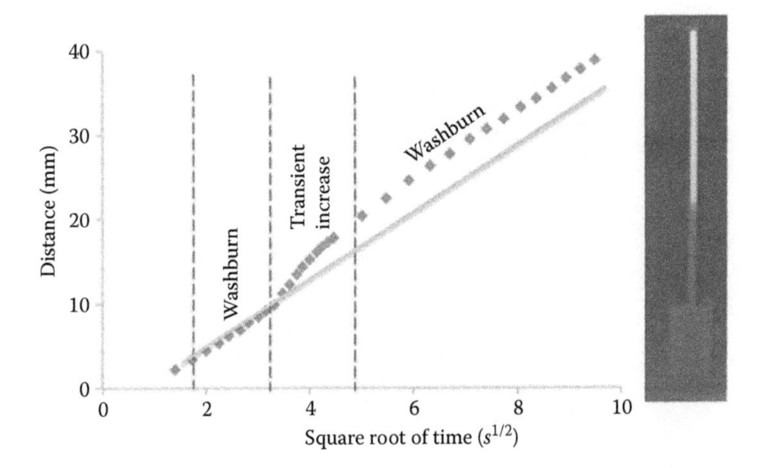

FIGURE 6.4 Transport of the fluid front in the case of a contraction geometry. The transport time of the fluid front is decreased relative to a constant width strip.

In the case of fully wetted flow (i.e., when the fluid front has reached the wicking pad), one can use the electrical circuit analogy to Darcy's law [28] for fluidic circuits. The pressure difference across the circuit is analogous to potential difference, the volumetric flow rate is analogous to current, and the fluidic resistance depends on the physical properties of the system and geometric factors describing the paper circuit. Fluidic resistances in series are summed, while fluidic resistances in parallel are added in reciprocal. Using this electrical circuit analogy for resistances in series, one can calculate the relative resistances of simple structures, such as those shown in Figure 6.5. Since resistance is proportional to length over cross-sectional area, the resistance of A is the greatest and the resistance of B is the smallest. For a uniform pressure difference across all three structures, the volumetric flow rate in A is the smallest and in B is the greatest. The transport time for flow through a strip with a multisegment geometry can be calculated from $t = V/Q$, where V is the volume of the geometry, and Q is the volumetric flow rate. Assuming that permeability and viscosity are constant, differences in the transport times of fluids in the strips will be solely due to geometric factors. The prediction is that the transport time will be fastest in the constant width strip A, and

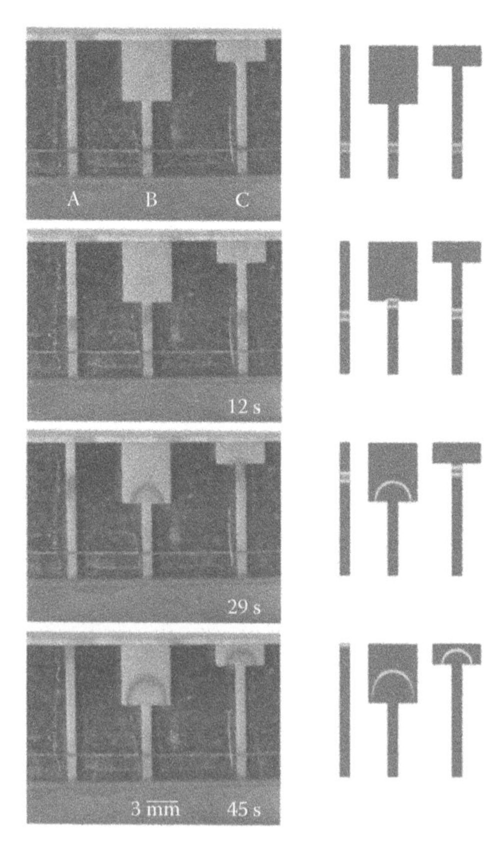

FIGURE 6.5 Comparison of experimental and model results for Darcy's law shows good agreement in both the shapes and the locations of the bands. (Adapted from Fu, E., Ramsey, S.A., Kauffman, P., Lutz, B., and Yager, P., *Microfluidics and Nanofluidics*, 10, 29–35, 2011.)

for the strips of varying widths, the transport time should be faster in strip C than in strip B. A time series comparison of experimental and simulation results (COMSOL Multiphysics) for flow in strips of different geometries is shown in Figure 6.5. The transport times show good quantitative agreement, demonstrating the ability to predict and control flow rates for simple changes in geometry [25].

Another key paper microfluidics tool is an on-switch for fluid flow. One type of on-switch for flow uses dissolvable barriers. Specifically, sugar barriers can be used to create delays in the transport of the fluid within a paper network. Both the extent of the sugar barrier and the concentration of the sugar solution used to form the barrier within the porous material can be used to control the delay time [24,29]. Delay times of up to 50 min were demonstrated when the latter were used [29]. These fluid delays can be critical tools in paper networks, in cases in which longer fluid delays are needed and it is no longer practical to use geometry alone.

A complementary tool for the manipulation of fluids in paper networks is an off-switch for fluid flow. One method to control the shutoff of multiple flows independently is the use of inlet legs that are submerged by varying distances into a common well [30]. The level of the fluid in the well drops as fluid wicks into the paper inlets. Fluid shuts off from the inlets at different times, in order of shortest to longest submerged lengths as shown in Figure 6.6. Thus, different volumes of multiple fluids are automatically input [30]. An alternative method to turn off flow from multiple inlets independently uses pads of varying fluid capacity that are prefilled to saturation [31]. Contact between the pads and inlets activates the flows. The release profiles from glass fiber source pads of different fluid capacities show that flow is Lucas–Washburn-like in that the distance traveled by the fluid front is proportional to the square root of time [31]. The properties of the pad, including the bed volume and the surface treatment, affect the release profile of fluid from the pad to the inlet [31].

Other interesting tools for controlling flow have been demonstrated by other groups, including modification of the wetting properties of the substrate and user-activated mechanical switches. In the context of a fluidic timer, the Phillips group has used wax to slow flow within the paper channel [32]. The Whitesides group has

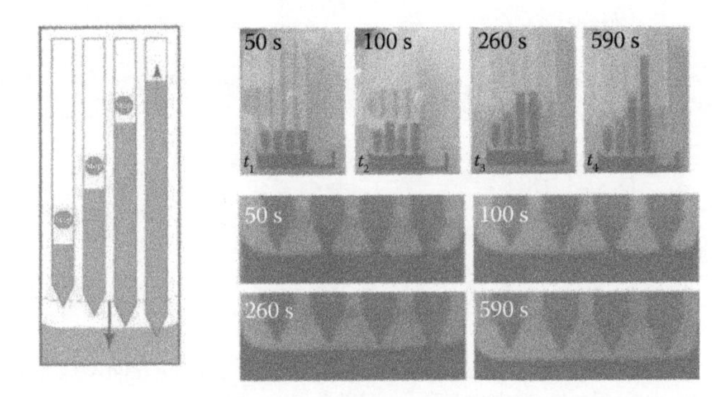

FIGURE 6.6 Inlet legs were immersed to different depths into the well as shown in the schematic. As fluid was wicked from the well, each leg broke contact with the fluid in a timed sequence to provide automated volume metering as shown in the image sequence. (Image data courtesy of P. Trinh.)

demonstrated the use of buttons that can be mechanically depressed by the user with a pen to activate flow between two initially disconnected fluidic paths in one of their 3D μPADs [33]. Finally, the Shen group demonstrated the analogous mechanical on-switch in a paper device composed of a single layer, using tabs that could be manipulated by the user to complete a fluidic pathway [34]. These are just a few examples of early work to create tools for the control of fluids in paper microfluidics.

6.2 APPLICATIONS OF 2DPNs

The ability to perform multiple automated processing steps enables a host of capabilities that can be implemented in 2DPNs for higher performance testing at the point of care. In this section, three examples—sample dilution [35], small molecule extraction [35], and signal amplification [31,36,37]—are described.

6.2.1 Sample Dilution and Mixing

Precise sample dilution is a particular type of mixing and often required for chemical reactions and binding-based assays. In conventional microfluidic systems, continuous dilution requires the combination of two fluid streams in a channel using expensive pumps and providing some means to mix the two fluids. A 2DPN can be used to create a paper dilution circuit that mixes two fluids and allows control over the dilution factor by simply changing the shape of the paper [35]. Figure 6.7 shows dilution of a fluid (top leg) with a buffer (right leg). The dilution factor is determined by the relative flow rates of the two fluids. In this case, the flow rates are not set by pumps, but rather by the relative fluidic resistances of the two inlet legs according to Darcy's law. For a given material, the resistance is simply proportional to the length of the leg and the viscosity of the fluid. As the length of the dilution arm increases, the volumetric flow rate of the diluent decreases, leading to a reduced dilution factor

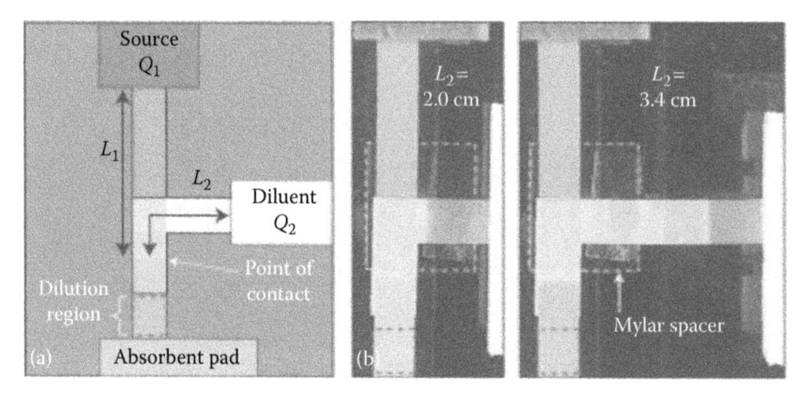

FIGURE 6.7 A sample can be diluted accurately by controlling the relative addition of sample and diluent in the network shown in (a). By modifying the relative resistance of two inlet channels (in this case, by increasing the length of the diluent inlet, L_2, while keeping the length of the source inlet, L_1, the same), their relative contributions can be controlled, allowing for precise sample dilution. (Reproduced from Osborn, J., Lutz, B., Fu, E., Kauffman, P., Stevens, D., and Yager, P., *Lab on a Chip*, 10, 2659–2665, 2010. With permission.)

in the common channel downstream. Serial dilutions are also possible by adding multiple dilution arms, allowing a wide range of dilutions without a single pump or pipetting step. Dilution and mixing of input reagents is one useful class of applications that can be automated in a 2DPN.

6.2.2 Small Molecule Extraction

Another sample pretreatment application that has been demonstrated in a 2DPN is extraction of small molecules from higher-molecular-weight components in a complex sample [35]. Previously, the Yager group developed a pump-driven microfluidic device called the H-filter that allowed extraction of small analytes from complex samples [38–40]. Separation of two species and subsequent extraction of a purified solution of the faster-diffusing species can be achieved when one inlet contains a mixed sample and the other a collection buffer. The efficiency of the extraction depends on the diffusion coefficient of each species, the contact time, and the dimensions of the common channel; no intervening membrane is required as long as the Reynolds number in the device is low. In conventional microfluidic devices, this requires a stable diffusion interface and multiple pumps. Figure 6.8 shows the classic H-filter recreated in a 2DPN. The significant advantages of the 2DPN H-filter are that it is pumpless, fully disposable, and much cheaper than any previous implementations using conventional microfluidics.

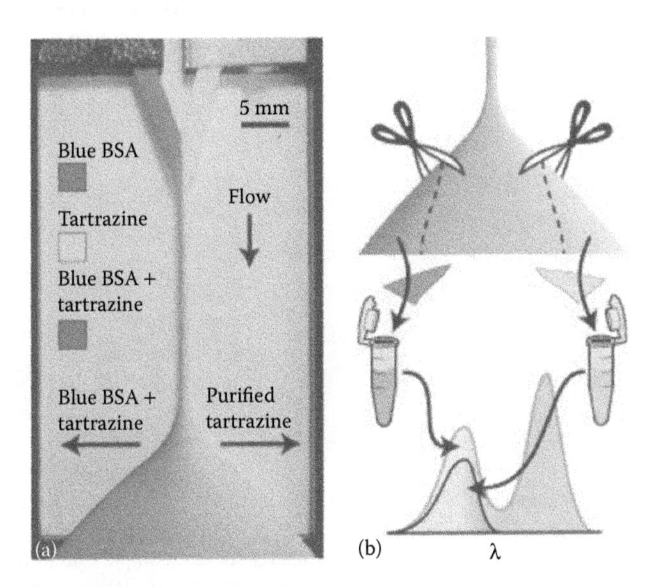

FIGURE 6.8 Small-molecule extraction in a 2DPN. In this proof-of-concept demonstration (a), small-molecule dye was separated from a larger component, dye labeled BSA. The extract was recovered by simply cutting out part of the 2DPN outlet. 2DPN H-filters could be used to extract analytes from complex samples for downstream analysis (b). (Reproduced from Osborn, J., Lutz, B., Fu, E., Kauffman, P., Stevens, D., and Yager, P., *Lab on a Chip*, 10, 2659–2665, 2010. With permission.)

6.2.3 Signal Amplification

Chemical signal amplification has been used in many systems to improve the sensitivity of an assay. The trade-off in the case of well-known laboratory ELISA is the requirement for many steps for labeling, washing, and amplifying the signal; these multiple critical steps are either performed by trained users or expensive laboratory robots. Using 2DPNs, we can program the structure of the paper device to automate the rinse and amplification steps [31,36,37]. Specifically, the simple three-inlet paper network of Figure 6.2 can be used to perform a basic three-step signal-amplified assay based on a conventional sandwich assay format.

An example of an amplified 2DPN assay for the detection of the malaria parasite protein *Pf*HRP2 is presented in Figure 6.9. The 2DPN card (shown on the left) was designed to automatically perform two additional processing steps—delivery of rinse and signal amplification reagents—for improved limit of detection (LOD) in an easy-to-use format. The three-inlet network and wicking pad were located on one side of the folding card, while the source pads were located on the opposite side of the folding card. The source pads contained the dry reagents, conjugate, buffer, and gold enhancement components, from left to right. The user steps were to add sample and water to appropriate pads on the card, then fold the device in a single activation step. This set of user steps is comparable in ease of use to commercially available conventional lateral flow tests and is much less complex than the many timed user steps required to operate alternate microfluidic formats proposed for performing signal amplification steps described in [31]. The schematic (shown on the right) outlines the capture sequence in the assay. After activation of the multiple flows in the 2DPN card, the sample plus antibody conjugated to a gold particle label was first delivered to the detection region. The signal produced at this stage is comparable to the signal

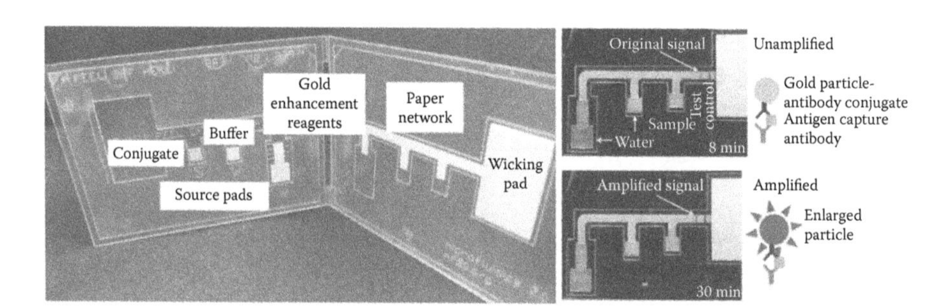

FIGURE 6.9 Easy-to-use 2DPN card format demonstrated for an amplified immunoassay. (*Left*) Conjugate, buffer, and the gold enhancement reagent components were stored dry on the card for rehydration at the time of use. The user steps are comparable to those required to run a conventional unamplified lateral flow strip test. (*Right*) 2DPN assay results for a high analyte concentration of 200 ng/mL. The top panel shows the original pink signal after 8 min due to formation of a conventional sandwich structure with a gold particle label in the detection region. The bottom panel shows the amplified, significantly darkened purple signal after 30 min. (Reproduced from Fu, E., Liang, T., Spicar-Mihalic, P., Houghtaling, J., Ramachandran, S., and Yager, P., *Analytical Chemistry*, 84, 4574–4579, 2012. With permission.)

from a conventional lateral flow test. Following this, a rinse buffer was delivered to the detection region to remove the nonspecifically bound label. Finally, the signal amplification reagent was delivered to the detection region to produce an amplified signal. In this case, application of a gold enhancement solution, consisting of gold salt and a reducer, resulted in the deposition of metallic gold onto the original gold particle labels. This enlargement of the gold particles produced a significant darkening of the original signal.

The detection region of the amplified assay for a concentration series of the analyte is shown in the image series of Figure 6.10. The signal-versus-concentration curves for the 2DPN amplified and unamplified assays are shown in the plot of Figure 6.10. The LOD of the amplified 2DPN malaria card using the gold enhancement reagent was 2.9 ± 1.2 ng/mL, an almost fourfold improvement over the unamplified case (10.4 ± 4.4). For context, the LOD of the amplified assay is similar to that reported for a *Pf*HRP2 ELISA of 4 ng/mL [31]. The commercially available gold enhancement system used here was chosen for ease of use, and has shown promise in other microfluidic formats (described in [31] and [37]). However, the 2DPN card format demonstrated here can also be used to implement other signal amplification

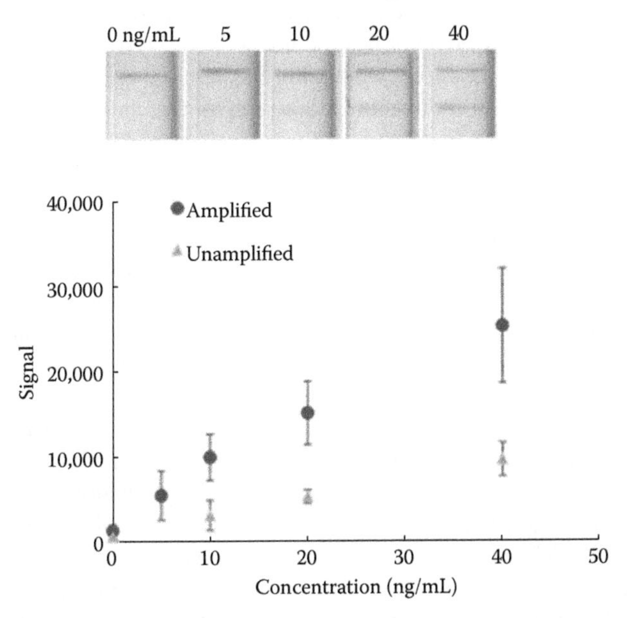

FIGURE 6.10 Sensitivity improvement in the amplified 2DPN assay [1]. (*Top*) Image series of the detection region for different concentrations of the protein *Pf*HRP2. (*Bottom*) Plot of the average signal for each concentration of the 2DPN card with rinse and amplification at 40 min (amplified assay, black circles). Also shown for comparison is a control case in which additional water rather than gold enhancement reagent was run in the 2DPN card format (unamplified assay, gray triangles). The error bars represent the standard deviation. The limit of detection of the amplified 2DPN malaria card was improved almost fourfold over the unamplified case. (Reproduced from Fu, E., Liang, T., Spicar-Mihalic, P., Houghtaling, J., Ramachandran, S., and Yager, P., *Analytical Chemistry*, 84, 4574–4579, 2012. With permission.)

methods. Other amplification methods reported to be useful in lateral flow and other formats include silver enhancement of gold nanoparticles [41,42] and the horseradish peroxidase/tetramethylbenzidine enzymatic system [43], which have the potential to further improve the assay LOD.

6.2.4 Selected Advances That Complement Paper Microfluidics Technology

Recently, there have been complementary advances to address the equipment-free challenge of the lowest-resource settings. We briefly highlight two lines of work: robust methods for electricity-free temperature control and cell phones for expanded detection capabilities with minimal dedicated instrumentation.

Recently, the Weigl group at PATH demonstrated the use of chemical heating, for example, hydration of CaO and phase-change materials to perform loop-mediated nucleic acid amplification [44]. Their device achieved a controlled elevated temperature of 65°C ± 1.5°C for over an hour [44]. The specific combination of exothermic reactants and the composition of the phase change material can be used to tune the thermal properties of the instrument-free heater for numerous applications including other isothermal nucleic acid amplification methods, cell lysis protocols, and sample concentration methods based on temperature-responsive polymers [44]. Building on this work of controlled chemical heating using phase change materials, the Bau group has demonstrated a disposable polymeric self-heating cartridge for performing nucleic acid amplification that is automatically activated using a water source traveling through a length of filter paper [45].

A second challenge to developing high-performance tests for low-resource settings is to achieve quantitative detection with minimal dedicated instrumentation. The use of a compact reader in conjunction with fluorescent or colorimetric labels has been a common strategy for providing quantitative readout of conventional lateral flow tests (e.g., from ESE GmbH/Qiagen) [46]. The Whitesides group has also demonstrated the use of a transmission-based reader for measurements in index-matched paper devices [47]. More recently, the use of cell phones for the acquisition, analysis, and transmission of assay data has become an area of active research and development. Challenges include the acquisition of high-quality image data given the expected wide range of lighting conditions and user variability of camera positioning [48]. The Whitesides group has demonstrated the use of a cell-phone camera for direct acquisition of end-point intensity measurements from a colorimetric paper assay [49], while the Shen group has demonstrated quantitative detection of chemiluminescence [50]. A related approach has been to develop an adapter module to interface with a standard cell phone. The Ozcan group has developed compact and lightweight adapters that couple to a smartphone camera for wide-field fluorescent and dark-field imaging capability [51], as well as for use with colorimetric conventional lateral flow tests to enable more robust output near the LOD of the assay [52]. Though this adapter model introduces a small piece of equipment to the system, it also has some significant positives. Since the adapter can be made to fit in lock and key fashion to both the smartphone and the disposable test, user variability in camera positioning and variability in lighting conditions can be well controlled.

6.3 SUMMARY

The overall challenge in point-of-care diagnostics development continues to be to create high-performance assays that are *appropriate* for the various multiconstraint settings relevant for global health applications. For the lowest-resource settings, the requirements for high performance with a rapid, instrument-free, easy-to-use, and very-low-cost device brings specific design and implementation challenges. Paper-based microfluidics is especially well suited to addressing these challenges. Specifically, the 2DPN is an enabling technology for implementing multistep assays, characteristic of gold-standard laboratory tests, in an automated disposable package. Coupled with advances to develop electricity-free methods for temperature control and cell phones for data acquisition, analysis, interpretation, and transmission, there is great potential for creating sophisticated assays appropriate for low-resource settings. The main challenges in the field of paper microfluidics are to develop a set of robust and precise paper fluidic tools for the automation of devices, to develop high-performance reagents and materials, and to create integrated systems for sample-in to result-out, including sample transfer and pretreatment. Meeting these challenges will enable the development of paper-based devices that can achieve the performance specifications needed for a positive impact in low-resource communities.

REFERENCES

1. Chin, C.D., V. Linder, and S.K. Sia, Lab-on-a-chip devices for global health: Past studies and future opportunities. *Lab on a Chip*, **7**: 41–57, 2007.
2. Yager, P., G.J. Domingo, and J. Gerdes, Point-of-care diagnostics for global health. *Annual Review of Biomedical Engineering*, **10**: 107–144, 2008.
3. Urdea, M., L.A. Penny, S.S. Olmsted, M.Y. Giovanni, P. Kaspar, A. Shepherd, P. Wilson, et al., Requirements for high impact diagnostics in the developing world. *Nature*, **444**(Suppl 1): 73–79, 2006.
4. Kettler, H., K. White, and S. Hawkes, *Mapping the Landscape of Diagnostics for Sexually Transmitted Infections*. Geneva: WHO/TDR, 2004.
5. Posthuma-Trumpie, G.A., J. Korf, and A. van Amerongen, Lateral flow (immuno) assay: Its strengths, weaknesses, opportunities and threats. A literature survey. *Analytical and Bioanalytical Chemistry*, **393**: 569–582, 2009.
6. O'Farrell, B., Evolution in lateral flow-based immunoassay systems, in R. Wong and H. Tse (eds), *Lateral Flow Immunoassay*, pp. 1–33. New York: Humana Press, 2009.
7. Martinez, A.W., S.T. Phillips, M.J. Butte, and G.M. Whitesides, Patterned paper as a platform for inexpensive, low-volume, portable bioassays. *Angewandte Chemie-International Edition*, **46**: 1318–1320, 2007.
8. Martinez, A.W., S.T. Phillips, and G.M. Whitesides, Three-dimensional microfluidic devices fabricated in layered paper and tape. *Proceedings of the National Academy of Sciences of the United States of America*, **105**: 19606–19611, 2008.
9. Martinez, A.W., S.T. Phillips, B.J. Wiley, M. Gupta, and G.M. Whitesides, FLASH: A rapid method for prototyping paper-based microfluidic devices. *Lab on a Chip*, **8**: 2146–2150, 2008.
10. Lu, Y., W.W. Shi, J.H. Qin, and B.C. Lin, Fabrication and characterization of paper-based microfluidics prepared in nitrocellulose membrane by wax printing. *Analytical Chemistry*, **82**: 329–335, 2010.
11. Carrilho, E., A.W. Martinez, and G.M. Whitesides, Understanding wax printing: A simple micropatterning process for paper-based microfluidics. *Analytical Chemistry*, **81**: 7091–7095, 2009.

12. Fenton, E.M., M.R. Mascarenas, G.P. Lopez, and S.S. Sibbett, Multiplex lateral-flow test strips fabricated by two-dimensional shaping. *ACS Applied Materials and Interfaces*, **1**: 124–129, 2009.
13. Abe, K., K. Kotera, K. Suzuki, and D. Citterio, Inkjet-printed paperfluidic immunochemical sensing device. *Analytical and Bioanalytical Chemistry*, **398**: 885–893, 2010.
14. Spicar-Mihalic, P., B. Toley, J. Houghtaling, T. Liang, P. Yager, and E. Fu, CO_2 laser cutting and ablative etching for the fabrication of paper-based devices. *Journal of Micromechanics and Microengineering*, **23**(6): 067003, 2013.
15. Li, X., J.F. Tian, and W. Shen, Quantitative biomarker assay with microfluidic paper-based analytical devices. *Analytical and Bioanalytical Chemistry*, **396**: 495–501, 2010.
16. Dungchai, W., O. Chailapakul, and C.S. Henry, Use of multiple colorimetric indicators for paper-based microfluidic devices. *Analytica Chimica Acta*, **674**: 227–233, 2010.
17. Wang, W., W.Y. Wu, W. Wang, and J.J. Zhu, Tree-shaped paper strip for semiquantitative colorimetric detection of protein with self-calibration. *Journal of Chromatography A*, **1217**: 3896–3899, 2010.
18. Drexler, J.F., A. Helmer, H. Kirberg, U. Reber, M. Panning, M. Muller, K. Hofling, B. Matz, C. Drosten, and A.M. Eis-Hubinger, Poor clinical sensitivity of rapid antigen test for influenza A pandemic (H1N1) 2009 virus. *Emerging Infectious Diseases*, **15**: 1662–1664, 2009.
19. Hurt, A.C., R. Alexander, J. Hibbert, N. Deed, and I.G. Barr, Performance of six influenza rapid tests in detecting human influenza in clinical specimens. *Journal of Clinical Virology*, **39**: 132–135, 2007.
20. Uyeki, T., Influenza diagnosis and treatment in children: A review of studies on clinically useful tests and antiviral treatment for influenza. *Pediatric Infectious Disease Journal*, **22**: 164–177, 2003.
21. Vasoo, S., J. Stevens, and K. Singh, Rapid antigen tests for diagnosis of pandemic (swine) influenza A/H1N1. *Clinical Infectious Diseases*, **49**: 1090–1093, 2009.
22. Center for Disease Control, Interim guidance for detection of novel influenza A virus using rapid influenza testing, 2009. http://www.cdc.gov/h1n1flu/guidance/rapid_testing.htm.
23. Bunce, R., G. Thorpe, J. Gibbons, L. Keen, and M. Walker, Liquid transfer devices, U.S. Patent US5198193, 1994.
24. Fu, E., B. Lutz, P. Kauffman, and P. Yager, Controlled reagent transport in disposable 2D paper networks. *Lab on a Chip*, **10**: 918–920, 2010.
25. Fu, E., S.A. Ramsey, P. Kauffman, B. Lutz, and P. Yager, Transport in two-dimensional paper networks. *Microfluidics and Nanofluidics*, **10**: 29–35, 2011.
26. Washburn, E.W., The dynamics of capillary flow. *Physical Review*, **17**: 273–283, 1921.
27. Lucas, R., Ueber das Zeitgesetz des Kapillaren Aufstiegs von Flussigkeiten. *Colloid and Polymer Science*, **23**: 15–22, 1918.
28. Darcy, H., *Les fontaines publiques de la ville de Dijon*. Paris: Dalmont, 1856.
29. Lutz, B., T. Liang, E. Fu, S. Ramachandran, P. Kauffman, and P. Yager, Dissolvable fluidic time delays for programming multi-step assays in instrument-free paper diagnostics. *Lab on a Chip*, **13**(14): 2840–2847, 2013.
30. Lutz, B.R., P. Trinh, C. Ball, E. Fu, and P. Yager, Two-dimensional paper networks: Programmable fluidic disconnects for multi-step processes in shaped paper. *Lab on a Chip*, **11**: 4274–4278, 2011.
31. Fu, E., T. Liang, P. Spicar-Mihalic, J. Houghtaling, S. Ramachandran, and P. Yager, Two-dimensional paper network format that enables simple multistep assays for use in low-resource settings in the context of malaria antigen detection. *Analytical Chemistry*, **84**: 4574–4579, 2012.
32. Noh, N. and S.T. Phillips, Metering the capillary-driven flow of fluids in paper-based microfluidic devices. *Analytical Chemistry*, **82**: 4181–4187, 2010.

33. Martinez, A.W., S.T. Phillips, Z.H. Nie, C.M. Cheng, E. Carrilho, B.J. Wiley, and G.M. Whitesides, Programmable diagnostic devices made from paper and tape. *Lab on a Chip*, **10**: 2499–2504, 2010.
34. Li, X., J.F. Tian, and W. Shen, Progress in patterned paper sizing for fabrication of paper-based microfluidic sensors. *Cellulose*, **17**: 649–659, 2010.
35. Osborn, J., B. Lutz, E. Fu, P. Kauffman, D. Stevens, and P. Yager, Microfluidics without pumps: Reinventing the T-sensor and H-filter in paper networks. *Lab on a Chip*, **10**: 2659–2665, 2010.
36. Fu, E., P. Kauffman, B. Lutz, and P. Yager, Chemical signal amplification in two-dimensional paper networks. *Sensors and Actuators B: Chemical*, **149**: 325–328, 2010.
37. Fu, E., T. Liang, J. Houghtaling, S. Ramachandran, S.A. Ramsey, B. Lutz, and P. Yager, Enhanced sensitivity of lateral flow tests using a two-dimensional paper network format. *Analytical Chemistry*, **18**: 7941–7946, 2011.
38. Hatch, A., E. Garcia, and P. Yager, Diffusion-based analysis of molecular interactions in microfluidic devices. *Proceedings of the IEEE*, **92**: 126–139, 2004.
39. Helton, K.L., K.E. Nelson, E. Fu, and P. Yager, Conditioning saliva for use in a microfluidic biosensor. *Lab on a Chip*, **8**: 1847–1851, 2008.
40. Helton, K.L. and P. Yager, Interfacial instabilities affect microfluidic extraction of small molecules from non-Newtonian fluids. *Lab on a Chip*, **7**: 1581–1588, 2007.
41. Yan, J., D. Pan, C.F. Zhu, L.H. Wang, S.P. Song, and C.H. Fan, A gold nanoparticle-based microfluidic protein chip for tumor markers. *Journal of Nanoscience and Nanotechnology*, **9**: 1194–1197, 2009.
42. Yeh, C.H., C.Y. Hung, T.C. Chang, H.P. Lin, and Y.C. Lin, An immunoassay using antibody-gold nanoparticle conjugate, silver enhancement and flatbed scanner. *Microfluidics and Nanofluidics*, **6**: 85–91, 2009.
43. Kolosova, A.Y., S. De Saeger, S.A. Eremin, and C. Van Peteghem, Investigation of several parameters influencing signal generation in flow-through membrane-based enzyme immunoassay. *Analytical and Bioanalytical Chemistry*, **387**: 1095–1104, 2007.
44. LaBarre, P., K. Hawkins, J. Gerlach, J. Wilmoth, A. Beddoe, J. Singleton, D. Boyle, and B. Weigl, A simple, inexpensive device for nucleic acid amplification without electricity—Toward instrument-free molecular diagnostics in low-resource settings. *Plos One*, **6**: e19738, 2011.
45. Liu, C.C., M.G. Mauk, R. Hart, X.B. Qiu, and H.H. Bau, A self-heating cartridge for molecular diagnostics. *Lab on a Chip*, **11**: 2686–2692, 2011.
46. Faulstich, K., R. Gruler, M. Eberhard, D. Lentzsch, and K. Haberstroh, Handheld and portable reader devices for lateral flow immunoassays, in R. Wong and H. Tse (eds), *Lateral Flow Immunoassay*, pp. 75–94. New York: Humana Press, 2009.
47. Ellerbee, A., S. Phillips, A. Siegel, K. Mirica, A. Martinez, P. Striehl, N. Jain, M. Prentiss, and G. Whitesides, Quantifying colorimetric assays in paper-based microfluidic devices by measuring the transmission of light through paper. *Analytical Chemistry*, **81**: 8447–8452, 2009.
48. Stevens, D., Development and optical analysis of a microfluidic point-of-care diagnostic device in bioengineering, PhD dissertation, University of Washington, Seattle, 2010.
49. Martinez, A.W., S.T. Phillips, E. Carrilho, S.W. Thomas, H. Sindi, and G.M. Whitesides, Simple telemedicine for developing regions: Camera phones and paper-based microfluidic devices for real-time, off-site diagnosis. *Analytical Chemistry*, **80**: 3699–3707, 2008.
50. Delaney, J.L., C.F. Hogan, J.F. Tian, and W. Shen, Electrogenerated chemiluminescence detection in paper-based microfluidic sensors. *Analytical Chemistry*, **83**: 1300–1306, 2011.
51. Zhu, H., O. Yaglidere, T. Su, D. Tseng, and A. Ozcan, Cost-effective and compact wide-field fluorescent imaging on a cell-phone. *Lab on a Chip*, **11**: 315–322, 2011.
52. Mudanyali, O., S. Dimitrov, U. Sikora, S. Padmanabhan, I. Navruz, and A. Ozcan, Integrated rapid-diagnostic-test reader platform on a cellphone. *Lab on a Chip*, **12**: 2678–2686, 2012.

7 Droplet-Based Digital Microfluidics for Single-Cell Genetic Analysis

Yong Zeng and Richard A. Mathies

CONTENTS

7.1 Introduction 172
7.2 High-Throughput, Programmable Microfluidic Droplet Generation 173
7.2.1 Experimental 174
7.2.1.1 MEGA Fabrication and Assembly 174
7.2.1.2 Droplet Generation 174
7.2.2 Results and Discussion 175
7.2.2.1 MEGA Design and Performance 175
7.2.2.2 Programmable Active Droplet Generation 176
7.3 Single-Cell Quantification and Genotyping of Bacterial Pathogens 180
7.3.1 Experimental 180
7.3.1.1 Cell Culture 180
7.3.1.2 Bead and PCR Preparation 181
7.3.1.3 PCR Reaction 181
7.3.1.4 Bead Recovery and Flow Cytometry Quantitation 182
7.3.2 Results 182
7.3.2.1 Digital Single-Cell Pathogen Detection at the Statistically Dilute Regime 182
7.3.2.2 High-Throughput and Sensitive SCGA 184
7.3.2.3 Limit of Detection and Dynamic Range 184
7.3.3 Discussion 186
7.4 Single-Cell Multiplex Gene Detection and Sequencing of Mammalian Cells 187
7.4.1 Methodology 188
7.4.2 Experimental 189
7.4.2.1 Cell Culture and Preparation 189
7.4.2.2 Agarose Droplet Generation 190
7.4.2.3 Single-Genome Purification 190
7.4.2.4 Emulsion PCR 190
7.4.2.5 Fluorescence Imaging 191
7.4.2.6 Single-Cell Sequencing 191

7.4.3 Results and Discussion 191
7.4.3.1 Single-Cell Encapsulation and Genome Purification 191
7.4.3.2 Multiplexed Single-Cell Droplet PCR 193
7.4.3.3 Single-Cell Sequencing 195
7.5 Conclusions 196
Acknowledgments 197
References 197

7.1 INTRODUCTION

Cellular heterogeneity and individuality are well documented and play essential roles in biological functions and in the development of disease.[1,2] For instance, recent results have revealed marked cellular heterogeneity in gene and protein expression,[1,3–5] genetic/genomic alterations,[2,6,7] and responsiveness to environmental and chemotherapeutic stimuli.[8,9] Circulating tumor cells (CTCs) are found at an estimated frequency of 1 per 10^6–10^7 nucleated blood cells, making their isolation, quantitative counting, and molecular characterization extremely difficult.[10] Traditional biological analyses probe large ensembles on the order of 10^3–10^6 cells, thereby revealing only the average genotypic and phenotypic characterization of the population. These average measurements from large population of cells cannot properly characterize the cellular heterogeneity of a population, detect the co-occurrence of various somatic mutations within single cells, or identify rare events (e.g., rare mutations and CTCs); these factors are important to elucidating the pathology, enabling early detection, and improving the treatment of diseases.[11] In particular, since initial mutagenesis occurs inherently at the single-cell level, investigation of carcinogenesis can be remarkably facilitated by molecular techniques with single-cell sensitivity and resolution. Identifying driver mutations that lead to carcinogenesis in a rare subset of cells is one key approach to the risk assessment, early detection, and treatment of cancer.[12,13] These somatic mutations can provide tumor-specific biomarkers and therapeutic targets for early detection, prognosis, and treatment of cancer. Clinical use of these genetic biomarkers relies on sensitive and quantitative measurement of these rare mutations in a vast excess of wild-type alleles.

Microfluidics has evolved from a scale-dependent technology that improves chemical analysis through miniaturization toward a versatile platform that enables the development of powerful new techniques to address challenges in many areas such as genetics, cell biology, and global health care.[14] Microfluidic technology offers the fundamentally new capabilities of manipulating fluids, molecules, and cells in space and time, thus opening new opportunities for developing high-throughput single-cell analysis technology.[15] Droplet-based digital microfluidics is an emerging paradigm in the field that takes advantages of microscale multiphase flow dynamics to create and manipulate uniform emulsion droplets of femtoliter to nanoliter volume for massively parallel and ultrasensitive assays.[16–18] For instance, emulsion polymerase chain reaction (ePCR) provides a powerful tool for high-throughput single-molecule amplification and counting of genetic targets.[19,20] However, traditional ePCR has inherent limitations due to the use of mechanical agitation for emulsion generation, which produces high shear forces and polydisperse droplet sizes. Droplet microfluidics overcomes these limitations due to its unique ability to generate monodisperse droplets with precise control over droplet size, which ensures equal population sampling and amplification efficiencies across all reaction compartments to enable accurate digital quantification of an absolute number of targets.[16,21–25] In addition, microfluidic encapsulation significantly mitigates the mechanical agitation and stress that can damage cells as it is normally performed with a shear stress lower than 10 dynes/cm^2.[29] Finally, microfluidic integration allows programmable *in situ* manipulation of droplets, such as droplet steering, trapping,[26] and fusion,[27] leading to a broader spectrum of applications, including real-time PCR,[21] protein expression studies,[25] and drug screening.[28]

We have developed a high-performance single-cell/copy genetic analysis (SCGA) technology that is based on high-throughput microfluidic emulsion generation and multiplexed droplet PCR assays.[29] This platform technology has now been shown to be useful in a variety of different applications. Here we will first describe the programmable active microfluidic droplet generation approach and devices that we have developed.[30] We will then discuss their applications to single-cell genetic analysis of pathogens[31] and mammalian cells.[32]

7.2 HIGH-THROUGHPUT, PROGRAMMABLE MICROFLUIDIC DROPLET GENERATION

Droplet microfluidics has been typically a passive process, consisting of droplet generation by flow focusing and cross-flow shearing,[33] and downstream manipulation based on hydrodynamic interactions of droplets.[34] Elucidation of complex and dynamic biological processes poses an urgent need for the ability to perform more complicated active operations, including reagent dosing, mixing, splitting, extraction, detection, and sorting, each of which requires precise and automated control in space and time. Precise and controllable droplet generation has been recognized as an important component in developing droplet-based platforms to interrogate complex biological systems. A variety of methods, such as piezoelectric actuators[35] and electrowetting,[36] have been investigated to achieve controllable and eventually programmable microfluidic droplet generation. Pneumatic valves fabricated by soft lithography are a particularly promising tool for controlled droplet formation, owing to the inherent compatibility of pneumatic valves with large-scale microfluidic integration and automation.[37] Current valve-based methods rely on external pumps to continuously flow the dispersed phase while actuating the integrated valves to modulate passive droplet formation in microchannels.[38] In contrast, we have developed an active microfluidic droplet generator (μDG) that uses an integrated valve-based diaphragm pump to control both fluidic transport and flow modulation of the dispersed phase.[23] Here, we will describe the microfluidic engineering approach to scale up our μDG to large-scale microfluidic emulsion generator array (MEGA) systems for high-throughput SCGA, as well as the systematic study of the on-demand droplet formation process driven by on-chip pumping.

7.2.1 Experimental

7.2.1.1 MEGA Fabrication and Assembly

MEGA chips shown schematically in Figure 7.1 were constructed from three 100 mm diameter glass wafers and a thin poly(dimethylsiloxane) (PDMS) membrane following a process described by Zeng et al.[31] Briefly, the valve structure was first lithographically fabricated on the top side of a Borofloat glass wafer. Using backside alignment lithography, the droplet generators were fabricated on the bottom of the same wafer in register with the pump structure. Fluidic and interlayer via holes were drilled and the wafer was thermally bonded to a clean glass substrate at 650°C for 6 h. Pneumatic control manifolds were microfabricated on another glass wafer. Prior to device assembly, the microchannels were coated with 0.1% octadecyltrichlorosilane in dry toluene for 10 min. A 250 μm thick PDMS membrane was sandwiched between a microchip and a pneumatic manifold piece immediately after 2 min UV-ozone activation of the interacting surfaces. A custom Plexiglas manifold was assembled with the MEGA stack to allow oil infusion and droplet collection.

7.2.1.2 Droplet Generation

The carrier oil contains 39.8% (w/w) Dow Corning 5225C Formulation Aid, 30% (w/w) Dow Corning 749 Fluid, 30% (w/w) AR20 Silicone Oil, and 0.2% (w/w) Triton X-100 surfactant. A syringe pump was used to continuously inject carrier oil into the device at various flow rates. The on-chip three-valve diaphragm pump was

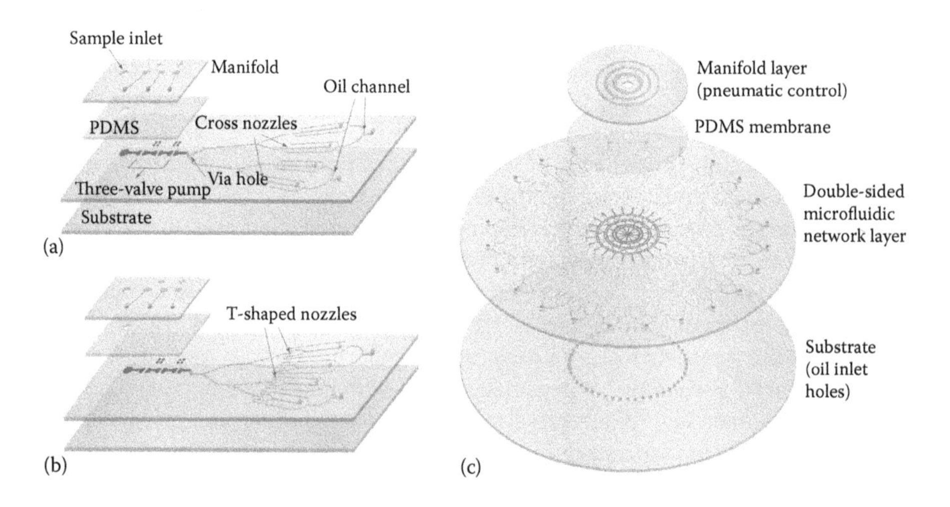

FIGURE 7.1 Microfluidic emulsion generator array (MEGA) devices. (a) Exploded view of a glass/PDMS/glass hybrid four-channel MEGA device with a pneumatically controlled three-valve micropump integrated to drive four nozzles for droplet generation. (b) Design of an eight-channel MEGA device that contains eight T-shaped droplet generators to increase device density. (c) Exploded view of the complete 96-channel MEGA device containing 96 parallel T-shaped droplet generators driven by a single coaxial three-ring valve pump.

pneumatically actuated by a homemade solenoid valve system to pump the aqueous solution through the channels. A Welch dry vacuum pump was used to supply constant vacuum (−80 kPa) for valve actuation while the pressure was varied as specified in the text. Pumping was conducted in either four- or three-step fashion under the control of a program written in LabVIEW.

7.2.2 Results and Discussion

7.2.2.1 MEGA Design and Performance

To enable detection of extremely low frequency events in a vast population, we have developed a series of MEGA systems with 4, 8, 32, and 96 parallel units (Figure 7.1). These multilayer devices consist of a bonded glass–glass microfluidic chip, a PDMS membrane, and a microfabricated manifold wafer. The main fluidic chip contains a pattern of valve seats microfabricated on the top surface, which is connected through a "via" hole to the enclosed microfluidic networks. The assembly is then completed by contact bonding a manifold wafer onto the fluidic chip with a PDMS membrane to form the micropump structure. The microfluidic networks consist of the symmetrically bifurcated channels for introduction of aqueous and oil phases, which form parallel droplet nozzles. Four- and eight-channel MEGAs are based on the single-pump design, but with different droplet generator nozzles: a cross-shaped nozzle in the four-channel MEGA and a T-shaped nozzle in eight-channel devices (Figure 7.1a and b). The four-channel array defines a basic unit that is used to build up multiplexed MEGAs, such as the 32-channel MEGA with eight such arrays integrated onto a 4″ wafer. Further multiplexing of the MEGA on a 4″ wafer is restricted by the number of individual pumps that can be symmetrically arranged in a circle. To achieve a higher density, we designed a new ring micropump composed of three pairs of coaxial ring-shaped valve seats connected by offset channels, as well as corresponding circular displacement trenches (Figure 7.1c). This compact micropump, along with the T-shaped nozzle design, enables the implementation of a 96-channel MEGA on a 4″ wafer. In the 96-channel MEGA, oil channels are connected to the oil inlet holes drilled on the bottom substrate of the device.

For droplet generation, a homemade manifold module is used to infuse oil phase and route generated droplets into PCR tubes, as sketched in Figure 7.2a. The bottom part of the manifold is designed to form a circular oil reservoir when sealed against the MEGA chip. This design yields uniform oil introduction across the whole channel network while minimizing the number of syringe pumps and tubing connections required. The compact ring-shaped micropump in the 96-channel MEGA confers sufficient power to drive multiple droplet generators in parallel. At the same time, the symmetrically designed microfluidic network ensures even fluidic transport, which is crucial for uniform droplet encapsulation. The symmetrical design of the MEGA assures uniform generation of monodisperse droplets in all channels. The microphotograph in Figure 7.2b demonstrates the generation of uniform droplets from a mock PCR mix containing agarose beads by flow shearing at the T-shaped injectors. The mean diameter of the droplets was determined to be 162 μm (2.2 nL) with a deviation of only 3.8% (Figure 7.2c). Such size uniformity is preserved across a range of droplet volumes (1–5 nL; RSD $<$ 5%).

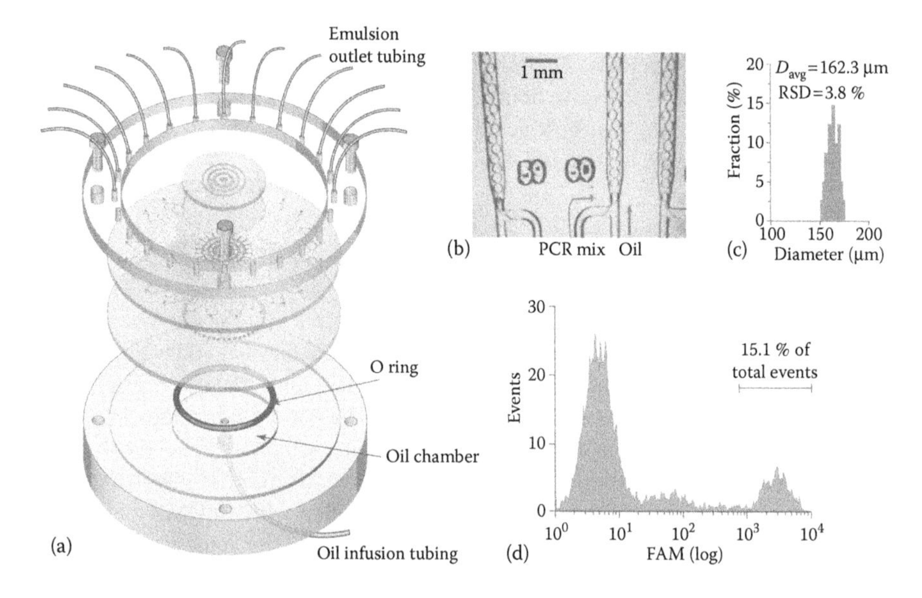

FIGURE 7.2 Characterization of droplet generation and PCR amplification using the 96-channel MEGA. (a) Exploded view of the complete four-layer MEGA device and the Plexiglass assembly module used to infuse oil and to collect the generated emulsion. (b) Image of droplet production at the T-shaped nozzles with a total throughput of 2.4×10^6 droplets per hour. (c) Size distribution of uniform ~2 nL droplets collected from 16 nozzles in a device. (d) A representative flow cytometric histogram of beads carrying the FAM-labeled PCR product from *E. coli* K12 at 0.2 cpd in 2.5 nL droplets. For droplet generation, the mock PCR mix containing ~100 beads per μL was used. (Modified with permission from Zeng, Y., Novak, R., Shuga, J., Smith, M.T., and Mathies, R.A., *Anal. Chem.*, 82, 3183–3190, Copyright 2010. American Chemical Society.)

To assess the encapsulation performance of the 96-channel MEGA, single-cell emulsion PCR was performed targeting the KI#128 island on the K12 genome. Bacterial *Escherichia coli* K12 cells and forward primer modified beads were introduced at a statistical dilution of 0.2 cells and 0.1 beads per 2.5 nL droplet, respectively. After PCR and isolation, flow cytometric analysis (Figure 7.2d) shows that 15% of the total bead population (4423 beads) is strongly fluorescent, corresponding well to a theoretical value of 18% predicted by the Poisson distribution. This good agreement indicates successful single-copy genetic amplification resulted from uniformly distributed cells and beads. A 96-channel MEGA can produce up to 3.4×10^6 droplets per hour (dph), which enables rapid detection of extremely low frequency events in a vast population.

7.2.2.2 Programmable Active Droplet Generation

We have systematically investigated the effects of pulsatile on-chip pumping and flow conditions on droplet formation using the eight-channel MEGA. A three-step actuation sequence was used here to increase the droplet formation rate. We first studied the effects of the oil flow rate, a critical factor in microfluidic droplet generation.

Figure 7.3a plots the droplet formation frequency ($f_{droplet}$) and droplet size ($R_{droplet}$) as a function of the oil flow rate (Q_{oil}) under an actuation sequence of 40, 50, 40 ms and a pressure of 35 kPa. At the low oil flow rate in regime (a), large uniform droplets were periodically formed, as seen in Figure 7.3b. Interestingly, $f_{droplet}$ is lower than the pumping frequency (f_{pump}) of 7.69 Hz and increases along with Q_{oil} while $R_{droplet}$ decreases. Microscopic visualization also observed a droplet break-up process, which resembles that of passive cross-flow droplet formation.[39] These results indicate that the droplet break-up within regime (a) is predominantly governed by the interaction of viscous shear stresses and interfacial tension between the two immiscible phases and that the force generated by mechanical pump actuation plays a minor role.[40] As Q_{oil} was further increased, a narrow transition regime (b) occurred in which droplet generation lacks both periodicity and size uniformity, as seen in Figure 7.3b. A sharp change of droplet generation behavior was observed when $f_{droplet}$ was tuned to a threshold value of 1.3 μL/min (Figure 7.3a). Within this regime of a wide range of Q_{oil}, the use of active on-chip pumping enables stable generation of uniform droplets, as exemplified in Figure 7.3b. Several unique phenomena were observed in this regime. First, the droplet generation synchronizes with the pumping frequency, which can be attributed to the pulsatile nature of on-chip pumping that modulates the process of droplet formation. Second, $f_{droplet}$ is independent of Q_{oil},

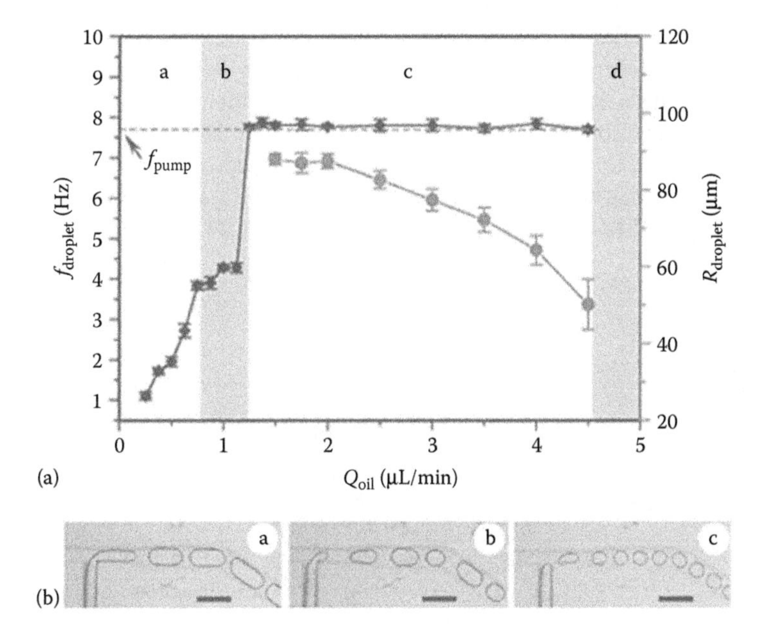

FIGURE 7.3 (a) Plot of droplet formation frequency ($f_{droplet}$) and droplet radius ($R_{droplet}$) as a function of the oil flow rate in each T-junction (Q_{oil}) under an actuation sequence of 40, 50, 40 ms and a pressure of 35 kPa. The *dashed line* indicates the pumping frequency (f_{pump}) of 7.69 Hz. (b) Images of different droplet formation behavior observed in the regimes identified in (a). The scale bars are 600 μm. (From Zeng, Y., Shin, M., and Wang, T., *Lab Chip*, 13, 267–273, 2013. Reproduced by permission of Royal Society of Chemistry.)

which indicates the dominance of the forces caused by pumping over the viscous shear stresses that depend on the flow parameters. Third, the droplet size can be independently tuned by varying Q_{oil} while droplet formation remains synchronized with the pumping. Increasing Q_{oil} leads to higher back pressure that suppresses the pneumatic actuation of the valves and thus reduces the flow rate of the aqueous phase being pumped into the channels.[41] Eventually, the droplet generators will stop working when the back pressure is larger than the pressure that the pump is able to generate (Figure 7.3a). These observations clearly demonstrate that on-chip pumping confers the ability to independently control frequency by adjusting pumping rate and droplet size by varying Q_{oil}. This unique property is in contrast to the passive droplet generation methods in which droplet generation frequency and droplet size are linked and sensitive to the flow conditions.[33] Figure 7.3 shows that a range of Q_{oil} needs to be experimentally defined to ensure precisely controlled droplet generation with excellent size uniformity. As expected, the Q_{oil} range and the droplet sizes were found to be also affected by the pumping conditions, such as f_{pump} and the pressure for pneumatic actuation.

Figure 7.4a shows that the threshold Q_{oil} at which droplet generation and pumping start to synchronize shows a roughly linear function of f_{pump}, especially for the high frequencies. At a fixed pressure (same as in Figure 7.3a) within the range that we commonly used, increasing pumping rate requires a higher oil flow rate to establish stable operation. Since the droplet size is also affected by Q_{oil} (Figure 7.3a), one should work at oil flow rates away from the boundary conditions such that the adjustment of droplet generation frequency by varying f_{pump} will require no changes of oil flow and thus have no effects on droplet sizes. In addition to oil flow rate, droplet sizes can be tuned by adjusting the actuation pressure. Droplet volume exhibits a peaking response to increasing actuation pressure under the conditions of pumping and oil flow that we commonly used for fast droplet generation. An example is shown in Figure 7.4b with a pumping sequence of 45, 45, 45 ms and Q_{oil} at 3 μL/min.

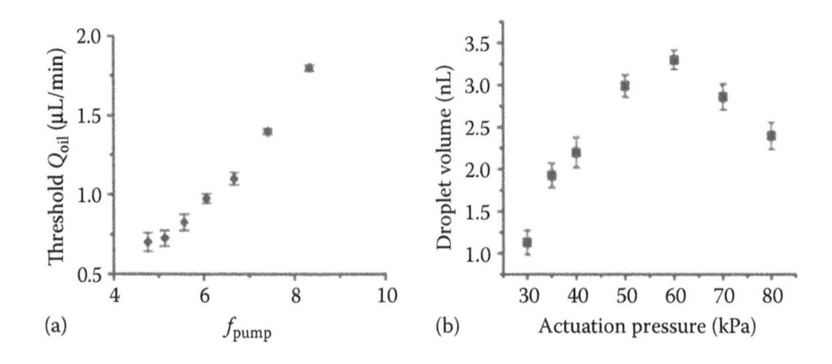

FIGURE 7.4 Effects of the pumping conditions on active droplet generation. (a) Plot of the threshold Q_{oil} as a function of pumping frequency (f_{pump}) obtained under an actuation pressure of 35 kPa. The duration for each actuation step was kept the same. (b) Plot of droplet volume as a function of actuation pressure. Q_{oil} was 3 μL/min and the pumping sequence was 45, 45, 45 ms. (From Zeng, Y., Shin, M., and Wang, T., *Lab Chip*, 13, 267–273, 2013. Reproduced by permission of Royal Society of Chemistry.)

The experimental observations in Figures 7.3 and 7.4 provide guidance for exploring proper conditions for independent and precise control of the droplet frequency and sizes.

It is important to note that these droplet generation behaviors are distinct from those of mechanical actuation-based methods that use PDMS or piezoelectric valves.[40,42] In a PDMS valve-based method, the dispersed phase is constantly driven by external pumps and a valve placed close to the droplet generation junction is periodically actuated to alter the instantaneous flow rate of the dispersed phase.[41] It was found that forced droplet generation synchronizes with natural droplet generation only if the actuation frequency is a multiple of the natural frequency, otherwise irregular droplet formation will occur. In our method, synchronized droplet formation can be obtained at any pumping frequency as long as the oil flow rate falls in regime (c) in Figures 7.3 and 7.4. Therefore, our studies suggest a different physical mechanism that should arise from the combination of pulsatile transport and modulation of the dispersed phase, in contrast to the valve-based devices that only modulate the flow being continuously injected. More experimental and theoretical studies are needed to fully elucidate the fluid dynamics of droplet formation driven by pulsing pumping.

Because the on-chip diaphragm pump directly delivers the dispersed phase in our device, the pumping frequency or period is expected to be an important factor in determining the flow rate of the dispersed phase and thus the droplet size. As seen in Figure 7.5a, the droplet size increases and then levels off as the pumping period is extended, consistent with the reported behavior of three-valve micropumps.[41] Adjusting the pumping period provides a means to control droplet generation that is

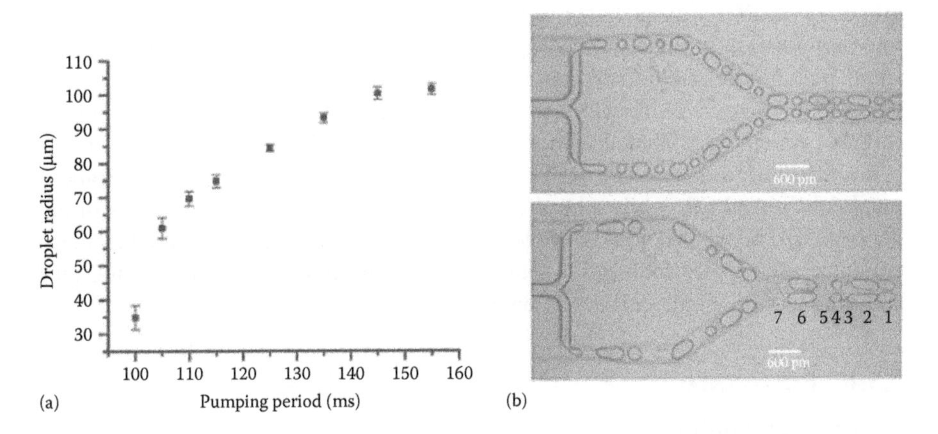

FIGURE 7.5 Programmable droplet generation. (a) Plot of droplet radius as a function of the pumping period. (b) The *top photo* shows a stream of two differently sized droplets formed by continuously alternating two pumping cycles of 150 and 105 ms. The *bottom photo* shows on-demand generation of more complex patterns. In this case, each repeating sequence consists of seven steps: (1,2) two consecutive pumping cycles of 130 and 165 ms, (3) 100 ms interval, (4) 110 ms pumping, (5) 300 ms interval, (6) 150 ms pumping, and (7) 500 ms interval. The actuation pressure was 45 kPa and Q_{oil} was 3.5 μL/min. (From Zeng, Y., Shin, M., and Wang, T., *Lab Chip*, 13, 267–273, 2013. Reproduced by permission of Royal Society of Chemistry.)

more compatible with computer programming. The top photo in Figure 7.5b shows a stream of two differently sized droplets that was formed by continuously alternating two pumping cycles of 150 and 105 ms. In addition, the droplets can be formed on demand to yield more complex patterns, as demonstrated in the bottom photo of Figure 7.5b. In this case, each repeating sequence consists of seven steps: two consecutive pumping cycles of 130 and 165 ms, 100 ms interval, 110 ms pumping, 300 ms interval, 150 ms pumping, and 500 ms interval. Clearly, the on-chip pumping method allows precise control of droplet generation frequency, size, and timing, which demonstrates its on-demand capability and programmability.

7.3 SINGLE-CELL QUANTIFICATION AND GENOTYPING OF BACTERIAL PATHOGENS

The integration of multiplex PCR with high-throughput droplet generation is valuable for the analysis of genomic deletion, forensic genotyping, mutation and polymorphism analysis, and the identification of pathogens. We will demonstrate high-throughput single-cell analysis of a major food-borne bacterial pathogen, *E. coli* O157:H7, which alone causes an estimated 73,000 infections and 61 deaths annually in the United States.[43,44] Sensitive pathogen detection is critical for many applications, such as food safety, where microbial pathogen detection needs to meet a zero tolerance policy for many foods.[45] To this end, we have developed a multiplex single-cell genetic analysis (SCGA) approach, as illustrated in Figure 7.6, which allows efficient high-throughput PCR amplification of multiple target genes specific to different cell types. In this process, primer-linked beads and cells are diluted in the PCR mix such that isolated individual beads or cells are encapsulated into individual uniform reaction droplets dispersed in the carrier oil. Statistically, a small fraction of droplets will contain both one bead and one or more cells. Beads are functionalized with forward primers for all targets, and the PCR mix contains reverse primers each labeled with a unique fluorescent dye. Each bead in a droplet containing only a single cell will carry one type of dye-labeled double-stranded amplicons after PCR, while a bead compartmentalized with different types of target cells will be labeled with multiple dyes. Post-PCR beads are recovered from the emulsion and rapidly analyzed by flow cytometry for multicolor fluorescent digital counting of each single-cell detection event.

7.3.1 Experimental

7.3.1.1 Cell Culture

All cell culture and preparation was performed in a class II biosafety cabinet to avoid contamination. Two types of cells, *E. coli* K12 and nontoxigenic *E. coli* O157, were grown separately in tryptic soy broth medium overnight at 37°C. *E. coli* K12 cells were transformed with a 3.9 kb pCR 2.1-TOPO vector to confer ampicillin resistance and grown with 1 mg/mL ampicillin added to the medium. *E. coli* O157 was untreated and frequently tested for contamination by PCR. Cells were washed three times in 1×PBS and the final cell density was determined using a hemacytometer.

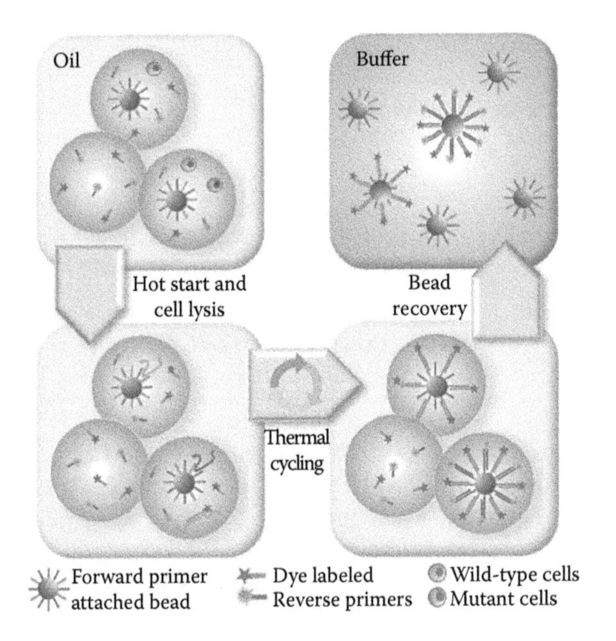

FIGURE 7.6 Multiplex single-cell genetic analysis (SCGA). Statistically dilute beads and templates are encapsulated into uniform nanoliter-volume PCR-mix-in-oil droplets, which are then thermally cycled for PCR amplification. Each bead is functionalized with forward primers for all targets. The PCR mix contains reverse primers, each labeled with a unique fluorescent dye. Each bead in a droplet containing only a single target will carry one type of fluorescent amplicon after PCR, while a bead compartmentalized with two different templates or cells will be linked with multiple dye-labeled products. Following emulsion PCR, the droplets are broken and the beads are recovered and analyzed by flow cytometry. (Reprinted with permission from Zeng, Y., Novak, R., Shuga, J., Smith, M.T., and Mathies, R.A., *Anal. Chem.*, 82, 3183–3190, Copyright 2010. American Chemical Society.)

7.3.1.2 Bead and PCR Preparation

Samples were handled in a UV-treated laminar flow hood. Primers specific to the KI#128 island on the K12 genome and the OI#43 island on the O157 genome were designed to prevent cross-amplification between strains. Reverse primers were labeled with 6-carboxyfluorescein (6-FAM) or Cy5 dye on the 5′ end. 5′ Amine-modified forward primers were linked to agarose beads (34 μm mean diameter) via amine-NHS conjugation chemistry. The coupling reaction was performed at a ratio of ~1.5 μmol oligos per gram beads for the pUC18 target. For *E. coli* cells, equimolar forward primers were used at a concentration of ~0.3 μmol oligo per gram beads. PCR mixes contained forward primer functionalized beads (40 beads/μL) and varied amounts of freshly prepared template DNA or cells.

7.3.1.3 PCR Reaction

The primer sequences specific for *E. coli* cells are listed in Table 7.1. For *E. coli* experiments with an average cell concentration of $\leq$1 cell per droplet, the PCR mix contains 1× AmpliTaq Gold buffer with 1.5 mM $MgCl_2$, 0.2 mM dNTPs,

TABLE 7.1
Primer Sequences for Gene Targets

Gene Target	T_m (°C)	Sequence	Size (bp)
***E. coli* K12#128 Island**			
Forward primer	60.2	5′-TTCGATTACACGGAGTGCTGGGAA-3′	262
Reverse primer	60.3	5′-FAM-CGTTGATTTGCCGTTCCATGTCGT-3′	
***E. coli* O157#43 Island**			
Forward primer	59.5	5′-GGGCAGGAAGAGAGTGACAGG -3′	192
Reverse primer	59.1	5′-Cy5-CGGCCTTACCCGTGAACAGTA-3′	

0.5 μg/μL heat-inactivated BSA, 0.01% Tween 80, 0.4 μM reverse and 0.04 μM forward primers for each cell target, 0.1 beads/2.5 nL droplet, 0.2 U/μL AmpliTaq Gold polymerase, and various amount of cells. At the average cell concentrations of 10 and 100 cells per droplet (cpd), the PCR mix was optimized to have 3 mM $MgCl_2$, 2 μM reverse and 0.2 μM forward primers for *E. coli* K12, with other ingredients kept the same. To minimize DNA/polymerase adsorption on glass and PDMS surfaces, the MEGA device was primed with a coating solution consisting of 1× AmpliTaq Gold buffer with 1.5 mM $MgCl_2$, 0.5 μg/μL heat-deactivated BSA, and 0.01% Tween 80. Droplets were thermocycled as follows: 10 min at 95°C; six cycles of 30 s at 94°C, 90 s at 65°C, and 60 s at 72°C; eight cycles of 30 s at 94°C, 90 s at 63°C, and 60 s at 72°C; 10 cycles of 30 s at 94°C, 90 s at 60°C, and 60 s at 72°C; 13 cycles of 30 s at 94°C, 90 s at 55°C, and 60 s at 72°C; and 7 min at 72°C.

7.3.1.4 Bead Recovery and Flow Cytometry Quantitation

The emulsion in the PCR tubes was vacuumed to pass a 15 μm mesh filter in a 13 mm diameter plastic filter holder. The droplets were broken and the beads were retained by the filter. After rinsing with isopropanol, 100% ethanol, and 1×Dulbecco's PBS (DPBS), the beads were recovered using a 5 mL syringe and stored in 1×DPBS for flow cytometry analysis. The bead suspension in PBS was analyzed using a multicolor flow cytometer (FC-500, Beckman-Coulter). FAM-labeled DNA product on the bead surface was quantified against standard beads with different known amounts of fluorophores (Bangs Laboratories Inc., Fishers, IN). The flow cytometry data were analyzed using WinMDI.

7.3.2 Results

7.3.2.1 Digital Single-Cell Pathogen Detection at the Statistically Dilute Regime

We targeted unique genes on the K12 genome (KI#128 island) and the O157 genome (OI#43 island) using primers labeled with different fluorophores. A four-channel MEGA device was operated for ~18 min to obtain ~3000 beads for flow cytometric analysis. A mixed *E. coli* bacterial sample containing 50% O157 cells was first analyzed at an average cell concentration (C_{avg}) of 0.2 cpd. As seen in Figure 7.7a, the flow cytometry profile shows four distinct bead populations: 212 FAM-positive beads due to the amplicons from K12 (7.93%, green), 198 Cy5-positive beads for O157 (7.40%, red), 45 double-positive beads due to coexistence of both cell types in a single droplet (1.68%, orange), and 2220 negative beads (82.99%, blue). The O157

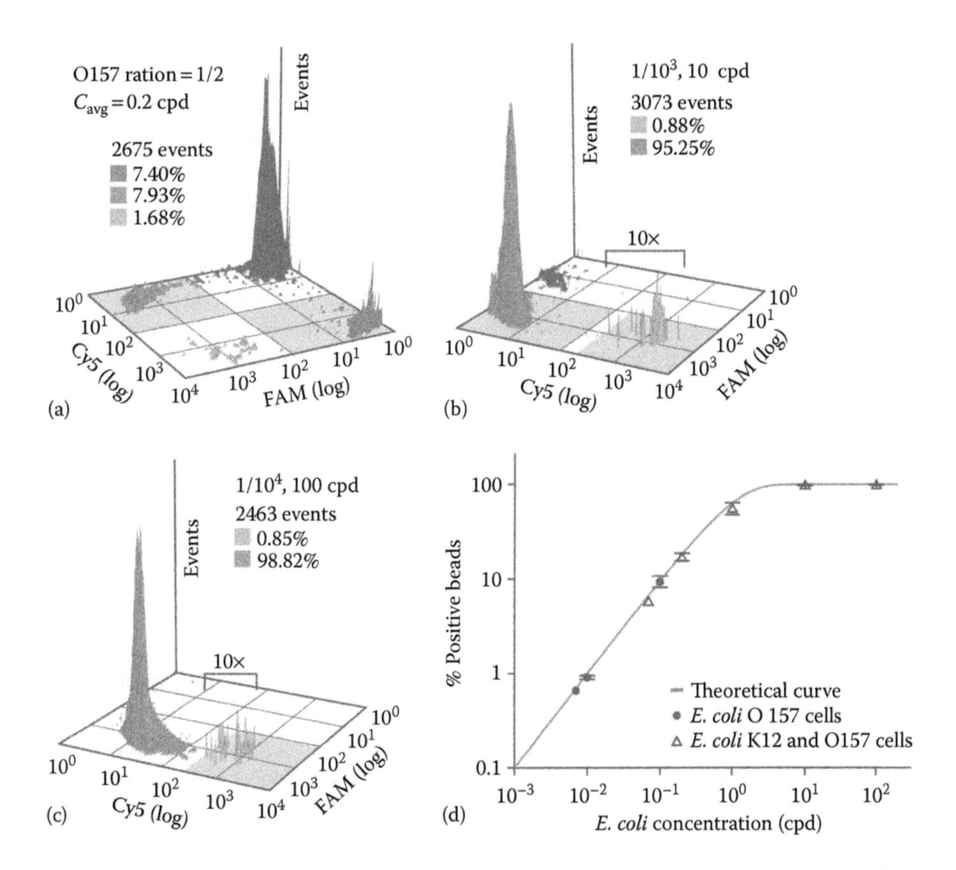

FIGURE 7.7 High-throughput digital multiplex detection of *E. coli* O157 in a background of *E. coli* K12. (a) With $C_{avg} = 0.2$ cpd, flow cytometry shows four distinct populations: negative, FAM-positive beads specific for K12, Cy5-positive beads specific for O157, and double-positive beads for both cells. Small populations are expanded along the event axis for better visualization, as indicated. The gray regions mark the population gating. (b) When $C_{avg} = 10$ cpd while keeping O157 cells at 0.01 cpd, the measured O157 ratio was $0.92/10^3$ (expected: $1/10^3$). Up to 3000 beads can be analyzed using a four-channel device for ~25 min run time. (c) O157 detection using a 96-channel MEGA at $C_{avg} = 100$ cpd shows the measured O157 ratios of $0.85/10^4$, consistent with the inputs of $1/10^4$. Up to 10^4 events were processed within a 5 min run time. (d) Comparison of the percentage of positive beads over total beads vs. starting cell concentration. The curve represents the value expected from Poisson statistics. The experimental data were obtained by detecting *E. coli* O157 cells in a background of *E. coli* K12 with overall cell concentrations of 0.07–100 cpd. Error bars represent standard deviation ($n \geq 3$). (Reprinted with permission from Zeng, Y., Novak, R., Shuga, J., Smith, M.T., and Mathies, R.A., *Anal. Chem.*, 82, 3183–3190, Copyright 2010. American Chemical Society.)

cell ratio (O157-positive beads/total positive beads) is then determined to be 0.48, in good accord with the input O157 cell fraction of 0.5.

7.3.2.2 High-Throughput and Sensitive SCGA

The detection of low frequency genetic variations requires high analysis throughput in order to obtain the statistically significant population for the target. The multiplex SCGA process discussed above uses droplets/beads inefficiently because cells are highly dilute so that most droplets are empty. One way to increase the process efficiency is to perform multiplex single-cell genetic analysis at elevated cell density while still keeping the cells of interest statistically dilute. This is realistic because the target pathogen or mutant cells are typically present at very low relative concentrations compared to normal cells. Figure 7.7b demonstrates the use of a four-channel MEGA device to detect *E. coli* O157 cells at a frequency of $1/10^3$. By increasing C_{avg} from 0.2 to 10 cpd, the effective density of O157 cells is raised from 0.0002 to 0.01 cpd, reducing the number of droplets/beads that must be processed by 50-fold. In this case, all beads should be FAM fluorescent due to coencapsulated K12 cells; the presence of O157 cells in the droplets will modify only a fraction of beads with the Cy5 dye. As expected, the vast majority of beads (1846 out of 1961 events, 96.13%) are FAM positive and a small fraction of double-positive beads (102 events, 0.88%) are observed. The measured O157 ratio is determined to be $0.92/10^3$ (the ratio of double-positive beads over all positive beads divided by C_{avg}), close to the $1/10^3$ input. Multiplex SCGA preserves the quantitative performance for pathogen detection even when the average concentration is up to 100 *E. coli* cells per 2.5 nL droplet. To further improve the detection sensitivity, we assessed the 96-channel MEGA which allows us to encapsulate up to 10^4 cells within 5 min when operating the device at ~7 Hz per channel. As shown in Figure 7.7c, the experiments at an input ratio of $1/10^4$ at 100 cpd detect 88 and 21 O157-positive events, giving an output fraction of $0.85/10^4$.

To verify that the observed performance is the result of digital quantification of each strain, we compare the percentage of positive beads obtained with various input ratios and average cell concentrations with that predicted by the Poisson distribution (Figure 7.7d). The multiplex detection is seen to follow Poisson statistics even when individual O157 cells were detected within a high background of 100 cpd. The good correspondence indicates successful single-cell emulsion PCR, which allows digital quantification of the absolute cell concentration. For instance, with $C_{avg} = 100$ cpd and O157 cells diluted to $1/10^4$ in a K12 background, the O157 cell density is 0.01 cpd and the detection resulted in $0.91\% \pm 0.04\%$ double-positive beads, consistent with the predicted ratio of 0.995%. Because of the presence of negative events, the average percentage of O157-positive beads is corrected to be $0.93\% \pm 0.05\%$ (double-positive beads divided by total positive beads), from which the O157 cell concentration is determined to be 3.7 ± 0.2 cells/μL (input 4 cells/μL).

7.3.2.3 Limit of Detection and Dynamic Range

To determine the limit of detection (LOD) of multiplex SCGA for *E. coli* O157 detection, the assay was carried out at lower pathogenic ratios and $C_{avg} = 100$ cpd using a 96-channel MEGA. Figure 7.8a presents a representative analysis at an input

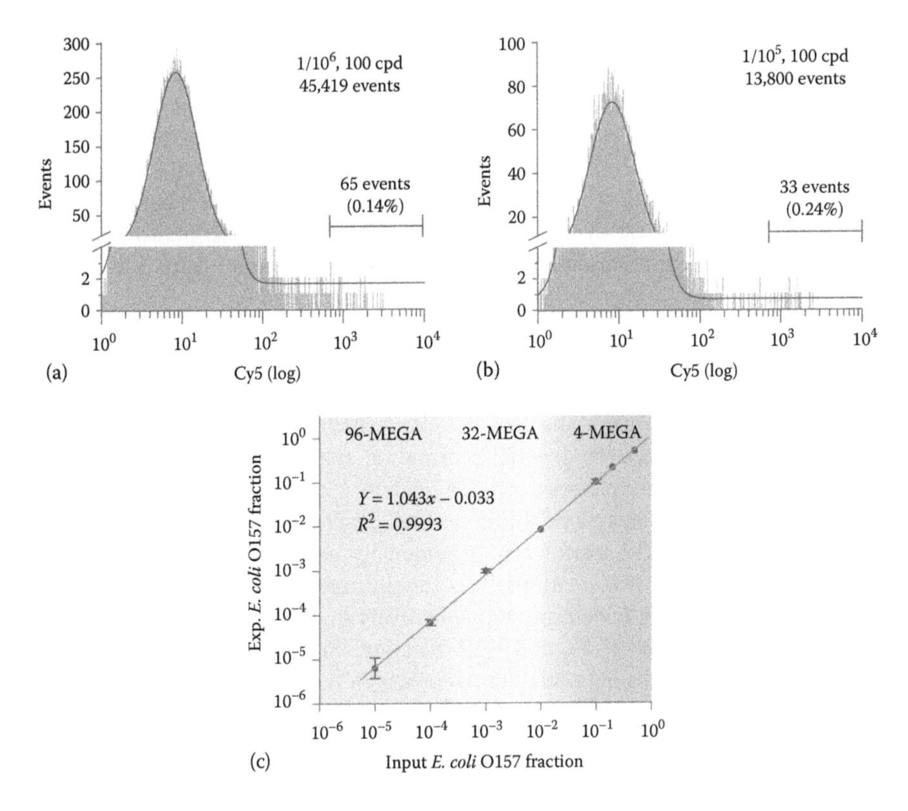

FIGURE 7.8 Detection limit and dynamic range of multiplex SCGA for *E. coli* O157 detection using a 96-channel MEGA. (a,b) Representative cytometric histograms (events vs. Cy5 signal) obtained with input O157 ratios of $1/10^6$ and $1/10^5$, respectively. The measured O157 ratios are: (a) $2.4/10^5$ and (b) $1.4/10^5$. The *solid lines* represent the Gaussian fittings for the Cy5-negative populations. (c) Plot of experimental readout vs. input O157 fraction showing a linear dynamic range for O157 fraction higher than $1/10^5$. The gray regions roughly mark the detection windows that meet the throughput offered by different MEGA devices within an experimentally accessible run time. Error bars represent standard deviation ($n \geq 3$). (Reprinted with permission from Zeng, Y., Novak, R., Shuga, J., Smith, M.T., and Mathies, R.A., *Anal. Chem.*, 82, 3183–3190, Copyright 2010. American Chemical Society.)

ratio of $1/10^6$ where the experimental output ratio ($1.4/10^5$) is one order of magnitude higher than that expected. In this case, to define a statistically significant population at the extremely low cell density (0.0001 cpd for O157 cells), a large volume of PCR mix (1500 μL) was used to produce ~8×10^5 droplets within 20 min, which leads to 45,419 total events analyzed and 65 Cy5-positive events detected. The experimental readout at the input of $1/10^6$ ($1.8/10^5 \pm 0.66/10^5$, $n = 4$) is significantly above the expected value ($p = .01$), indicating that there is significant background signal. It is seen in Figure 7.8a that the peak for K12 cells (FAM-positive only) tails off into the Cy5-positive region, suggesting that the background could be attributed to misamplification caused by the nonspecific binding of O157 primers to the K12 templates. Figure 7.8b is the result obtained with an input ratio of $1/10^5$ that detects 33 Cy5-positive beads. At this pathogenic ratio, the experimental readouts ($2.4/10^5 \pm 0.91/10^5$, $n=4$) are in the same order as the expected value and significantly different from that obtained at $1/10^6$ at $p=.05$ using the two-sample Student's *t*-test. By subtracting the background, the corrected value at the input ratio of $1/10^5$ is determined to be $0.69/10^5 \pm 0.34/10^5$ ($n=4$). Figure 7.8c summarizes the background corrected calibration of *E. coli* O157 detection as a function of the input O157 fraction, which shows a linear dynamic range for O157 fraction higher than $1/10^5$. From this plot, the concentration detection limit is determined to be $3/10^5$ with a 99% confidence.

7.3.3 Discussion

Multiplex single-cell PCR is the cornerstone of the high-throughput SCGA technique, which maximizes the genetic information extracted from each single-cell detection event. In this process, it is critical to compartmentalize single cells and beads into monodisperse nanoliter-volume droplets. Our results clearly demonstrate that the droplet uniformity provided by our active droplet generators enables strict Poisson statistical analysis to calibrate the performance of digital single-cell PCR, which is determined collectively by the statistical encapsulation of beads and cells, cell lysis, and PCR reaction. Such analysis permits digital quantification of the absolute number of targets in the initial sample. Other digital PCR techniques using agitation-based emulsification fundamentally lack this capability because of the extremely polydisperse droplets produced.[46] In addition, the uniform droplets contain the same amount of reactants, enabling a quantitative comparison of the PCR products. Thus, one should be able to perform large-scale gene expression profiling at the single-cell level by transforming reverse transcription-PCR (RT-PCR) assays to the SCGA format.[47] The multiplex SCGA displays great tolerance to PCR inhibition as the cell lysate and debris are significantly diluted in the large nanoliter droplets generated by MEGA, and each reaction is independent. Efficient and specific multiplex PCR amplification can be achieved even when each droplet is loaded with up to 100 *E. coli* cells on average, which greatly increases the analysis throughput and hence the detection sensitivity, without excessively extending the droplet production time. This result indicates the feasibility of SCGA for the large-scale genetic analysis of larger and more complex mammalian cells. For instance, the analysis of cancer development and progression, CTCs, and stem cell differentiation, where single-cell resolution may facilitate a deeper understanding of the biological mechanisms involved, can be envisioned.

Most PCR-based microdevices reported the detection of only one bacterial strain by PCR, with detection limits ranging from a few to 10^4 bacterial cells.[48–50] Recent work in our group showed that PCR reactions in a 250 nL microreactor can detect *E. coli* O157:H7 in a mixture with the K12 strain down to a ratio of 1:500,[51] and that the detection limit of the microscale PCR can be as low as 1:1000 by using on-chip affinity cell preconcentration.[52] Our quantitative digital format remarkably outperforms previous low-throughput microsystems with its capability to identify and quantify both *E. coli* O157 and K12 cells down to a pathogen-to-background ratio in the order of 1:10^5 (Figure 7.8). Such sensitivity makes the technique a promising candidate to achieve the level of detection and speed required for a zero-pathogen-tolerance policy. Droplet generation using the 96-channel MEGA requires <30 min of run time to achieve a LOD of 1:10^5. The entire procedure, including PCR thermal cycling, post-PCR cleanup, and flow cytometry takes approximately 4 h and compares favorably with standard PCR-based detection assays while providing better sensitivity. In contrast, most methods commonly used today require at least 2 days to achieve this level of sensitivity since they rely on pathogen culturing.[45] The SCGA technique can be further extended to detect multiple pathogenic microorganisms in one sample. Compared to the small beads used in the BEAMing technique, the large surface area of the microsphere used here allows multiple different primers to be conjugated with a density that supports efficient on-bead PCR reaction for each target. Thus, our detection multiplicity is limited primarily by the number of fluorescent dyes that a flow cytometer can detect (commonly five colors) since that is a relatively standard degree of PCR multiplicity.

We believe that the detection limit of our multiplex SCGA can be further improved to beyond one in a million. Some nonspecific amplification was observed when the 2.5 nL droplets were loaded with 100 *E. coli* cells. Although the multiplex PCR conditions have been optimized to minimize nonspecific amplification, it still can contribute to the false-positive scores that limit the low-abundance detection to the level of 1:10^5 in this proof-of-concept work. In practical applications, we can perform droplet generation for longer periods of time at lower cell concentrations to minimize the effect of nonspecific PCR amplification, and thus lower the detection limit. On the other hand, trace exogenous contamination from microorganisms and/or DNA present in air, reagents, and instruments may cause considerable interference at these extremely low detection limits. We did see false-positive events occasionally in negative control experiments performed at a cell concentration of < 1 cpd. Sample preparation and experimental operation with more stringent environmental controls should prevent contaminations and enable us to lower the detection limit to 1:10^6 or better.

7.4 SINGLE-CELL MULTIPLEX GENE DETECTION AND SEQUENCING OF MAMMALIAN CELLS

Cytometric sorting, limiting dilution, and micromanipulation have been previously used to perform single-cell PCR assays in 96-well PCR plates, but these approaches are not ideal for large-scale screening applications.[53] Microfluidic droplet technology is particularly advantageous for single-cell analysis because it facilitates rapid statistical compartmentalization of targets for massively parallel pico- to nanoliter-scale assays.[29,54] To date most single-cell genomic analyses have been reported on bacterial samples.[55,56] For mammalian cells, droplet-based genetic analyses have predominantly implemented reverse transcription PCR for phenotypic profiling.[57] A difficulty in single-cell PCR is the persistent technical challenge of integrating a robust and scalable DNA release/extraction method.[55,58] The relative lack of suitable single-cell genomic analysis technologies combined with the significant genetic heterogeneity associated with cancer, underscores the importance of developing new microdroplet methodologies that integrate robust single-cell genome preparation with multiplexed PCR.

To address the fundamental technical challenge of single-genome extraction, we have adapted our SCGA methodology and the emulsion generator array technology to facilitate a novel agarose droplet-based method. In this method, individual cells are confined in gelled agarose droplets for integration with a standard DNA extraction protocol consisting of detergent membrane solubilization and enzymatic protein digestion to release and protect genomic DNA within the droplet. We explored multilocus single-cell sequencing of the control gene β-actin and the chromosomal translocation t(14;18) to validate this technology. The *BCL-2/IgH* translocation t(14;18) is highly prevalent in many blood cancers, including ~80% of follicular lymphoma (FL) cases and ~25% of large-cell B-cell lymphoma cases.[59] This translocation brings the B-cell lymphoma-2 (*BCL-2*) gene from 18q21 under the control of the strong enhancers of the *IgH* locus, ultimately disrupting *BCL-2*'s normal pattern of expression in B cells.[60] *BCL-2* is an antiapoptotic protein and its overexpression can be intimately involved in the pathogenesis of B-cell neoplasms.[61] t(14;18) is also found in healthy individuals at very low levels and may be an early biomarker for lymphoma.[62–64] A high-throughput technique that can sequence and quantify t(14;18) could provide insight into the molecular pathology and clinical importance of t(14;18).

7.4.1 Methodology

The underlying concept of our highly parallel single-cell purification and genetic analysis method is the microfluidic encapsulation of cells in agarose droplets that serves two important roles: first, it enables cell lysis and genomic DNA purification with single-genome integrity; and second, it defines monodispersed water-in-oil droplet reactors for efficient multiplex emulsion PCR target amplification and subsequent genetic analysis. The workflow of the agarose droplet-based SCGA method for the genetic analysis and multilocus sequencing of single mammalian cells is illustrated in Figure 7.9. Briefly, cells and primer-functionalized beads suspended in a molten agarose solution are injected into a MEGA device that is kept at an elevated temperature. Single cells are microfluidically encapsulated together with primer-functionalized beads in uniform agarose droplets. When the droplets are cooled, the agarose droplet forms a rigid but porous gel bead to protect the individual cell and its genomic DNA while allowing fluidic access to enzymes and chemicals for cell lysis, washing, and PCR. The genomes of single cells are released and trapped in the gel beads upon SDS lysis and digestion with proteinase K according to a standard protocol (Figure 7.9b). For genetic analysis, the agarose droplets are equilibrated in PCR buffer containing fluorescent forward primers and then emulsified with oil by mechanical agitation to form monodispersed nanoliter emulsion reactors for massively parallel single-cell amplification (Figure 7.9c). Following multiplex PCR amplification, primer beads are recovered by breaking the emulsion and melting the agarose. The fluorescent amplicon-decorated beads are then rapidly quantified by flow cytometry or further subjected to PCR amplification for DNA sequencing of target genes (Figure 7.9d).

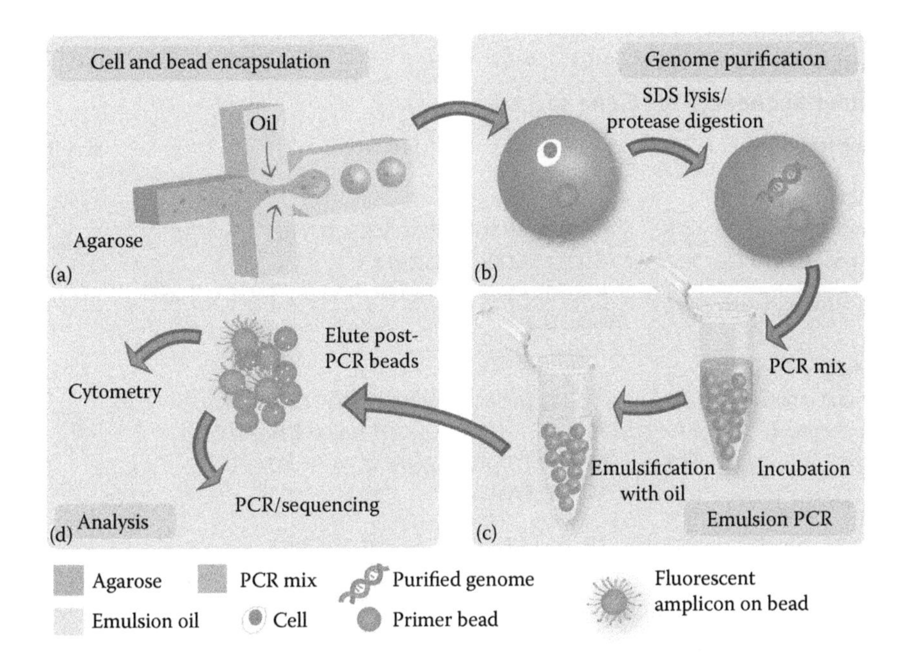

FIGURE 7.9 Workflow diagram demonstrating the agarose-based SCGA technology for genetic detection and multilocus sequencing of single mammalian cells. (a) A MEGA chip is used to statistically encapsulate single cells and primer-functionalized beads in uniform agarose droplets. (b) The genomes of single cells are released in the gelled droplets using a standard SDS lysis/proteinase K digestion protocol. (c) The agarose droplets are equilibrated in PCR buffer, mechanically emulsified in oil, and thermally cycled. (d) After PCR amplification, amplicon beads are released by breaking the emulsion and melting the agarose. The post-PCR beads will be quantified using flow cytometry or further PCR-amplified for DNA sequencing. (Reprinted with permission from Novak, R. et al., *Angew. Chem. Int. Ed. Engl.*, 50, 390–395, Copyright 2011. Wiley-VCH Verlag GmbH & Co. KGaA.)

7.4.2 Experimental

Device fabrication, bead functionalization, and bead recovery and flow cytometry followed the same protocols as described before. The primer sequences specific for *E. coli* cells are listed in Table 7.2.

7.4.2.1 Cell Culture and Preparation

Cell lines of t(14;18)$^-$ TK6 (CRL 8015) and t(14;18)$^+$ RL (CRL 2261) lymphoblasts were cultured in RPMI 1640 media supplemented with 10% FBS at 37°C in a 5% CO_2 atmosphere. Cell preparation was performed in a class II biosafety cabinet. Final cell density was determined to be between 2×10^5 and 2×10^6 using a hemacytometer. Cells were washed with PBS, combined with primer-conjugated beads, and mixed with 2% low-melt agarose to a final concentration of 400 beads/μL and 80–120 cells/μL in 1.5% agarose at 40°C.

TABLE 7.2
Primer Sequences for Gene Targets

Gene Target	T_m (°C)	Sequence	Size (bp)
Actin			
Forward primer	60.5	5′-Cy5-TACGTTGCTATCCAGGCTGTGC-3′	200
Reverse primer	58.7	5′-CTGTAGCCGCGCTCGGT-3′	
Probe	73.9	5′-JOE-ATCCCTGTACGCCTCTGGCCGTACCACT-Iowa Black FQ-3′	
t(14;18)			
Forward primer	60.2	5′TTCGATTACACGGAGTGCTGGGAA-3′	296
Reverse primer	60.3	5′-FAM-CGTTGATTTGCCGTTCCATGTCGT-3′	
Probe	74.6	5′-FAM-TTTCAACACAGACCCACCCAGAGCC CTCCTG-TAMRA-3′	

Source: Novak, R. et al., *Angew. Chem. Int. Ed. Engl.*, 50, 390–395, 2011.

7.4.2.2 Agarose Droplet Generation

The preparation of the carrier oil and the device operation have been described previously.[29] Molten agarose solution mixed with cells and beads was injected into the channels by the on-chip three-valve diaphragm pump. The three valves form the input, displacement, and output of a diaphragm pump and are cycled in a four-step fashion under the control of a LabVIEW graphical interface. The whole MEGA assembly was continually heated using a heated air stream (42°C–45°C) to avoid agarose gelling in the device. Pumping parameters were optimized to produce uniform ~3 nL 1.5% agarose droplets. The sample reservoir was constantly replenished with agarose suspension to sustain droplet generation. Agarose droplets were collected and immediately gelled in 0.5 mL PCR tubes placed in freezer trays.

7.4.2.3 Single-Genome Purification

Agarose droplets were extracted from carrier oil by washing with isopropanol, 100% ethanol, and PBS (10 mL each) and resuspended in 1 mL PBS. To lyse cells, 1 mL 2× SDS lysis buffer (1% SDS, 200 mM EDTA, 20 mM Tris–HCl) and 100 mg/mL proteinase K were added, and samples were incubated overnight at 52°C. Agarose droplets were washed with 2% (w/v) Tween 20 to remove PCR-inhibiting SDS, with 100% ethanol to inactivate any remaining proteinase K, with 0.02% (w/v) Tween 20 five times (10 mL each), and finally 10 mL H_2O. The supernatant was removed prior to emulsion PCR. Samples not immediately used were stored at 4°C in 100% ethanol.

7.4.2.4 Emulsion PCR

The PCR mix contained 1× AmpliTaq Gold buffer with 1.5 mM $MgCl_2$, 0.2 mM dNTPs, 4 μg/μL heat-inactivated BSA, 0.01% Tween 80, 0.45 μM forward and 0.045 μM reverse primers for each target gene, 0.25 U/μL AmpliTaq Gold polymerase, and 12.5 μL agarose droplets containing purified single-cell genomes in each 50 μL PCR reaction. Agarose droplets were equilibrated in PCR mix in standard 0.5 mL PCR tubes for 30 min at 4°C with occasional agitation. To reemulsify the agarose droplets, 160 μL carrier oil were added and mechanically agitated. Carrier oil was identical to the oil used for initial agarose droplet formation but contained 0.8% (w/w) Triton X-100 surfactant for increased stability. Emulsions were generated by mechanical agitation. Thermal cycling, carried out in a PTC200 thermocycler (MJ Research, Waltham, MA), involved ramping to 95°C (0.1°C/s) in order to melt agarose droplets without merging, followed by a 10 min hot start at 95°C, and 33 cycles of 95°C for 30 s, 60°C for 60 s, 72°C for 90 s, and a final 72°C extension for 5 min. The reactions were cooled to 4°C.

7.4.2.5 Fluorescence Imaging

Spinning disk confocal microscopy was performed with 491 and 561 nm excitation lasers through a 63 × 1.4 NA oil immersion objective (Carl Zeiss, Inc., Thornwood, NY). Images were captured using a Cascade II charge-coupled device. Samples were stained with (50 g/mL) acridine orange or (500 g/mL) propidium iodide. Epifluorescence imaging of amplicon-labeled beads was performed using a 10 × 0.25 NA Nikon Plan objective. Images were contrast adjusted and cropped if needed for optimal viewing.

7.4.2.6 Single-Cell Sequencing

Amplicon-bound beads from emulsion PCR were reamplified in 50 μL TaqMan qPCR reaction in reactions in standard 96-well plates. Beads were counted using a hemocytometer and diluted to approximately one bead per reaction. PCR mix contained 1× Ampli buffer with 5 mM $MgCl_2$, 0.2 mM dNTPs, 0.025% DMSO, 1 μM ROX reference dye, 0.15 μM forward and 0.15 μM reverse primers and 0.15 μM TaqMan probe for each target gene, and 0.035 U/μL AmpliTaq Gold polymerase. Thermal cycling in an ABI7300 cycler consisted of a 10 min hot start at 95°C, and 20 cycles of 95°C for 15 s, 60°C for 30 s, and 72°C for 30 s. Reactions were loaded on a 1.5% agarose gel and the two amplicons were separated and then excised, purified using a QIAquick Gel Extraction Kit (QIAGEN, Valencia, CA), and sequenced on an ABI3730 at the UC Berkeley Core Sequencing Facility.

7.4.3 Results and Discussion

7.4.3.1 Single-Cell Encapsulation and Genome Purification

Single lymphoblast cells were encapsulated along with primer-functionalized beads in 1.5% low-melting-point agarose by using a four-channel MEGA. Micropump actuation was optimized to account for the increased viscosity of molten agarose without modification of the microfluidic design. Droplet generation at approximately 40°C for 30 min resulted in the encapsulation of approximately 18,000 cells at up to 0.3 cpd on average. Figure 7.10a shows the generation of uniform 3 nL agarose droplets containing primer-functionalized beads. The inset highlights an example of cell and bead coencapsulation in a single droplet. Encapsulation in agarose droplets enables reproducible single-cell DNA extraction and isolation. Figure 7.10b shows

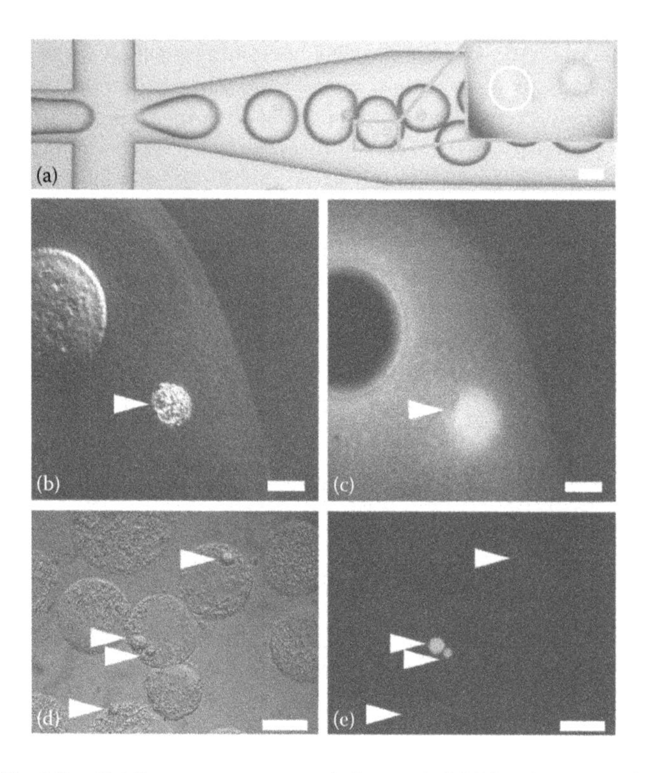

FIGURE 7.10 Microfluidic agarose encapsulation and SCGA of mammalian cells. (a) Microfluidic agarose emulsion generation showing single cells encapsulated along with primer-labeled beads in uniform 3 nL droplets. Inset highlights a cell (*circled*) and a bead in one droplet. (b) DIC image of a cell (*arrow*) at the edge of an agarose droplet after protease digestion. (c) Z-projection of confocal micrograph of the previous image showing acridine-orange-stained DNA confined in the agarose after release from the cell (*arrow*). (d) Following PCR amplification of the chromosomal translocation t(14;18), agarose droplets containing beads (*arrows*) remain intact and do not merge. (e) Epifluorescent micrograph of the previous image demonstrates primer-labeled beads (*arrows*) becoming fluorescent if a t(14;18)$^+$ RL cell is present in the same droplet but otherwise remaining dark. Scale bars are 100 μm in a, d, and e and 10 μm in b and c. (Reprinted with permission from Novak, R. et al., *Angew. Chem. Int. Ed. Engl.*, 50, 390–395, Copyright 2011. Wiley-VCH Verlag GmbH & Co. KGaA.)

an example of an agarose droplet in which a bead and a cell were coencapsulated (the bead is larger than the cell). Cell lysis and digestion of the DNA-binding histone proteins upon overnight incubation of the gel droplets in SDS lysis buffer containing proteinase K led to complete release of genomic DNA in a cell. The void left by the cell in the agarose was occupied by brightly fluorescent genomic DNA, which exhibited minimal diffusion into the surrounding gel (Figure 7.10c) as a result of the relatively small pore size of 1.5% agarose (~130 nm). An incubation temperature of 52°C facilitated enzymatic protein digestion while preserving the integrity of the agarose droplets. By staining with propidium iodide, we were able to visualize single high-molecular-weight DNA strands protruding from the relatively small agarose droplets. This result indicated that the majority of nuclear proteins were removed by the combination of proteinase K and SDS. The agarose droplets were stable for at least 1 week when stored in ethanol at 4°C, as determined by confocal imaging of DNA diffusion radii.

7.4.3.2 Multiplexed Single-Cell Droplet PCR

A key benefit of agarose encapsulation is the ability to mechanically manipulate the isolated genomic DNA without mixing the genetic content of different cells. The agarose matrix also protects the genomic DNA from physical damage caused by flow shearing during liquid sample preparation. Without going through another microfluidic droplet generation, agarose droplets equilibrated with PCR mix can be easily reemulsified by mechanical agitation in dispersing oil to produce uniform nanoliter-droplet "reactors" for massively parallel single-cell PCR analysis. Excess PCR mix produces microfines (emulsions <1 μm in diameter), which enhance emulsion stability during thermal cycling.[65] The agarose droplets melt during the hot start phase of PCR and remain liquid throughout the amplification process, maximizing reagent and amplicon diffusion rates. We varied the initial heating rate and tested various concentrations of Triton X-100 as well as combinations of Abil em90 and Span 80 detergents in oil. A slow temperature ramp profile (0.1°C/s) resulted in improved short-term stability, and the addition of BSA (4 mg/mL) to the PCR mix and 0.8% Triton X-100 to the emulsion oil minimized droplet merger over the course of PCR thermal cycling (Figure 7.10d). Fluorescently labeled amplicons bound to the primer-functionalized beads could be seen inside the agarose droplets following PCR (Figure 7.10e). The 34 μm cross-linked beads were selected for their ability to amplify targets exceeding 1 kb in amounts of at least 100 amol per bead.[23] The absence of fluorescence from beads in droplets without cellular genomic DNA indicated that genomic targets were not transferred between agarose droplets.

We demonstrated highly parallel genotyping with single-cell resolution by performing a multiplex PCR assay of cancer cells harboring the t(14;18) translocation at various mutant-to-wild type (RL/TK6) cell ratios. Labeling of the control gene product, β-actin, with the cyanine dye Cy5 enabled the quantification of total cell frequency, whereas the *BCL-2/IgH* translocation t(14;18) product labeled with FAM spanned *BCL-2* and *IgH* genes across their breakpoint regions and could be used to determine mutation frequency. A representative flow cytometric profile of beads following multiplex PCR amplification with 50% RL cells at an average cell frequency of 0.3 cpd (Figure 7.11a) demonstrated distinct populations of negative beads, Cy5-positive beads with β-actin only, and double-positive beads with FAM-labeled t(14;18) and Cy5-labeled β-actin. Beads containing t(14;18) only were never observed, which further indicates the conservation of single-genome integrity during cell lysis and PCR. By maintaining a constant total cell density in the single-cell regime (0.1–0.3 cpd on average) and varying the relative concentrations of mutant RL and wild-type TK6 cells, we generated a standard curve (Figure 7.11b) to confirm that amplification originated from single cells. In this stochastic regime, the linearity ($r = 0.993$) of the measured concentration of mutant RL cells with respect to the input in the 0%–100% RL-cell-frequency range tested indicated successful genetic analysis of single cells. Importantly, in the subset of samples further tested, the ratio of total amplicon-positive bead frequency determined by flow cytometry

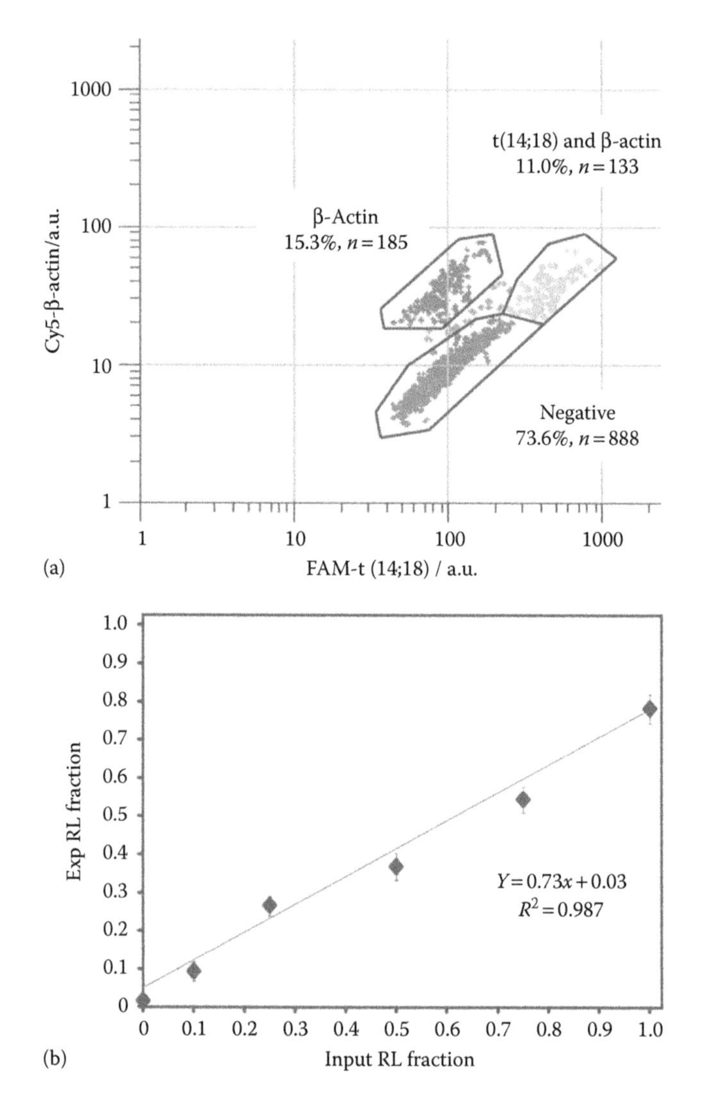

FIGURE 7.11 High-throughput digital genetic analysis of cancer cells. (a) A representative flow cytometry plot and gated populations of beads from a sample containing 50% RL lymphoblast cells ($N = 1206$) at a frequency of 0.3 cpd. The observed fraction of total positive events (26.3%) compares favorably with an expected frequency of 25.9% based on the Poisson distribution. (b) Standard curve of $t(14;18)^+$ RL cell detection from mixtures of mutant RL and wild-type TK6 cells. Experimental quantification by multiplex emulsion PCR and flow cytometry exhibits a linear response between the 10% and 100% mutant cell frequencies tested. (Reprinted with permission from Novak, R. et al., *Angew. Chem. Int. Ed. Engl.*, 50, 390–395, Copyright 2011. Wiley-VCH Verlag GmbH & Co. KGaA.)

to the microscopically observed cell-encapsulation frequency indicated high PCR efficiency (113% ± 24%).

7.4.3.3 Single-Cell Sequencing

For the single-cell sequencing of both target gene loci, the post-PCR beads were diluted to the stochastic limit and reamplified in a standard 96-well plate, and the products from individual wells were separated by gel electrophoresis. Gel analysis further confirmed amplification from single genomes, and frequencies of β-actin single-positive to double-positive events matched flow cytometry frequency results in the samples tested. Fluorescence-based sorting can be applied to amplicon-labeled beads to sequence only populations of interest and thereby increase efficiency. Size-separated products from single beads were excised from the gel and reamplified for Sanger sequencing. The correct sequences were recovered for both the t(14;18) target and the β-actin control (Figure 7.12). The random nucleotide insertion sequence in t(14;18) matched the unique translocation "fingerprint" determined by sequencing of RL cells in bulk.[66] The reamplification step bridges single-cell PCR amplification with standard molecular-biology protocols for the sequencing of multiple genetic loci with single-cell resolution at a rate required for meaningful population analysis.

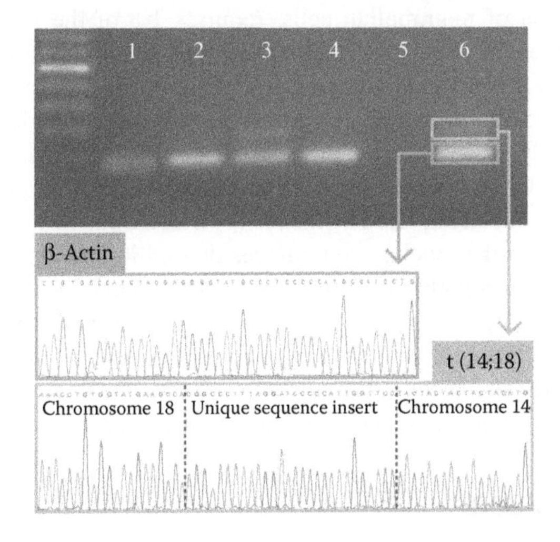

FIGURE 7.12 Multiplex single-cell sequencing. Two loci from DNA products isolated from single cells were sequenced by the reamplification of individual amplicon-labeled beads in separate PCR reactions and the separation of the two products by gel electrophoresis. This representative image of a section of the gel from a 50% RL cell sample shows amplified t(14;18) and β-actin from single beads (lanes 3 and 6), β-actin originating from TK6 cells (lanes 1, 2, and 4), the lack of β-actin and t(14;18) from negative beads (lane 5), and a 100 bp DNA ladder. A lack of t(14;18) amplification without β-actin indicates true single-cell analysis. Both products were excised and sequenced to give sequence data from two genes colocalized in a single cell. The breakpoint locations on chromosomes 14 and 18 as well as the unique insertion sequence confirmed the recovery of the expected RL lymphoblast sequence. (Reprinted with permission from Novak, R. et al., *Angew. Chem. Int. Ed. Engl.*, 50, 390–395, Copyright 2011. Wiley-VCH Verlag GmbH & Co. KGaA.)

Using this approach, it is possible to sequence multiple target genes from hundreds of single cells in a single experiment and to perform statistical analysis of gene-sequence variation of a cell population. Our agarose cell-encapsulation and emulsion PCR method has the potential to combine with next-generation sequencing technologies to provide additional gains in throughput for single-cell sequencing. In the case of follicular lymphoma, identification of the breakpoint sequence along *IgH* and *BCL-2* genes can help in the elucidation of the mechanisms of erroneous genetic recombination that cause the translocation t(14;18), whereas the random insertion sequence between the two chromosomes is a unique identifier of distinct mutation events.[66] Alternative single-cell analysis approaches, such as fluorescence *in situ* hybridization and *in situ* PCR, are utilized for investigating cell-mutation progression, their throughput is limited, and the inability to recover amplicons prevents the identification of unknown mutations.

Agarose droplet encapsulation offers a robust means for parallel cell lysis and DNA purification of cell types that are difficult to screen using other single-cell PCR techniques. The method can be adapted to encapsulate live single cells for cell culture and subsequent analysis of clonal populations. Although a similar approach was demonstrated recently for the screening and amplification of alginate-encapsulated *E. coli* colonies containing plasmid libraries,[67] our approach enables the robust genome purification of mammalian cells, requires 1% of the reagent volume for emulsion PCR, does not require droplet sorting, and maintains single-cell segregation throughout all downstream analyses as a result of the incorporation of primer-functionalized beads as amplicon substrates. Furthermore, single-cell sequencing techniques generally involve cell types that are amenable to lysis during PCR and the sorting of single cells into 96-well PCR plates. However, PCR amplification directly from single cells has been hindered by the lack of a DNA-purification step that would remove histones and other nuclear components that inhibit polymerase activity. Our approach incorporates a highly parallel single-cell-lysis and DNA-purification step, resulting in near 100% amplification efficiency. Finally, the ability to purify genomic DNA from single cells in a supporting matrix opens up the possibility of single-cell epigenetic analysis, such as the methylation-specific PCR of gene regulatory sequences following conversion with bisulfite for sequencing.

7.5 CONCLUSIONS

Cancer is an evolving disease driven by genetic instability, which results in constant clonal divergence of the tumor cell population. The accumulation of mutations in the genes coding for cellular pathways plays an important role in carcinogenesis, metastasis, and therapeutic resistance. Significant cellular heterogeneity may therefore exist in tumors; this heterogeneity changes dynamically at different stages of disease progression.[68] Droplet microfluidics is uniquely positioned for high-throughput, sensitive single-molecule and single-cell analyses as it offers significant advantages compared to conventional robotic laboratory systems, such as speed, throughput, sensitivity, and ultralow sample consumption. We have developed a MEGA technology as a core droplet-based digital microfluidic platform for high-throughput single-cell genetic analysis. We have characterized a unique droplet formation mechanism underlying the active pump-driven droplet formation, which allows for programmable, on-demand formation of complex droplet patterns. Based on the MEGA technology, we developed a multiplexed SCGA method that is used to address multiple problems in biology. We first demonstrated the feasibility of ultrasensitive pathogen detection and genotyping with high speed with sensitivity extending to one pathogen in a 10^6 background. The digital microfluidic technology was then adapted to multilocus detection and sequencing of genes in single cells for high-throughput analysis of variable regions as well as mutation colocalization. The ability to sequence multiple targets within single cells with high throughput will facilitate investigation of the synergistic effects of mutation co-occurrence and its impact on disease progression and treatment. Furthermore, the screening of large cell populations will uncover potential tumor heterogeneity at the single-nucleotide level that is otherwise obscured by the ensemble average. Overall, our high-performance microfluidic single-cell genetic analysis technology provides a powerful tool for probing the stochastic mechanisms of carcinogenesis, progression, and response to chemotherapy.

ACKNOWLEDGMENTS

The authors gratefully acknowledge the contributions of Richard Novak, Dr. Joe Shuga, Professor Martyn T. Smith from the University of California at Berkeley, Mimi Shin and Dr. Tanyu Wang from the University of Kansas, and all the other collaborators who contributed to the work described in this chapter. This work was supported by the trans-NIH Genes, Environment and Health Initiative, Biological Response Indicators of Environmental Systems Center Grant U54 ES016115-01, by the Superfund Basic Research Program NIEHS Grant P42 ES004705, and the new faculty start-up funds from the University of Kansas.

REFERENCES

1. Cai, L., Friedman, N. and Xie, X.S. Stochastic protein expression in individual cells at the single molecule level. *Nature* **440**, 358–362 (2006).
2. Hanahan, D. and Weinberg, R.A. The hallmarks of cancer. *Cell* **100**, 57–70 (2000).
3. Elowitz, M.B., Levine, A.J., Siggia, E.D. and Swain, P.S. Stochastic gene expression in a single cell. *Science* **297**, 1183–1186 (2002).
4. Toriello, N.M. et al. Integrated microfluidic bioprocessor for single-cell gene expression analysis. *Proc. Natl Acad. Sci. USA* **105**, 20173–20178 (2008).
5. King, K.R. et al. A high-throughput microfluidic real-time gene expression living cell array. *Lab Chip* **7**, 77–85 (2007).
6. Klein, C.A. et al. Genetic heterogeneity of single disseminated tumour cells in minimal residual cancer. *Lancet* **360**, 683–689 (2002).
7. Fuhrmann, C. et al. High-resolution array comparative genomic hybridization of single micrometastatic tumor cells. *Nucleic Acids Res.* **36**, e39 (2008).
8. Batchelor, E., Loewer, A. and Lahav, G. The ups and downs of p53: Understanding protein dynamics in single cells. *Nat. Rev. Cancer* **9**, 371–377 (2009).
9. Li, X.J., Ling, V. and Li, P.C.H. Same-single-cell analysis for the study of drug efflux modulation of multidrug resistant cells using a microfluidic chip. *Anal. Chem.* **80**, 4095–4102 (2008).

10. Pantel, K., Brakenhoff, R.H. and Brandt, B. Detection, clinical relevance and specific biological properties of disseminating tumour cells. *Nat. Rev. Cancer* **8**, 329–340 (2008).
11. d'Amore, F. et al. Clonal evolution in t(14;18)-positive follicular lymphoma, evidence for multiple common pathways, and frequent parallel clonal evolution. *Clin. Cancer Res.* **14**, 7180–7187 (2008).
12. Carter, H. et al. Cancer-specific high-throughput annotation of somatic mutations: Computational prediction of driver missense mutations. *Cancer Res.* **69**, 6660–6667 (2009).
13. Chiu, B.C.H. et al. Agricultural pesticide use and risk of t(14;18)-defined subtypes of non-Hodgkin lymphoma. *Blood* **108**, 1363–1369 (2006).
14. Arora, A., Simone, G., Salieb-Beugelaar, G.B., Kim, J.T. and Manz, A. Latest developments in micro total analysis systems. *Anal. Chem.* **82**, 4830–4847 (2010).
15. Zeng, Y. and Wang, T. Quantitative microfluidic biomolecular analysis for systems biology and medicine. *Anal. Bioanal. Chem.* **405**, 5743–5758 (2013).
16. Teh, S.Y., Lin, R., Hung, L.H. and Lee, A.P. Droplet microfluidics. *Lab Chip* **8**, 198–220 (2008).
17. Niu, X. and Demello, A.J. Building droplet-based microfluidic systems for biological analysis. *Biochem. Soc. Trans.* **40**, 615–623 (2012).
18. Guo, M.T., Rotem, A., Heyman, J.A. and Weitz, D.A. Droplet microfluidics for high-throughput biological assays. *Lab Chip* **12**, 2146–2155 (2012).
19. Nakano, M. et al. Single-molecule PCR using water-in-oil emulsion. *J. Biotechnol.* **102**, 117–124 (2003).
20. Dressman, D., Yan, H., Traverso, G., Kinzler, K.W. and Vogelstein, B. Transforming single DNA molecules into fluorescent magnetic particles for detection and enumeration of genetic variations. *Proc. Natl Acad. Sci. USA* **100**, 8817–8822 (2003).
21. Beer, N.R. et al. On-chip, real-time, single-copy polymerase chain reaction in picoliter droplets. *Anal. Chem.* **79**, 8471–8475 (2007).
22. Beer, N.R. et al. On-chip single-copy real-time reverse-transcription PCR in isolated picoliter droplets. *Anal. Chem.* **80**, 1854–1858 (2008).
23. Kumaresan, P., Yang, C.J., Cronier, S.A., Blazei, R.G. and Mathies, R.A. High-throughput single copy DNA amplification and cell analysis in engineered nanoliter droplets. *Anal. Chem.* **80**, 3522–3529 (2008).
24. Mazutis, L. et al. Droplet-based microfluidic systems for high-throughput single DNA molecule isothermal amplification and analysis. *Anal. Chem.* **81**, 4813–4821 (2009).
25. Huebner, A. et al. Quantitative detection of protein expression in single cells using droplet microfluidics. *Chem. Commun.* (12), 1218–1220 (2007).
26. Huebner, A. et al. Static microdroplet arrays: A microfluidic device for droplet trapping, incubation and release for enzymatic and cell-based assays. *Lab Chip* **9**, 692–698 (2009).
27. Mazutis, L., Baret, J.C. and Griffiths, A.D. A fast and efficient microfluidic system for highly selective one-to-one droplet fusion. *Lab Chip* **9**, 2665–2672 (2009).
28. Brouzes, E. et al. Droplet microfluidic technology for single-cell high-throughput screening. *Proc. Natl Acad. Sci. USA* **106**, 14195–14200 (2009).
29. Zeng, Y., Novak, R., Shuga, J., Smith, M.T. and Mathies, R.A. High-performance single cell genetic analysis using microfluidic emulsion generator arrays. *Anal. Chem.* **82**, 3183–3190 (2010).
30. Zeng, Y., Shin, M. and Wang, T. Programmable active droplet generation enabled by integrated pneumatic micropumps. *Lab Chip* **13**, 267–273 (2013).
31. Zeng, Y., Novak, R., Shuga, J., Smith, M.T. and Mathies, R.A. High-performance single cell genetic analysis using microfluidic emulsion generator arrays. *Anal. Chem.* **82**, 3183–3190 (2010).

32. Novak, R. et al. Single-cell multiplex gene detection and sequencing with microfluidically generated agarose emulsions. *Angew. Chem. Int. Ed. Engl.* **50**, 390–395 (2011).
33. Christopher, G.F. and Anna, S.L. Microfluidic methods for generating continuous droplet streams. *J. Phys. D Appl. Phys.* **40**, R319–R336 (2007).
34. Prakash, M. and Gershenfeld, N. Microfluidic bubble logic. *Science* **315**, 832–835 (2007).
35. Shemesh, J., Nir, A., Bransky, A. and Levenberg, S. Coalescence-assisted generation of single nanoliter droplets with predefined composition. *Lab Chip* **11**, 3225–3230 (2011).
36. Gu, H., Murade, C.U., Duits, M.H. and Mugele, F. A microfluidic platform for on-demand formation and merging of microdroplets using electric control. *Biomicrofluidics* **5**, 11101 (2011).
37. Zeng, S., Li, B., Su, X., Qin, J. and Lin, B. Microvalve-actuated precise control of individual droplets in microfluidic devices. *Lab Chip* **9**, 1340–1343 (2009).
38. Lin, R., Fisher, J.S., Simon, M.G. and Lee, A.P. Novel on-demand droplet generation for selective fluid sample extraction. *Biomicrofluidics* **6**, 24103 (2012).
39. Christopher, G.F., Noharuddin, N.N., Taylor, J.A. and Anna, S.L. Experimental observations of the squeezing-to-dripping transition in T-shaped microfluidic junctions. *Phys. Rev. E* **78**, 036317 (2008).
40. Willaime, H., Barbier, V., Kloul, L., Maine, S. and Tabeling, P. Arnold tongues in a microfluidic drop emitter. *Phys. Rev. Lett.* **96**, 054501 (2006).
41. Grover, W.H., Skelley, A.M., Liu, C.N., Lagally, E.T. and Mathies, R.A. Monolithic membrane valves and diaphragm pumps for practical large-scale integration into glass microfluidic devices. *Sens. Actuators B Chem.* **89**, 315–323 (2003).
42. Bransky, A., Korin, N., Khoury, M. and Levenberg, S. A microfluidic droplet generator based on a piezoelectric actuator. *Lab Chip* **9**, 516–520 (2009).
43. Mead, P.S. et al. Food-related illness and death in the United States. *Emerg. Infect. Dis.* **5**, 607–625 (1999).
44. Nugen, S.R. and Baeumner, A.J. Trends and opportunities in food pathogen detection. *Anal. Bioanal. Chem.* **391**, 451–454 (2008).
45. Batt, C.A. Materials science: Food pathogen detection. *Science* **316**, 1579–1580 (2007).
46. Diehl, F. et al. Detection and quantification of mutations in the plasma of patients with colorectal tumors. *Proc. Natl Acad. Sci. USA* **102**, 16368–16373 (2005).
47. Zhong, J.F. et al. A microfluidic processor for gene expression profiling of single human embryonic stem cells. *Lab Chip* **8**, 68–74 (2008).
48. Lagally, E.T. et al. Integrated portable genetic analysis microsystem for pathogen/infectious disease detection. *Anal. Chem.* **76**, 3162–3170 (2004).
49. Liu, R.H., Yang, J.N., Lenigk, R., Bonanno, J. and Grodzinski, P. Self-contained, fully integrated biochip for sample preparation, polymerase chain reaction amplification, and DNA microarray detection. *Anal. Chem.* **76**, 1824–1831 (2004).
50. Cady, N.C., Stelick, S., Kunnavakkam, M.V. and Batt, C.A. Real-time PCR detection of *Listeria monocytogenes* using an integrated microfluidics platform. *Sens. Actuators B Chem.* **107**, 332–341 (2005).
51. Thaitrong, N., Toriello, N.M., Del Bueno, N. and Mathies, R.A. Polymerase chain reaction-capillary electrophoresis genetic analysis microdevice with in-line affinity capture sample injection. *Anal. Chem.* **81**, 1371–1377 (2009).
52. Beyor, N., Yi, L.N., Seo, T.S. and Mathies, R.A. Integrated capture, concentration, polymerase chain reaction, and capillary electrophoretic analysis of pathogens on a chip. *Anal. Chem.* **81**, 3523–3528 (2009).
53. Vogelstein, B. and Kinzler, K.W. Digital PCR. *Proc. Natl Acad. Sci. USA* **96**, 9236–9241 (1999).
54. Beer, N.R. et al. On-chip, real-time, single-copy polymerase chain reaction in picoliter droplets. *Anal. Chem.* **79**, 8471–8475 (2007).

55. Chao, T.C. and Ros, A. Microfluidic single-cell analysis of intracellular compounds. *J. R. Soc. Interface* **5**(Suppl 2), S139–S150 (2008).
56. Ottesen, E.A., Hong, J.W., Quake, S.R. and Leadbetter, J.R. Microfluidic digital PCR enables multigene analysis of individual environmental bacteria. *Science* **314**, 1464–1467 (2006).
57. Zhang, H., Jenkins, G., Zou, Y., Zhu, Z. and Yang, C.J. Massively parallel single-molecule and single-cell emulsion reverse transcription polymerase chain reaction using agarose droplet microfluidics. *Anal. Chem.* **84**, 3599–3606 (2012).
58. Zare, R.N. and Kim, S. Microfluidic platforms for single-cell analysis. *Annu. Rev. Biomed. Eng.* **12**, 187–201 (2010).
59. Aster, J.C. and Longtine, J.A. Detection of BCL-2 rearrangements in follicular lymphoma. *Am. J. Pathol.* **160**, 759–763 (2002).
60. Cleary, M.L. and Sklar, J. Nucleotide sequence of a t(14;18) chromosomal breakpoint in follicular lymphoma and demonstration of a breakpoint-cluster region near a transcriptionally active locus on chromosome 18. *Proc. Natl Acad. Sci. USA* **82**, 7439–7443 (1985).
61. Korsmeyer, S.J. BCL-2 gene family and the regulation of programmed cell death. *Cancer Res.* **59**, 1693s–1700s (1999).
62. Rabkin, C.S., Hirt, C., Janz, S. and Dolken, G. t(14;18) Translocations and risk of follicular lymphoma. *J. Natl Cancer Inst. Monogr.* (39), 48–51 (2008).
63. Biagi, J.J. and Seymour, J.F. Insights into the molecular pathogenesis of follicular lymphoma arising from analysis of geographic variation. *Blood* **99**, 4265–4275 (2002).
64. Bretherick, K.L. et al. Elevated circulating t(14;18) translocation levels prior to diagnosis of follicular lymphoma. *Blood* **116**, 6146–6147 (2010).
65. Margulies, M. et al. Genome sequencing in microfabricated high-density picolitre reactors. *Nature* **437**, 376–380 (2005).
66. McHale, C.M. et al. Chromosome translocations in workers exposed to benzene. *J. Natl Cancer Inst. Monogr.* (39), 74–77 (2008).
67. Walser, M. et al. Novel method for high-throughput colony PCR screening in nanoliter-reactors. *Nucleic Acids Res.* **37**, e57 (2009).
68. Navin, N. et al. Inferring tumor progression from genomic heterogeneity. *Genome Res.* **20**, 68–80 (2010).

8 Droplet-Based Microfluidics for Biological Sample Preparation and Analysis

Xuefei Sun and Ryan T. Kelly

CONTENTS

8.1 Introduction 201
8.2 Droplet-Based Operations 203
 8.2.1 Droplet Generation 203
 8.2.2 In-Droplet Reagent Combination and Mixing 205
 8.2.3 Droplet Incubation 208
 8.2.4 Droplet Readout Strategies 208
8.3 Perspectives for Droplet-Based Microfluidics 211
 8.3.1 Enhanced LC/MS-Based Proteomic Analysis 211
 8.3.2 Single-Cell Chemical Analysis 212
8.4 Conclusions 215
References 216

8.1 INTRODUCTION

Modern biological research often requires massively parallel experiments to analyze a large number of samples in order to find biomarkers, screen drugs, or elucidate complex cellular pathways. These processes frequently involve time-consuming sample preparation and expensive biochemical measurements. Another constraint frequently encountered in bioanalysis is limited amounts of available sample. Microfluidics or lab-on-a-chip platforms offer promise for addressing the challenges encountered in biological research because a large number of small samples can be handled and processed with different functional elements in an automated fashion.

Droplet-based microfluidics, in which reagents of interest are compartmentalized within femtoliter-to-nanoliter-sized aqueous droplets or plugs that are encapsulated and dispersed in an immiscible oil phase, has emerged as an attractive platform for small-volume bioanalysis [1–9]. This new platform elegantly addresses challenges encountered with conventional continuous flow systems by, for example, limiting reagent dilution caused by diffusion and Taylor dispersion and minimizing cross-contamination and surface-related adsorptive losses [10]. The microdroplets isolated by the immiscible liquid can serve as microreactors, allowing for high-throughput chemical reaction screening and extensive biological research [1]. Droplet-based microfluidics also offers great promise for reliable quantitative analysis because monodisperse microdroplets can be generated with controlled sizes and preserve temporal information that is easily lost to dispersion in continuous flow systems [11,12].

Biological analysis begins with sample selection and preparation. The initial sampling can comprise cell sorting, tissue dissection, or extraction of protein or other analytes of interest from cells or tissues [13]. The biological samples are then prepared by, for example, combining reagents, mixing, incubating, purifying, and enriching. Depending on the complexity of the sample, the subsequent analytical measurements can be very simple, employing, for example, laser-induced fluorescence (LIF) to detect a single labeled analyte. With more complex samples having multiple analytes of interest, chemical separation techniques including capillary electrophoresis (CE) and liquid chromatography (LC), and information-rich detection methods such as mass spectrometry (MS) become necessary. To date, many operational components for microdroplets have been well-developed to perform most of these basic operations. For example, stable aqueous droplets dispersed in an oil phase can be generated using various droplet generator designs for sampling in a confined small volume, the most common of which are the T-junction [14,15] and flow focusing [16,17] geometries. Addition of reagents to existing droplets can be realized by fusion with other droplets, enabling the initiation and termination of the compartmentalized reactions confined in the microdroplets [18,19]. Rapid mixing of fluids within droplets enables a homogeneous reactive environment to be achieved and can be enhanced by means of chaotic advection [20]. In addition, droplets can be incubated in delay lines [21] or stored in reservoirs [22,23] or traps [24,25] for extended periods of time to complete reactions or facilitate biological processes.

Droplet-based microfluidic platforms have been successfully applied in a variety of chemical and biological research areas. For example, a droplet-based platform for polymerase chain reaction (PCR) amplification has proven able to significantly improve amplification efficiency over conventional microfluidic formats [26], which is mainly due to the elimination of both reagent dilution and adsorption on the channel surfaces. Droplets have also been employed to encapsulate, sort, and assay single cells [12,27] or microorganisms [24], study enzyme kinetics [11] and protein crystallization [28], and synthesize small molecules and polymeric micro- and nanoparticles.

Although droplet-based microfluidic technology has developed to a degree where droplets can be generated and manipulated with speed, precision, and control, some real challenges still exist that limit the widespread use of these systems. One challenge is how to extract and acquire the enormous amount of chemical information that can be contained in the picoliter-sized droplets. Detection of droplet contents has historically been limited to optical methods such as LIF, while coupling with chemical separations and nonoptical detection has proven difficult. Combining the advantages of droplet-based platforms with more information-rich analytical techniques including LC, CE, and MS can greatly extend their reach. This requires extraction of the droplets from the oil phase for downstream analysis and detection.

This chapter focuses primarily on the integrated droplet-based microsystems having the ability to couple with chemical separations and nonoptical detection, allowing for *ex situ* analysis and identification of the biochemical components contained in the microdroplets. Some unit operations for microdroplets will be briefly introduced, including droplet generation, fusion, and incubation. All approaches and techniques developed for droplet detection, droplet extraction, coupling CE separation, and electrospray ionization (ESI)-MS detection will be reviewed. An example of integrated droplet-based microfluidics, including on-demand droplet generation and fusion, robust and efficient droplet extraction, and a monolithically integrated nano-ESI emitter, will be given to demonstrate its potential for chemical and biological research.

8.2 DROPLET-BASED OPERATIONS

8.2.1 Droplet Generation

Currently, most planar microfluidic droplet generators are designed using T-junction [14,15] and flow focusing [16,17] geometries, in which small droplets are spontaneously formed at an intersection taking advantage of the interface instability between oil and aqueous streams. Using these approaches, droplets can be generated over a broad range of frequencies ranging from ~0.1 Hz to 10 kHz and using flow rates on the order of 0.1–100 μL/min [29]. Droplet volume and generation frequency depend on several factors, including the physical properties of the immiscible phases, flow rates, intersection geometry, and so on. For a given geometry and solvent composition, flow focusing and T-junction interfaces exhibit interdependence between flow rate and droplet generation frequency and cannot be easily modulated over short timescales.

For lower frequencies and applications for which the ability to rapidly change droplet size and generation frequency is desirable, on-demand droplet generation strategies become more favorable, as they ensure precise control and fine manipulation of individual droplets. Various approaches have been developed to generate droplets on demand, for example, by carefully balancing the pressure and flow in the system [27], as well as electrical [30] or laser pulsing [31] and piezoelectric actuation [32]. Pneumatic valving has also been explored and has been found to provide easy, independent control over both droplet size and generation frequency [33–36]. Galas et al. utilized a single pneumatic valve that was embedded in an active connector and assembled close to a T-junction to regulate the flow of the dispersed phase [33]. Constant pressures were applied on the inlets of two immiscible liquids to drive the flow in the microchannel. Individual droplets were created by briefly opening the valve. The aqueous droplet size depended on the valve actuation time and frequency, as well as the pressure applied at the oil inlet. Therefore, the droplet volume, spacing, and speed could be controlled accurately and independently. This device not only generated periodic sequences of identical droplets, but also enabled the production of nonperiodic droplet trains with different droplet sizes or spacing. Lin and coworkers also reported a similar platform for pneumatic valve-assisted on-demand droplet generation [34]. Negative pressure was applied at the outlet of the device to drive the flow of the two immiscible liquids through the microchannel. The dependence of droplet size on the valve actuation time and applied pressure was investigated. In addition, they utilized several aqueous flow channels, each with independently controlled microvalves, to generate arrays of droplets containing different compositions by alternately actuating the valves.

We have also investigated valve-controlled on-demand droplet generation. To minimize the dead volume and control droplet volume precisely, the pneumatic valve was placed over the side channel exactly at the T-junction (Figure 8.1a). Carrier oil flow was driven by a syringe pump, and the dispersed aqueous phase was injected by finely controlled air pressure. Figure 8.1a shows the generation process of an individual droplet. The valve is initially closed and the aqueous fluorescein solution is confined in the side channel. When the valve is opened briefly, a small volume of aqueous solution is dispensed into the oil channel to form a droplet, which is then flushed downstream by the carrier oil flow. Compared with conventional microfluidic droplet generation techniques based on a T-junction or flow focusing, the valve-integrated system can generate droplets with precise control over droplet volume, generation frequency, and velocity. Droplet velocity is determined by the syringe pump driving the oil stream, while the droplet generation rate is controlled by the

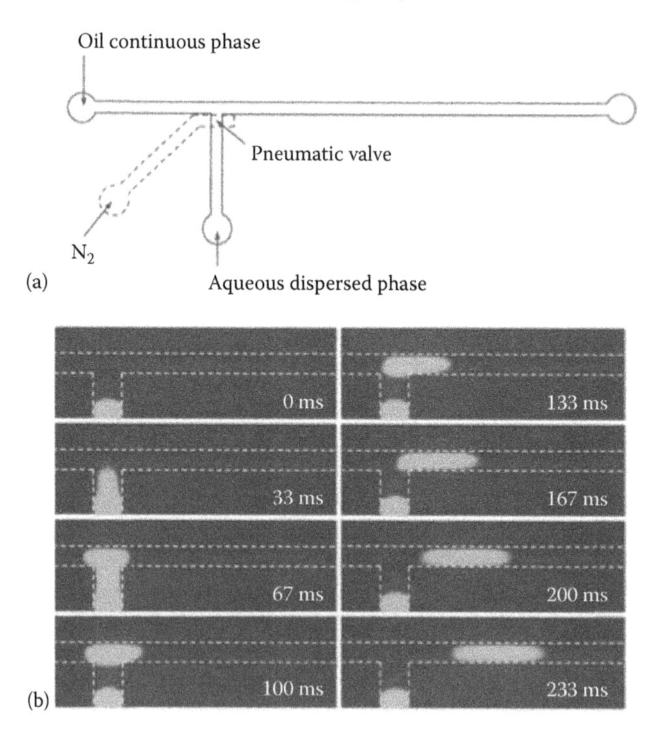

FIGURE 8.1 (a) Schematic depiction of a T-junction droplet generator controlled by a pneumatic valve. (b) Micrograph sequences depicting pneumatic valve-controlled generation of an individual fluorescein droplet. The width of the oil flow channel was 100 μm, the oil flow rate was 0.5 μL/min, and the sample injection pressure was 8 psi. Valve actuation time and pressure were 33 ms and 25 psi, respectively.

valve actuation frequency as defined in the software. The droplet spacing is determined by the interval between valve openings and the oil flow velocity. Droplet volume depends on several parameters including the valve actuation time, the pressure of the aqueous solution, the oil phase flow rate, and the valve control pressure (Figure 8.2).

8.2.2 In-Droplet Reagent Combination and Mixing

Besides controlled droplet generation, droplet fusion is of crucial importance in relation to the development of microreactors because it allows precise and reproducible mixing of reagents at well-defined points to initiate, modify, and terminate reactions [6]. Ismagilov and coworkers carried out the pioneering work to combine different reagents into individual droplets by causing two reagent solutions to flow in a microchannel as two laminar streams [37]. To prevent the prior contact of two reagents before droplet generation, an inert center stream was used to separate them. Thus, three streams were continuously injected into an immiscible carrier oil phase to form droplets. The gradient droplets in the reagent concentrations were achieved by varying the relative flow rates of the three streams [11,38]. A subsequent winding channel

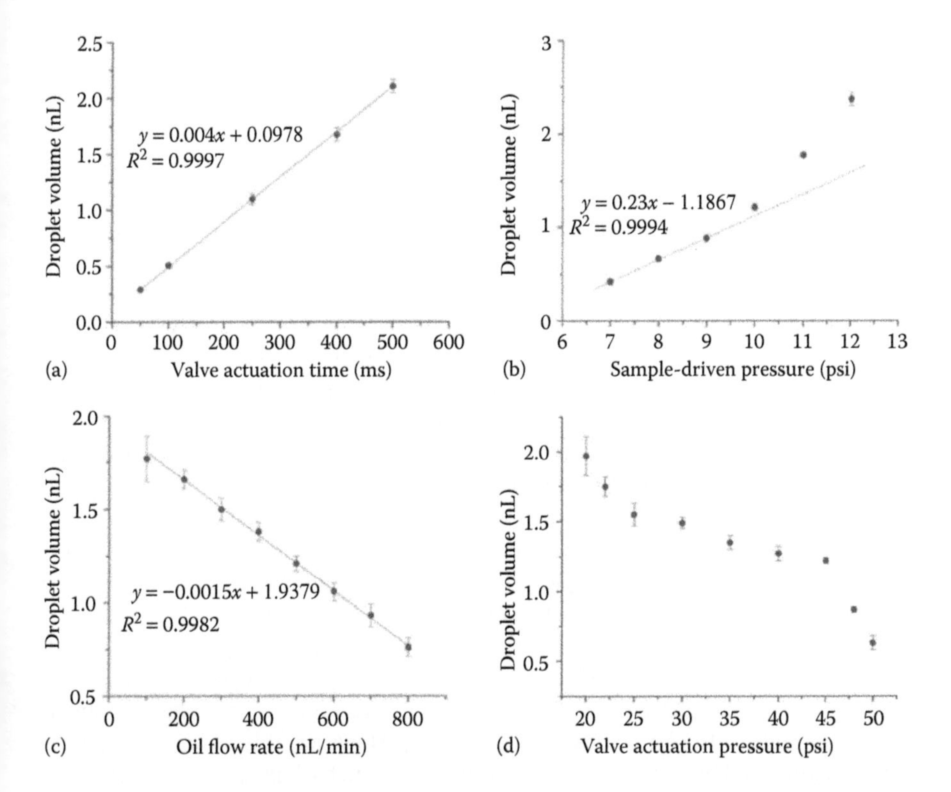

FIGURE 8.2 Plots of droplet volume dependence on (a) valve actuation time, (b) sample driven pressure, (c) oil flow rate, and (d) valve actuation pressure. Channel dimensions were the same as for Figure 8.1.

was designed to accelerate mixing by chaotic advection [20,37]. This approach has been widely employed to control networks of chemical reactions [37], study reaction kinetics [11], screen protein crystallization conditions [38], and investigate single-cell-based enzyme assays [39] and protein expression [12].

Recently, Weitz and colleagues presented a robust picoinjector to add reagents to droplets in microfluidic systems [40]. The picoinjector was controlled by an electric field to trigger the injection of a controlled volume of reagents into each droplet. The injection volume was precisely controlled by adjusting the droplet velocity and injection pressure. Selective injection was realized by switching the electric field on and off at kilohertz frequencies.

In-channel droplet fusion is another attractive approach to combine different reagents in individual droplets to initiate or terminate the confined reactions. The process of droplet merging introduces convective flows into the system, resulting in far more rapid mixing than relying on diffusion alone [41]. In-channel droplet fusion is readily achieved by bringing two or more surfactant-free droplets into contact. Both passive and active methods have been developed to control droplet fusion. For passive fusion devices, droplet coalescence is usually initiated by utilizing specially designed fusion elements in the channel. For example, Bremond et al. incorporated an expanded coalescence chamber in the channel network in which two droplets were brought into close proximity and merged together before they entered a narrow channel [42]. Fidalgo et al. reported a method for droplet fusion based on a surface energy pattern inside a microfluidic channel where the segmented flow was disrupted and the droplets were trapped and fused together [43]. In this case, full control of droplet fusion could be achieved by varying channel and pattern dimensions, as well as the fluid flow. This surface-induced droplet fusion method enabled the merging of multiple droplets containing different reagents to form a large droplet. However, this approach could potentially cause cross-contamination between droplets from the patterned surface. Niu et al. developed a pillar-induced droplet merging device, in which rows of pillars were constructed in the channel network serving as passive fusion elements or chambers [44]. The pillar array trapped droplets and drained the carrier oil phase through the apertures between pillars. The first trapped droplet was suspended and merged with succeeding droplets until the surface tension was overwhelmed by the hydraulic pressure. The merging process depended on the droplet size, and the number of droplets that could be merged relied on the mass flow rate and volume ratio between the droplets and merging chamber.

Active fusion methods that can be controlled externally and selectively have also been developed using, for example, electric fields [45–48] and laser pulses [49] to trigger coalescence. To perform active droplet fusion effectively, synchronization of droplets is a key factor because fusion efficiency relies on the droplets being in very close proximity [50]. Currently, special designs are often employed to synchronize droplets in two parallel channels, which then merge into a single channel downstream to realize droplet coalescence [34,42,51]. However, this system can potentially be disturbed by a few factors such as flow rate and back pressure in the channel, which may reduce the fusion efficiency. Recently, Jambovane et al. used valve-based droplet generation for multiple reagents to perform controlled reactions and establish chemical gradients among arrays of droplets [52]. Droplets were generated at a valve-controlled side channel and then different reagents were added to the droplets as they passed by similar side channels downstream.

An efficient method for reagent combination that we have recently developed employed two pneumatic valves integrated at a double-T intersection (Figure 8.3a). Reagents were introduced through the different side channels, each controlled by a separate valve, and simultaneous opening of the valves resulted in the creation of an aqueous plug containing both reagents. Upon actuation, the oil between the two side channels is quickly displaced, and the two aqueous streams collide and combine. No sample cross-contamination was observed because of the applied pressures, the rapid valve actuation, and the offset between the two side channels. The two liquids mixed together by a combination of diffusion and convection caused by an equalization of internal pressures following the combining of the aqueous streams. The linear dependence of droplet volume on valve actuation time and the independent valving for the two aqueous streams provide a high degree of control over droplet composition. Figure 8.3b shows arrays of six droplets containing different ratios of two dyes that were created by controlling the operation of two valves. Such control should be useful for optimizing or screening reactions and for studying reaction kinetics.

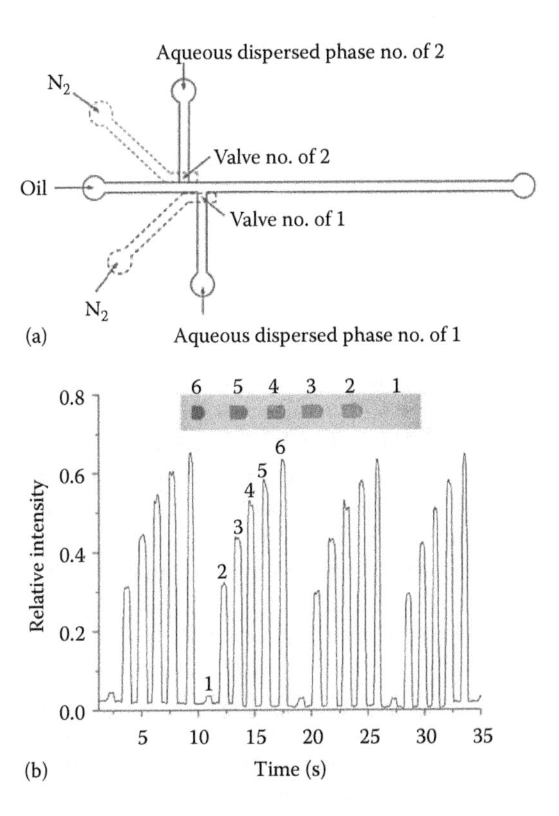

FIGURE 8.3 (a) Schematic depiction of the droplet generation, fusion, and mixing portion of the device. (b) Relative intensity of an array of six droplets containing different volume ratios of colored dyes.

8.2.3 Droplet Incubation

Many biological assays involving, for example, enzymatic reactions, have relatively slow kinetics, requiring microdroplets to be incubated for minutes to hours for efficient reaction. Similarly, studies involving cell incubation or protein expression require extended incubations. A straightforward method for microdroplet incubation is to simply increase the channel length following droplet generation [53,54], but increased back pressure and disruption of droplet formation can quickly become an issue. Frenz et al. incorporated deeper and wider delay lines following the droplet generation section, which enabled the length of times for reactions in the droplets to increase from 1 min to >1 h [21]. Similarly, Kennedy and coworkers have interfaced capillaries or Teflon tubing with the droplet generation devices to collect and store sample plugs for 1–3 h [55]. For longer-term online incubation, droplets can be stored in reservoirs, traps, or dropspot arrays. For example, Courtois et al. fabricated a large reservoir for storing droplets for periods of up to 20 h to study the retention of small molecules in droplets [23]. Huebner et al. designed a droplet trapping array to store and incubate picoliter-sized droplets for extended periods of time to investigate the encapsulated cells and enzymatic reactions [25]. Weitz and colleagues introduced a "Dropspots" device to immobilize and store thousands of individual droplets in a round chamber array over a 15 h incubation period [56]. Droplets can also be incubated off-chip for periods ranging from minutes to several days when appropriate surfactants are used to stabilize the droplets [57,58]. The incubated droplets can then be reinjected into microfluidic devices for further processing and detection.

8.2.4 Droplet Readout Strategies

To date, in-droplet fluorescence detection remains the most widely used method for analyzing the contents of droplets due to its ability to measure in real time and with high sensitivity. Fluorescence detection has been implemented to study enzyme kinetics within droplets [11,59,60], characterize the behavior of encapsulated single cells [12,27], detect PCR products [61,62], and investigate the interactions between biological samples [63]. Fluorescence detection is ideally suited to rapid, sensitive detection of a small number of distinct species. For cases in which a large number of analytes need to be detected and identified (e.g., proteomics and metabolomics) and where fluorescent labeling is not desirable, alternative measurement strategies are needed.

In-droplet Raman spectroscopy has recently been used to detect and analyze droplet contents [64,65]. It is a nondestructive and label-free detection approach with high molecular selectivity which can track the droplets in real time to determine fundamental droplet properties and chemical contents, including droplet sizes, encapsulated species, structures, and concentrations. Surface-enhanced Raman spectroscopy (SERS) can offer higher sensitivity and reproducible quantitative analysis of the droplets due to the enhancement of the Raman signal intensity [66]. Electrochemical detection is an inexpensive and label-free approach to collect information on the physical and chemical properties of droplets and can monitor droplet production and measure droplet length, frequency, and velocity [67]. It can provide chemical information when the reaction within the droplets involves an electrochemically active reactant or product [68]. Another advantage of electrochemical measurement is its compatibility with alternative chip materials, including opaque substrates, which are difficult to implement for conventional optical detection strategies such as fluorescence. Nuclear magnetic resonance (NMR) has also been used for droplets or segmented flow analysis. Karger and coworkers developed a microcoil NMR probe for high-throughput analysis of sample plugs in dimethyl sulfoxide (DMSO) [69].

While the above detection strategies can be employed *in situ*, others require the contents to be removed from the droplet for subsequent analysis. Once extracted to an aqueous stream, the droplet contents can be analyzed using more information-rich techniques including LC, CE, and MS. MS is an especially attractive technique for in-depth, label-free biological analysis because of its ability to identify and provide structural information for hundreds or more unique species in a given analysis [70]. Below we detail methods used for droplet extraction and subsequent analysis.

Ismagilov and colleagues used a microfluidic system to screen and optimize organic reaction conditions in microdroplets detected using matrix-assisted laser desorption ionization mass spectrometry (MALDI-MS) [71]. The incubated reaction plugs were deposited onto a sample plate for MALDI-MS analysis. Kennedy and coworkers directly pumped nanoliter plugs of sample into a mass spectrometer for analysis through a metal-coated capillary nanospray emitter, separating the analyte from the carrier at the emitter itself [72,73]. Teflon tubing was placed close to the emitter tip to siphon the accumulated oil away from the tip, and could maintain stable electrospray at flow rates as high as 2000 nL/min. However, it is generally necessary to extract the aqueous droplet from the oil phase for further separation or on-line MS analysis to avoid contamination of the mass spectra with peaks from the oil and to maintain the electrospray Taylor cone in the most efficient cone-jet mode of operation.

Edgar et al. first reported the extraction of aqueous droplet contents into a channel for CE separation [74]. A femtoliter-volume aqueous droplet was directly delivered to fuse with the aqueous phase in the separation channel for CE separation. Niu et al. employed a similar method to inject the droplets in which the LC eluent was fractionated into a CE channel for comprehensive two-dimensional separations in both time and space [75]. A pillar array was constructed at the interface to evacuate the carrier oil phase prior to loading samples into the separation channel. In these two cases, it was very difficult to maintain a robust extraction because the segmented flow was perpendicular to the CE separation channel.

Kennedy and colleagues exploited a surface modification method to form a stable interface at the junction between two immiscible phases in the microchannel [76–78]. They selectively patterned glass surfaces in the segmented flow channel to be hydrophobic in order to stabilize the oil–water interface and facilitate droplet extraction. But in some cases only part of each droplet was extracted due to the presence of a "virtual wall," and was not suitable for quantitative analysis because of irreproducibility and loss of information [76]. Fang and coworkers employed a similar surface modification technique to obtain a hydrophilic tongue-based droplet extraction interface that could control the droplet extraction by regulating the waste reservoir height [79]. The extracted droplet contents were then detected by MS through an integrated ESI emitter. More recently, Filla et al. used a corona treatment to hydrophilize a portion of a polydimethylsiloxane (PDMS) chip to establish an extraction interface [80]. Aqueous droplets were transferred into the hydrophilic channel when the segmented flow encountered the interface. The droplet contents were subsequently analyzed by electrochemistry or microchip-based electrophoresis with electrochemical detection.

Huck and coworkers employed electrocoalescence to control droplet extraction [81,82]. The segmented flow and continuous aqueous flow met at a rectangular-shaped chamber where an interface between two immiscible phases was built up. A pulsed electric field was applied over the chamber to force droplets to coalesce with the continuous aqueous stream, which then delivered the droplet contents to a capillary emitter for ESI-MS detection [82]. This droplet extraction approach required careful adjustments of the flow of the two immiscible phases to maintain a stable interface in the extraction chamber and avoid cross-contamination of the aqueous and oil streams. In addition, the severe dilution of the droplet contents resulted in high detection limits (~500 μM bradykinin). Lin and coworkers used an electricity-based method to control the droplet breaking and extraction at the stable oil–water interface [83]. One reported issue in this case was the difficulty of achieving complete extraction with high efficiency, which limited its compatibility with quantitative analysis.

Kelly et al. invented a droplet extraction interface that was constructed with an array of cylindrical posts to separate the segmented flow channel and the continuous aqueous phase channel [84]. When the aqueous stream and carrier oil phase flow rates were well controlled to balance the pressure at the junction, a stable oil–aqueous interface based on interfacial tension alone was formed to prevent bulk crossover of the two immiscible streams. The droplets could be transferred through the apertures to the continuous aqueous stream and finally detected by ESI-MS with virtually no dilution, enabling nanomolar detection limits.

Most of the reported methods and techniques for droplet extraction, as mentioned above, need to adjust two immiscible liquid flow rates to stabilize the interface and extract entire droplets. It is desirable to perform effective and complete droplet extraction independent of the flow rates, which would provide added flexibility for device operation. Recently, we have developed a robust interface for reliable and efficient droplet extraction, which was integrated in a droplet-based PDMS microfluidic assembly. The droplet extraction interface consisted of an array of cylindrical posts (Figure 8.4a), the same as was previously reported [84], but the aqueous stream microchannel surface was selectively treated by corona discharge to be hydrophilic. The combination of different surface energies and small flow-through apertures (~3 × 25 μm) enabled a very stable liquid interface between two immiscible steams to be established over a broad range of aqueous and oil flow rates. All aqueous droplets were entirely transferred to the aqueous stream (Figure 8.4b) and detected by MS following ionization at a monolithically integrated nanoelectrospray emitter.

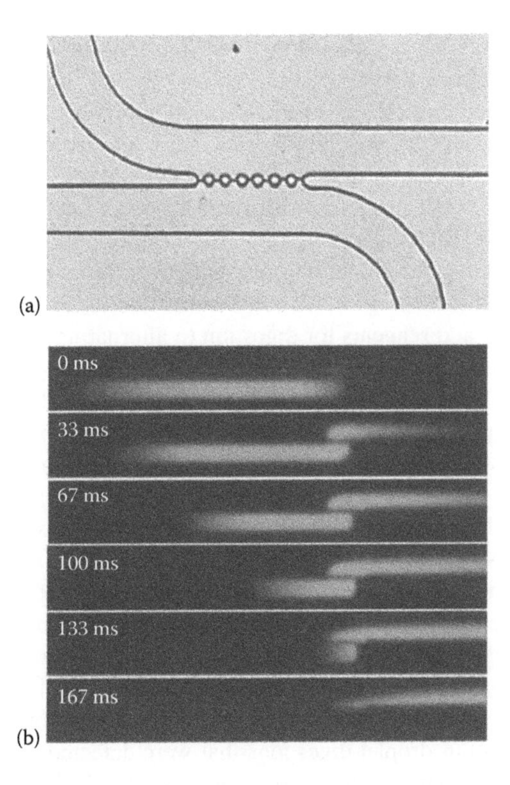

FIGURE 8.4 (a) Photograph of the droplet extraction region of the device. Water and oil fill the top and bottom channels, respectively, and the interface for the two liquids can be seen between the circular posts. (b) Micrograph sequences depicting the extraction of an individual fluorescein droplet. The flow rate in both channels was 400 nL/min.

8.3 PERSPECTIVES FOR DROPLET-BASED MICROFLUIDICS

As mentioned above, droplet-based microfluidics have been employed for a wide range of analyses and due to their unique advantages, their use will undoubtedly grow. The following subsections outline a few promising applications that will leverage the strengths of the platform.

8.3.1 Enhanced LC/MS-Based Proteomic Analysis

MS-based proteomics studies are vital for biomarker discovery, identification of drug targets, and fundamental biological research. In a typical "bottom-up" proteomics workflow [85], proteins are extracted from a sample, purified, and enzymatically digested into peptides. The peptides are then separated by LC, ionized by ESI, identified by MS, and those identified peptides are then matched to their corresponding proteins based on genomic information. Alternatively, for "top-down" proteomics [86], intact proteins are separated and identified directly by MS, providing potentially more complete sequence information and the ability to characterize posttranslational modifications. However, MS identification of intact proteins is far more challenging and has a lower throughput, limiting the widespread use of top-down approaches at present.

As top-down and bottom-up proteomics approaches each have unique and complementary advantages, it would be especially attractive to obtain both intact protein and peptide-level information from a single analysis. We propose that this could be achieved by encapsulating separated proteins into droplets as they elute from an LC column, thus preserving temporal information and separation resolution while enabling further processing. For example, using our droplet-on-demand and droplet merging technologies, we could encapsulate eluting proteins into droplets, and selectively add reagents for digestion to alternating droplets. The droplets could then be incubated in a delay line to allow sufficient reaction time prior to extracting and ionizing the droplets. The result would be that each droplet containing unreacted protein would be followed by a droplet containing digested peptide such that conventional bottom-up MS would be complemented with the intact molecular mass.

To this end, we have begun combining proteins with proteases in droplets to evaluate the conditions needed for digestion. The platform incorporated our integrated droplet-on-demand interface that enabled controlled in-droplet reactions, incubation in the oil stream, extraction from the aqueous stream, and ionization of the droplet contents at an integrated nanoelectrospray (nano-ESI) emitter [87] for MS analysis (Figures 8.5 and 8.6). This integrated microfluidic platform has been successfully utilized to combine myoglobin and pepsin from separate aqueous streams into droplets to perform rapid in-droplet digestions that were detected and identified on-line by nano-ESI-MS following droplet extraction (Figure 8.7). Given the short incubation time (18 s), the digestion did not go to completion such that peaks from the intact protein are still evident in the mass spectrum, but numerous peptides are confidently identified based on their m/z ratio as well (Table 8.1). We expect that simply extending the incubation time will dramatically improve digestion efficiency and enable the application of the platform to combined top-down/bottom-up proteomic analyses.

8.3.2 Single-Cell Chemical Analysis

Sensitivity limitations on biochemical measurements typically dictate that large samples are required comprising populations of cells. These ensemble measurements average over important cell-to-cell differences. Direct chemical analysis at the single-cell level will enable the heterogeneity that is currently obscured to be better understood. The sensitivity of MS instrumentation used for proteomic and metabolomic studies has increased to the point that such single-cell measurements are now feasible. For example, while inefficient ionization and transmission of ions generated at atmospheric pressure to the high vacuum region of the mass spectrometer previously were serious problems, recent improvements have produced combined efficiencies that can exceed 50% in some cases [88]. Indeed, around 50 proteins have been identified from samples containing just 50 pg of protein [89], which is as much protein as is contained in an average eukaryotic cell [90]. However, despite having adequate analytical sensitivity, existing methods for sample preparation, involving manual pipetting and multiple reaction vessels, are incompatible with single cells.

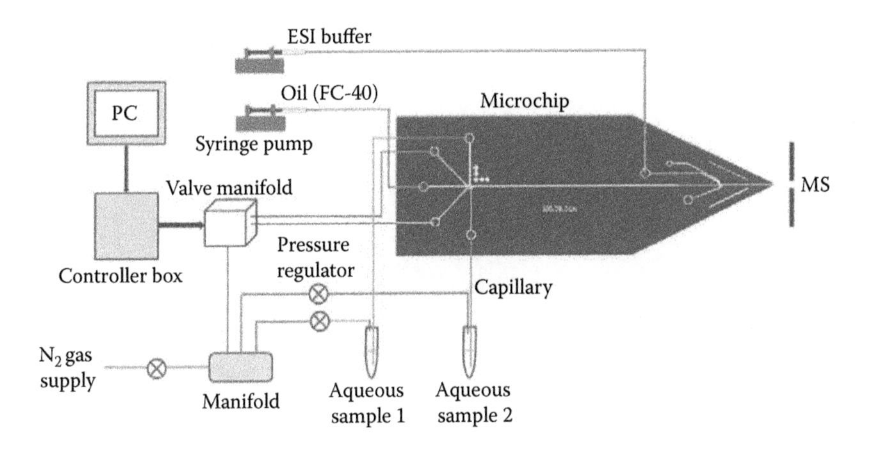

FIGURE 8.5 Schematic of the experimental set-up for droplet generation, fusion, mixing, extraction, and MS detection.

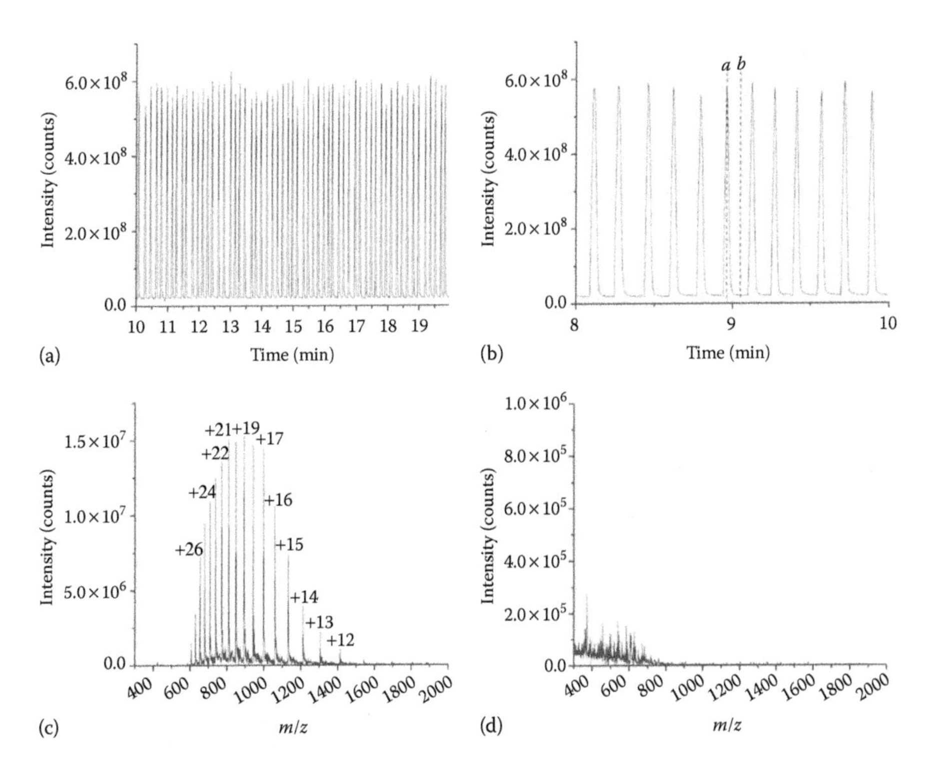

FIGURE 8.6 (a) MS detection of the extracted 1 μg/μL apomyoglobin droplets. Oil phase flow rate was 100 nL/min, ESI buffer flow rate was 400 nL/min, and droplet generation frequency was 0.1 Hz. (b) Detailed view of the MS-detected extracted apomyoglobin droplets. (c,d) Mass spectra obtained from the peak and baseline indicated as *a* and *b* in (b), respectively.

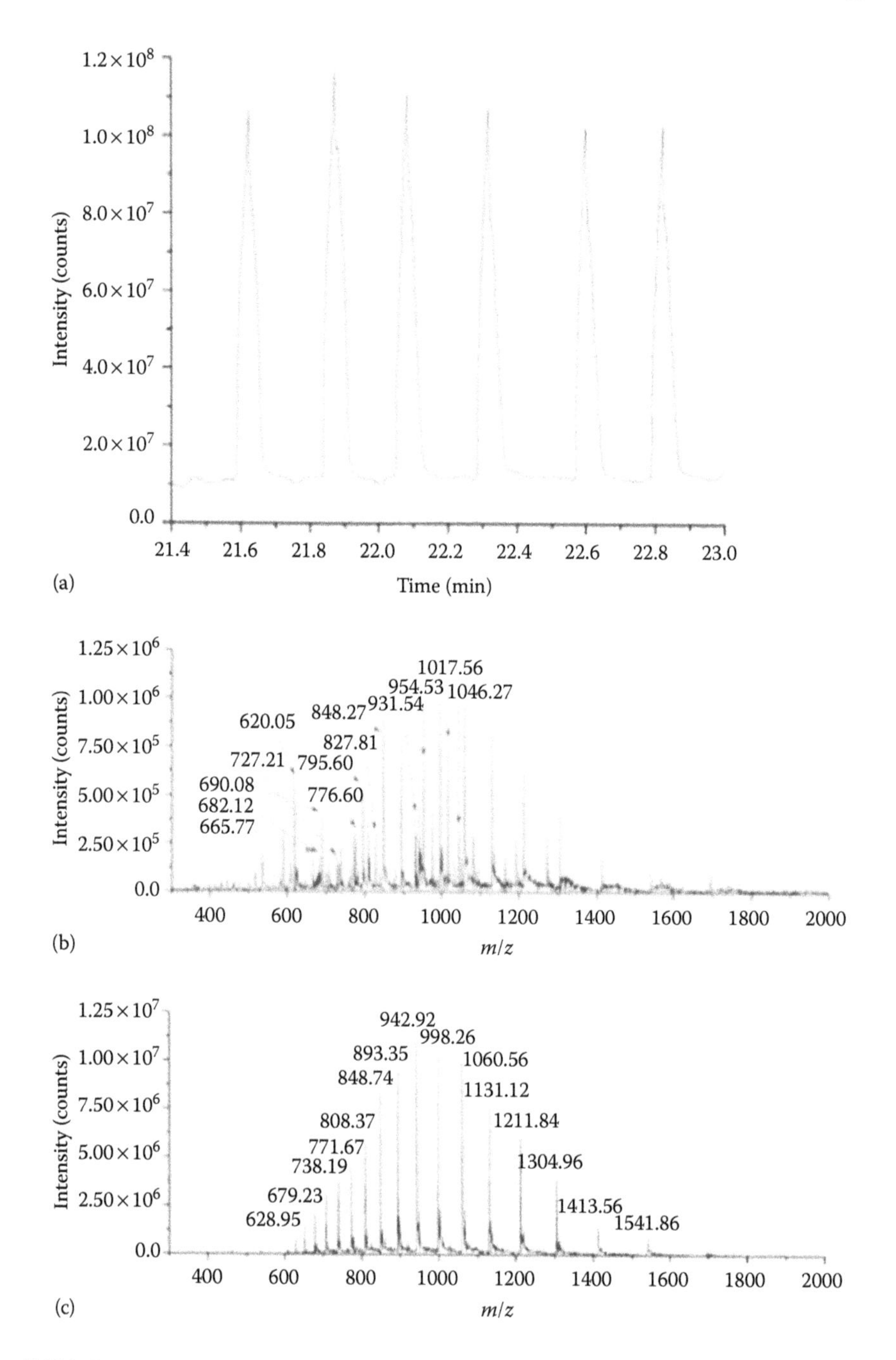

FIGURE 8.7 (a) MS detection of the fused droplets mixing 1 μg/μL apomyoglobin with 1 μg/μL pepsin in water containing 0.1% formic acid (pH ~3). The flow rates of oil and ESI buffer streams were 0.1 and 0.4 μL/min, respectively. (b) and (c) MS spectra of the fused droplet and 1 μg/μL apomyoglobin, respectively. The sequences of some digested peptide fragments labeled in (b) are listed in Table 8.1.

TABLE 8.1
Sequence of Apomyoglobin and Identification of Peptide Fragments from In-Droplet Digested Apomyoglobin Shown in Figure 8.7b.

m/z	Mass	*z*	Position	Sequence
620.05	1856.0	3+	138–153	FRNDIAAKYKELGFQG
690.08	4133.9	6+	70–106	TALGGILKKKGHHEAELKPLAQSHATKHKIPIKYLEF
827.81		5+		
665.77	4653.4	7+	30–69	IRLFTGHPETLEKFDKFKHLKTEAEMKASEDL
776.60		6+		KKHGTVVL
931.54		5+		
682.12	4767.5	7+	110–153	AIIHVLHSKHPGDFGADAQGAMTKALE
795.60		6+		LFRNDIAAKYKELGFQG
954.53		5+		
727.21	5082.8	7+	107–153	ISDAIIHVLHSKHPGDFGADAQGAMTK
848.27		6+		ALELFRNDIAAKYKELGFQG
1017.56		5+		
1046.27	3133.6	3+	1–29	GLSDGEWQQVLNVWGKVEADIAGHGQEVL

Note: Apomyoglobin sequence:
GLSDGEWQQVLNVWGKVEADIAGHGQEVLIRLFTGHPETLEKFDKFKHLKTEAEM
KASEDLKKHGTVVLTALGGILKKKGHHEAELKPLAQSHATKHKIPIKYLEFISDAIIHV
LHSKH PGDF GADAQGAMTKALELFRNDIAAKYKELGFQG.

This is another area where droplet-based microfluidics should be able to meet the need. Droplets have been previously used for single-cell encapsulation, and cells have also been lysed within droplets, with the surrounding oil preventing further dilution of the contents. Using such technologies for encapsulation and lysis in combination with our approaches for reagent mixing and droplet compatibility with ultrasensitive MS should enable us to dig deeper into the proteome and metabolome of single cells than has been accomplished previously.

8.4 CONCLUSIONS

Droplet-based microfluidics has developed substantially as a technology and will likely assume a higher-profile role in biological analyses in the future. Not only are much smaller amounts of reagents and samples consumed, but also thousands of reactions and screening experiments can be performed within droplets simultaneously. Perhaps more importantly, droplet-based microfluidics is a promising tool to help us understand some fundamental biological processes such as enzymatic reactions in a confined and crowding environment, protein–protein and protein–ligand interactions, interfacial functions in biological systems, and single-cell proteomics and metabolomics. A number of operational units have been well developed for droplet-based microfluidics, including droplet generation, fusion, and incubation. Others, such as droplet extraction for subsequent analysis of the contents, have been developed recently and promise to add versatility to the platform. Robust integration of multiple functions to create a true "lab-on-a-chip" continues to be a challenge, but the unique advantages of droplets for sample-limited biological analyses will undoubtedly spawn further development and we anticipate significant growth in the number of applications that rely on this technology in coming years.

REFERENCES

1. Song, H., D.L. Chen, and R.F. Ismagilov, Reactions in droplets in microfluidic channels. *Angew. Chem. Int. Ed.*, **45**: 7336–7356, 2006.
2. Huebner, A., et al., Microdroplets: A sea of applications? *Lab Chip*, **8**: 1244–1254, 2008.
3. Teh, S.-Y., et al., Droplet microfluidics. *Lab Chip*, **8**: 198–220, 2008.
4. Chiu, D.T., R.M. Lorenz, and G.D.M. Jeffries, Droplets for ultrasmall-volume analysis. *Anal. Chem.*, **81**: 5111–5118, 2009.
5. Chiu, D.T. and R.M. Lorenz, Chemistry and biology in femtoliter and picoliter volume droplets. *Accounts Chem. Res.*, **42**(5): 649–658, 2009.
6. Theberge, A.B., et al., Microdroplets in microfluidics: An evolving platform for discoveries in chemistry and biology. *Angew. Chem. Int. Ed.*, **49**(34): 5846–5868, 2010.
7. Yang, C.-G., Z.-R. Xu, and J.-H. Wang, Manipulation of droplets in microfluidic systems. *Trends Anal. Chem.*, **29**: 141–157, 2010.
8. Kintses, B., et al., Microfluidic droplets: New integrated workflows for biological experiments. *Curr. Opin. Chem. Biol.*, **14**: 548–555, 2010.
9. Casadevall i Solvas, X. and A.J. deMello, Droplet microfluidics: Recent developments and future applications. *Chem. Commun.*, **47**: 1936–1942, 2011.
10. Roach, L.S., H. Song, and R.F. Ismagilov, Controlling nonspecific protein adsorption in a plug-based microfluidic system by controlling interfacial chemistry using fluorous-phase surfactants. *Anal. Chem.*, **77**: 785–796, 2005.
11. Song, H. and R.F. Ismagilov, Millisecond kinetics on a microfluidic chip using nanoliters of reagents. *J. Am. Chem. Soc.*, **125**: 14613–14619, 2003.
12. Huebner, A., et al., Quantitative detection of protein expression in single cells using droplet microfluidics. *Chem. Commun.*, **28**: 1218–1220, 2007.
13. Aebersold, R. and M. Mann, Mass spectrometry-based proteomics. *Nature*, **422**(6928): 198–207, 2003.
14. Thorsen, T., et al., Dynamic pattern formation in a vesicle-generating microfluidic device. *Phys. Rev. Lett.*, **86**: 4162–4166, 2001.
15. Garstecki, P., et al., Formation of droplets and bubbles in a microfludic T-junction-scaling and mechanism of break-up. *Lab Chip*, **6**: 437–446, 2006.
16. Anna, S.L., N. Bontoux, and H.A. Stone, Formation of dispersions using "flow focusing" in microchannels. *Appl. Phys. Lett.*, **82**: 364–366, 2003.
17. Ward, T., et al., Microfluidic flow focusing: Drop size and scaling in pressure versus flow rate driven pumping. *Electrophoresis*, **26**: 3716–3724, 2005.
18. Baroud, C.N., F. Gallaire, and R. Dangla, Dynamics of microfluidic droplets. *Lab Chip*, **10**(16): 2032–2045, 2010.
19. Gu, H., M.H.G. Duits, and F. Mugele, Droplets formation and merging in two-phase flow microfluidics. *Int. J. Mol. Sci.*, **12**(4): 2572–2597, 2011.
20. Song, H., et al., Experimental test of scaling of mixing by chaotic advection in droplets moving through microfluidic channels. *Appl. Phys. Lett.*, **83**: 4664–4666, 2003.
21. Frenz, L., et al., Reliable microfluidic on-chip incubation of droplets in delay lines. *Lab Chip*, **9**: 1344–1348, 2009.
22. Courtois, F., et al., An integrated device for monitoring time-dependent in vitro expression from single genes in picolitre droplets. *ChemBioChem*, **9**: 439–446, 2008.

23. Courtois, F., et al., Controlling the retention of small molecules in emulsion microdroplets for use in cell-based assays. *Anal. Chem.*, **81**: 3008–3016, 2009.
24. Shi, W., et al., Droplet-based microfluidic system for individual *Caenorhabditis elegans* assay. *Lab Chip*, **8**: 1432–1435, 2008.
25. Huebner, A., et al., Static microdroplet arrays: A microfluidic device for droplet trapping, incubation and release for enzymatic and cell-based assays. *Lab Chip*, **9**: 692–698, 2009.
26. Schaerli, Y., et al., Continuous flow polymerase chain reaction of single copy DNA in microfluidic microdroplets. *Anal. Chem.*, **81**: 302–306, 2009.
27. He, M., et al., Selective encapsulation of single cells and subcellular organelles into picoliter- and femtoliter-volume droplets. *Anal. Chem.*, **77**: 1539–1544, 2005.
28. Lau, B.T.C., et al., A complete microfluidic screening platform for rational protein crystallization. *J. Am. Chem. Soc.*, **129**: 454–455, 2007.
29. Yobas, L., et al., High performance flow focusing geometry for spontaneous generation of monodispersed droplets. *Lab Chip*, **6**: 1073–1079, 2006.
30. He, M., J.S. Kuo, and D.T. Chiu, Electro-generation of single femtoliter- and picoliter-volume aqueous droplets in microfluidic systems. *Appl. Phys. Lett.*, **87**: 031916, 2005.
31. Park, S.-Y., et al., High-speed droplet generation on demand driven by pulse laser-induced cavitation. *Lab Chip*, **11**: 1010–1012, 2011.
32. Bransky, A., et al., A microfluidic droplet generator based on a piezoelectric actuator. *Lab Chip*, **9**: 516–520, 2009.
33. Galas, J.C., D. Bartolo, and V. Studer, Active connectors for microfluidic drops on demand. *New J. Phys.*, **11**: 075027, 2009.
34. Zeng, S., et al., Microvalve-actuated precise control of individual droplets in microfluidic devices. *Lab Chip*, **9**: 1340–1343, 2009.
35. Choi, J.-H., et al., Designed pneumatic valve actuators for controlled droplet breakup and generation. *Lab Chip*, **10**: 456–461, 2010.
36. Abate, A.R., et al., Valve-based flow focusing for drop formation. *Appl. Phys. Lett.*, **94**: 023503, 2009.
37. Song, H., J.D. Tice, and R.F. Ismagilov, A microfluidic system for controlling reaction networks in time. *Angew. Chem. Int. Ed.*, **42**: 768–772, 2003.
38. Zheng, B., L.S. Roach, and R.F. Ismagilov, Screening of protein crystallization conditions on a microfluidic chip using nanoliter size droplets. *J. Am. Chem. Soc.*, **125**: 11170–11171, 2003.
39. Huebner, A., et al., Development of quantitative cell-based enzyme assays in microdroplets. *Anal. Chem.*, **80**: 3890–3896, 2008.
40. Abate, A.R., et al., High throughput injection with microfluidics using picoinjectors. *Proc. Natl. Acad. Sci. USA*, **107**: 19163–19166, 2010.
41. Rhee, M. and M.A. Burns, Drop mixing in a microchannel for lab on a chip platforms. *Langmuir*, **24**: 590–601, 2008.
42. Bremond, N., A.R. Thiam, and J. Bibette, Decompressing emulsion droplets favors coalescence. *Phys. Rev. Lett.*, **100**: 024501, 2008.
43. Fidalgo, L.M., C. Abell, and W.T.S. Huck, Surface-induced droplet fusion in microfluidic devices. *Lab Chip*, **7**: 984–986, 2007.
44. Niu, X., et al., Pillar-induced droplet merging in microfluidic circuits. *Lab Chip*, **8**: 1837–1841, 2008.
45. Priest, C., S. Herminghaus, and R. Seemann, Controlled electrocoalescence in microfluidics: Targeting a single lamella. *Appl. Phys. Lett.*, **89**: 134101, 2006.
46. Link, D.R., et al., Electric control of droplets in microfluidic devices. *Angew. Chem. Int. Ed.*, **45**: 2556–2560, 2006.
47. Zagnoni, M. and J.M. Cooper, On-chip electrocoalescence of microdroplets as a function of voltage, frequency and droplet size. *Lab Chip*, **9**: 2652–2658, 2009.

48. Niu, X., et al., Electro-coalescence of digitally controlled droplets. *Anal. Chem.*, **81**: 7321–7325, 2009.
49. Baroud, C.N., M.R. de Saint Vincent, and J.P. Delville, An optical toolbox for total control of droplet microfluidics. *Lab Chip*, **7**: 1029–1033, 2007.
50. Thiam, A.R., N. Bremond, and J. Bibette, Breaking of an emulsion under an ac electric field. *Phys. Rev. Lett.*, **102**: 188304, 2009.
51. Frenz, L., et al., Microfluidic production of droplet pairs. *Langmuir*, **24**: 12073–12076, 2008.
52. Jambovane, S., et al., Creation of stepwise concentration gradient in picoliter droplets for parallel reactions of matrix metalloproteinase II and IX. *Anal. Chem.*, **83**: 3358–3364, 2011.
53. Agresti, J.J., et al., Ultrahigh throughtput screening in drop based microfluidics for directed evolution. *Proc. Natl. Acad. Sci. USA*, **107**: 4004–4009, 2010.
54. Brouzes, E., et al., Droplet microfluidic technology for single-cell high throughput screening. *Proc. Natl. Acad. Sci. USA*, **106**: 14195–14200, 2009.
55. Slaney, T.R., et al., Push-pull perfusion sampling with segmented flow for high temporal and spatial resolution in vivo chemical monitoring. *Anal. Chem.*, **83**: 5207–5213, 2011.
56. Schmitz, C.H.J., et al., Dropspots: A picoliter array in a microfluidic device. *Lab Chip*, **9**: 44–49, 2009.
57. Mazutis, L., et al., Multi-step microfluidic droplet processing: Kinetic analysis of an in vitro translated enzyme. *Lab Chip*, **9**: 2902–2908, 2009.
58. Clausell-Tormos, J., et al., Droplet based microfluidic platforms for the encapsulation and screening of mammalian cells and multicellular organisms. *Chem. Bio.*, **15**: 427–437, 2008.
59. Damean, N., et al., Simultaneous measurements of reactions in microdroplets filled by concentration gradients. *Lab Chip*, **9**: 1707–1713, 2009.
60. Bui, M.P.N., et al., Enzyme kinetic measurements using a droplet based microfluidic system with a concentration gradient. *Anal. Chem.*, **83**: 1603–1608, 2011.
61. Beer, N.R., et al., On chip, real time, single copy polymerase chain reaction in picoliter droplets. *Anal. Chem.*, **79**: 8471–8475, 2007.
62. Beer, N.R., et al., On chip single copy real time reverse transcription PCR in isolated picoliter droplets. *Anal. Chem.*, **80**: 1854–1858, 2008.
63. Srisa-Art, M., et al., Monitoring of real time streptavidin biotin binding kinetics using droplet microfluidics. *Anal. Chem.*, **80**: 7063–7067, 2008.
64. Marz, A., et al., Droplet formation via flow through microdevices in Raman and surface enhanced Raman spectroscopy: Concepts and applications. *Lab Chip*, **11**: 3584–3592, 2011.
65. Cristobal, G., et al., On line laser Raman spectroscopic probing of droplets engineered in microfluidic devices. *Lab Chip*, **6**: 1140–1146, 2006.
66. Strehle, K.R., et al., A reproducible surface enhanced Raman spectroscopy approach. Online SERS measurements in a segmented microfluidic system. *Anal. Chem.*, **79**: 1542–1547, 2007.
67. Liu, S., et al., The electrochemical detection of droplets in microfluidic devices. *Lab Chip*, **8**: 1937–1942, 2008.
68. Han, Z., et al., Measuring rapid enzymatic kinetics by electrochemical method in droplet based microfluidic devices with pneumatic valves. *Anal. Chem.*, **81**: 5840–5845, 2009.
69. Kautz, R.A., W.K. Goetzinger, and B.L. Karger, High throughput microcoil NMR of compound libraries using zero-dispersion segmented flow analysis. *J. Comb. Chem.*, **7**: 14–20, 2005.
70. Liu, T., et al., Accurate mass measurements in proteomics. *Chem. Rev.*, **107**(8): 3621–3653, 2007.

71. Hatakeyama, T., D.L. Chen, and R.F. Ismagilov, Microgram-scale testing of reaction conditions in solution using nanoliter plugs in microfluidics with detection by MALDI-MS. *J. Am. Chem. Soc.*, **128**: 2518–2519, 2006.
72. Pei, J., et al., Analysis of samples stored as individual plugs in a capillary by electrospray ionization mass spectrometry. *Anal. Chem.*, **81**: 6558–6561, 2009.
73. Li, Q., et al., Fraction collection from capillary liquid chromatography and off-line electrospray ionization mass spectrometry using oil segmented flow. *Anal. Chem.*, **82**: 5260–5267, 2010.
74. Edgar, J.S., et al., Capillary electrophoresis separation in the presence of an immiscible boundary for droplet analysis. *Anal. Chem.*, **78**(19): 6948–6954, 2006.
75. Niu, X.Z., et al., Droplet based compartmentalization of chemically separated components in two dimensional separations. *Chem. Commun.*, (41): 6159–6161, 2009.
76. Roman, G.T., et al., Sampling and electrophoretic analysis of segmented flow streams using virtual walls in a microfluidic device. *Anal. Chem.*, **80**: 8231–8238, 2008.
77. Wang, M., et al., Microfluidic chip for high efficiency electrophoretic analysis of segmented flow from a microdialysis probe and in vivo chemical monitoring. *Anal. Chem.*, **81**: 9072–9078, 2009.
78. Pei, J., J. Nie, and R.T. Kennedy, Parallel electrophoretic analysis of segmented samples on chip for high-throughput determination of enzyme activities. *Anal. Chem.*, **82**: 9261–9267, 2010.
79. Zhu, Y. and Q. Fang, Integrated droplet analysis system with electrospray ionization-mass spectrometry using a hydrophilic tongue-based droplet extraction interface. *Anal. Chem.*, **82**: 8361–8366, 2010.
80. Filla, L.A., D.C. Kirkpatrick, and R.S. Martin, Use of a corona discharge to selectively pattern a hydrophilic/hydrophobic interface for integrating segmented flow with microchip electrophoresis and electrochemical detection. *Anal. Chem.*, **83**: 5996–6003, 2011.
81. Fidalgo, L.M., et al., From microdroplets to microfluidics: Selective emulsion separation in microfluidic devices. *Angew. Chem. Int. Ed.*, **47**: 2042–2045, 2008.
82. Fidalgo, L.M., et al., Coupling microdroplet microreactors with mass spectrometry: Reading the contents of single droplets online. *Angew. Chem. Int. Ed.*, **48**(20): 3665–3668, 2009.
83. Zeng, S., et al., Electric control of individual droplet breaking and droplet contents extraction. *Anal. Chem.*, **83**: 2083–2089, 2011.
84. Kelly, R.T., et al., Dilution-free analysis from picoliter droplets by nano-electrospray ionization mass spectrometry. *Angew. Chem. Int. Ed.*, **48**(37): 6832–6835, 2009.
85. Swanson, S.K. and M.P. Washburn, The continuing evolution of shotgun proteomics. *Drug Discov. Today*, **10**(10): 719–725, 2005.
86. Zhou, H., et al., Advancements in top-down proteomics. *Anal. Chem.*, **84**(2): 720–734, 2012.
87. Sun, X., et al., Ultrasensitive nanoelectrospray ionization-mass spectrometry using poly(dimethylsiloxane) microchips with monolithically integrated emitters. *Analyst*, **135**: 2296–2302, 2010.
88. Marginean, I., et al., Achieving 50% ionization efficiency in subambient pressure ionization with nanoelectrospray. *Anal. Chem.*, **82**(22): 9344–9349, 2010.
89. Shen, Y., et al., Ultrasensitive proteomics using high-efficiency on-line micro-SPE-nanoLC-nanoESI MS and MS/MS. *Anal. Chem.*, **76**(1): 144–154, 2004.
90. Zhang, Z.R., et al., One-dimensional protein analysis of an HT29 human colon adenocarcinoma cell. *Anal. Chem.*, **72**(2): 318–322, 2000.

9 A Review of Tubeless Microfluidic Devices

Pedro J. Resto, David J. Beebe, and Justin C. Williams

CONTENTS

9.1 Introduction 221
9.2 Flow Rate Analysis 223
9.3 Evaporation 224
9.4 Sample Concentration 230
9.5 Concentration Gradients 232
9.6 Backflow Mechanisms 235
9.7 High Throughput 239
9.8 Passive Washing and Logic Circuitry 245
9.9 Biology 247
9.10 Integrated Electrodes 254
9.11 Other Types of Tubeless Devices 260
References 261

9.1 INTRODUCTION

Microfluidics is a field of science driven by the vision of entire biological and chemical laboratories developed on a miniaturized silicon or polymer chip. Microfluidic technologies offer advantages such as reduced sample size, decreased assay times, and minimized reagent consumption, reducing experimental costs and speeds while increasing experimental precision over their macroscale counterparts. Micromachining technologies have evolved to the point where valves, mixers, and pumps can all be built. The ultimate goal has been to integrate multiple components to form 'lab-on-a-chip' devices, where complex tasks are performed in a miniaturized chip instead of in large laboratories that require individual systems for specific tasks. The physics that dominate at the microscale are different from the physics that dominate at the macroscale, allowing for added functionalities not possible at the macroscale.[1]

Surface tension, van der Waals' forces, electrostatic force, capillary action, viscosity, laminar flow, diffusion, fluidic resistance, and surface area to volume ratios are all examples of dominant forces acting at the microscale. The macroscale is largely dominated by, for example, momentum, turbulence, and gravity. By understanding and using the unique physics at the microscale, microfluidic technologies can perform tasks not possible at the macroscale and permit new methods, ways of thinking, and solutions to emerge.[1]

Researchers are shifting their attention toward simplifying the way they control fluid flow inside microfluidic devices. Many different passive microfluidic devices have been successfully demonstrated in the literature, including passive valves, mixers, extractors, filters, and actuation schemes. A review of these methods can be found elsewhere.[2] There are also many types of active pumps used for moving flow in microscale devices.[3] These include reciprocating pumps, rotary pumps, centrifugal pumps, dynamic micropumps based on electromagnetic fields, electroosmotic pumps that use surface charge phenomena, and magnetohydrodynamic pumps. Although each type of pump has an advantage over others for any given application, most require complex manufacturing steps, while others rely on large and expensive external equipment to work.

Passive microfluidic devices have advantages, for instance they do not require external power sources and are easy to integrate into existing laboratory infrastructure. A category of microfluidic devices called tubeless microfluidic devices brings the added advantage of removing the dead volume that would be present in tubing. Examples of tubeless microfluidic devices include those used in surface tension passive pumping and in electroosmotic flow. Surface tension passive pumping is both passive and tubeless. With this technique, the surface tension inherent to small drops provides a means of moving fluid without using external actuation mechanisms. Surface tension

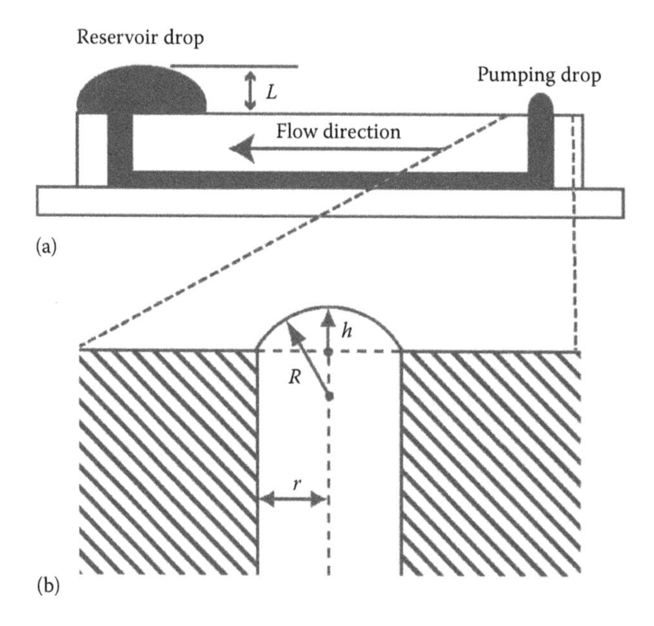

FIGURE 9.1 (a) Side view of a microchannel. An outlet port with a large drop and inlet port with a smaller drop are required to create fluid flow. (b) Description of inlet drop with height (h), drop radius (R), and inlet port radius (r). (From Walker, G. and Beebe, D.J., *Lab Chip*, 2, 131–134, 2002.)

passive pumping was first described by Glenn Walker and David Beebe in 2002.[4] Surface tension passive pumping, from here on referred to as passive pumping, is an actuation mechanism where the surface tension in small drops is used to create flow within a microfluidic device. The Laplace law, $\Delta P = \gamma(1/R_1 + 1/R_2)$, where ΔP is a pressure differential across a droplet interface, γ is the liquid surface tension, and R_1 and R_2 are the principal radii of curvature, is used to describe how a small droplet contains a higher inner pressure than a larger droplet. When a small drop is connected via a microfluidic device to a larger drop, flow moves from the small to the large drop (Figure 9.1). The characterization of this flow is the focus of many works that aim to apply passive pumping to a wide range of biological and biochemical studies.

9.2 FLOW RATE ANALYSIS

Biological protocols require understanding of the flow rates being used during experiments or an understanding of the pressure necessary to move flow across a microfluidic device. Berthier et al. developed a mathematical model that characterizes the collapse of a drop of water placed in the inlet of a microfluidic passive pumping device.[5] The drop's collapse into the inlet of the device is composed of two phases (Figure 9.2). The first phase is characterized by a decrease in the drop's contact angle with respect to the surface while a constant contact area with the surface is maintained. The contact angle diminishes until it reaches a critical contact angle value. The second phase starts once this critical contact angle is reached. In the second phase, the contact area between the drop and the liquid diminishes while the contact angle remains constant. Using the calculations available in this paper, one can calculate the time it takes for a drop of a given volume to collapse into the microfluidic device. Flow rates and liquid velocities can then be estimated as a function of channel dimensions and inlet drop volume. Chen et al. provided a more

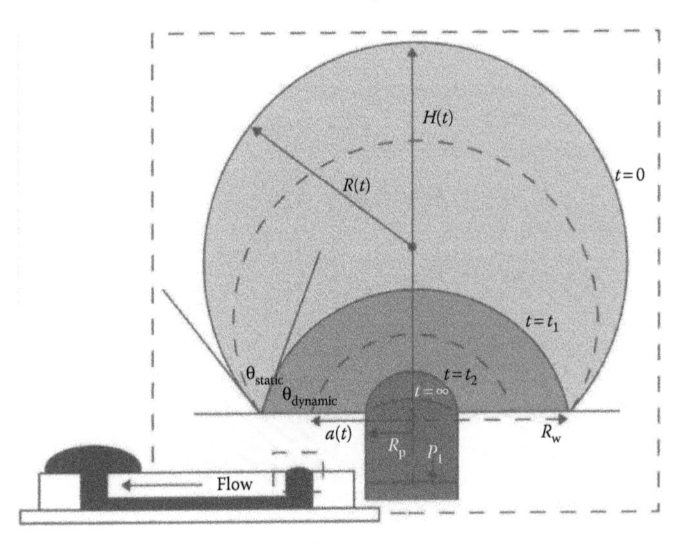

FIGURE 9.2 Two phase inlet drop collapse. (From Berthier, E. and Beebe, D.J., *Lab Chip*, 7, 1475–1478, 2007.)

general numerical model of volumetric flow rate as a function of time under various experimental conditions for surface tension passive pumping.[6] In their modeling, the authors consider the surface energy of small droplets, the hydrostatic pressure difference between the input and output ports of the microfluidic device, and the hydrodynamic resistance of the microfluidic channel.

9.3 EVAPORATION

The prelude to surface tension passive pumping was the use of a polydimethylsiloxane (PDMS) microfluidic device filled with water to show how evaporation can be used to create flow and concentrate beads.[7] The PDMS device connects a reservoir port to a collection port. A large drop is placed in the reservoir port. Liquid from the collection port evaporates creating flow from the reservoir to the collection port. Fluorescent beads are carried with the flow and are concentrated at the collection port (Figure 9.3). A bead concentration gradient forms as beads are collected. The gradient becomes shorter as time progresses ($t_c > t_b > t_a$).

Evaporation is an important topic in tubeless microfluidics. The open air–liquid interface and the potential for large surface area to volume ratios present in tubeless microfluidic devices make evaporation a major issue. Methods to mitigate evaporation include replacing the evaporated liquid, covering the open air–liquid interface with a solid or oil, and adding a high-boiling-point component to lower the liquid's vapor pressure. However, these methods may be too invasive or may require constant monitoring of the microfluidic device to ensure it is in working order. Berthier et al.

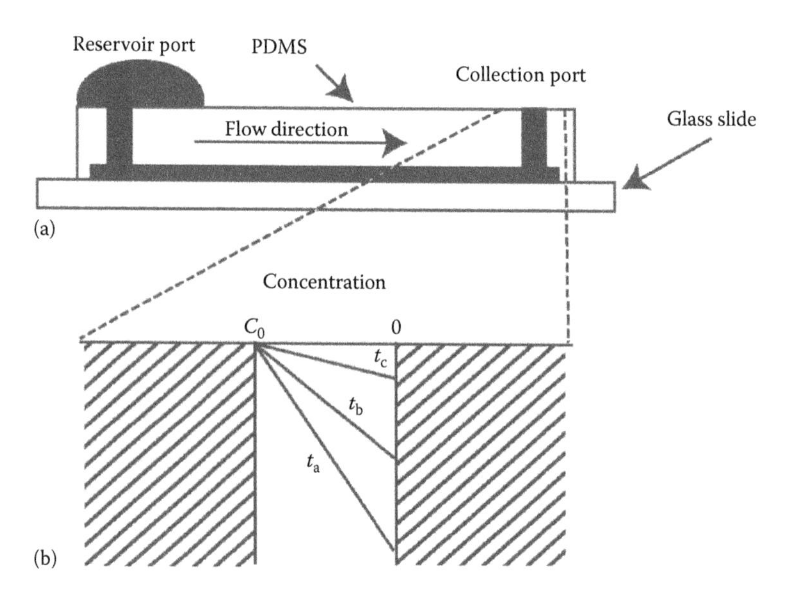

FIGURE 9.3 (a) Side view of microfluidic device. (b) Air–water interface at collection port modeled as a horizontal surface. Lines (t_a, t_b, t_c) depict the concentration gradient of particles as evaporation progresses. (From Walker, G.M. and Beebe, D.J., *Lab Chip*, 2, 57–61, 2002.)

analyzed a less invasive method where they increased the humidity in the environment by placing sacrificial drops around the device of interest (Figure 9.4).[8]

Berthier et al. followed this original work with a study of the interplay between convection and diffusion inside a microfluidic passive pumping channel.[9] Passive pumping occurs from a small drop to a large drop; however, evaporation will cause flow from the large drop to the small drop (in the opposite direction to the passive pumping flow) during storage of the device (Figure 9.5).

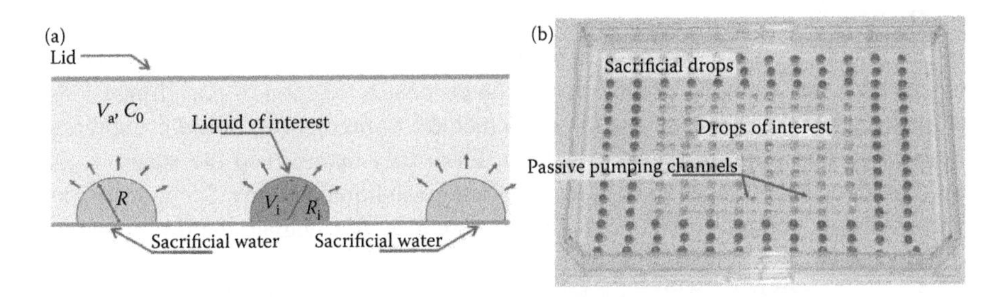

FIGURE 9.4 (a) A drop with volume V_i and radius R_i is placed in the environment of volume V_a and humidity C_0, surrounded by sacrificial drops of radius R. (b) Passive pumping–based assay in an Omnitray containing drops of interest and sacrificial drops for reducing volume loss on the drops of interest via evaporation. (From Berthier, E., Warrick, J., Yu, H., and Beebe, D.J., *Lab Chip*, 8, 852–859, 2008.)

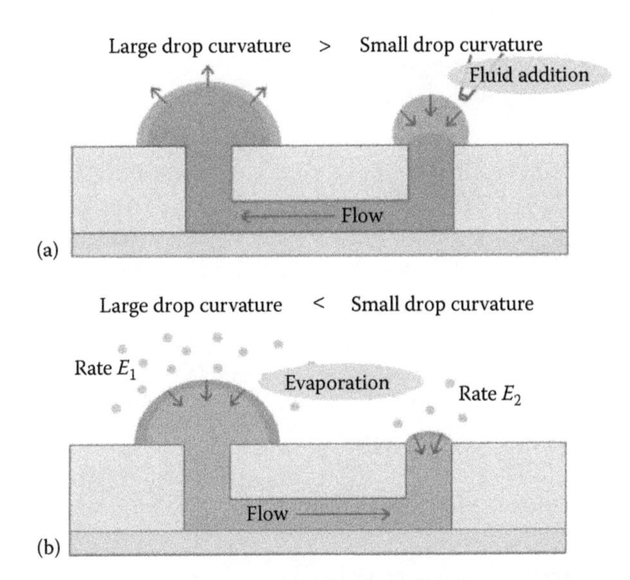

FIGURE 9.5 (a) Flow occurs from a small drop to a large drop via surface tension passive pumping. (b) Evaporation occurs at both the large and small drop; however, relative volume loss is greater in the small drop, causing flow from the large to the small drop. (From Berthier, E., Warrick, J., Yu, H., and Beebe, D.J., *Lab Chip*, 8, 860–864, 2008.)

The volume loss associated with evaporation is important in cell biology as it leads to changes in osmolarity and has a direct impact on cellular mechanisms. The convection caused by flow due to evaporation competes with the diffusion of soluble proteins in cell–cell interactions (Figure 9.6). The less humid the environment, the more effect evaporation will have on the liquid. Ways to mitigate convection–diffusion effects include increasing the cross-sectional area, adding alternate flow paths to reduce convection in the areas of interest, and reducing the size of the air–liquid interface at the ports. Adding sacrificial water around the device of interest will increase the humidity in the environment and decrease evaporation.

Zhu et al.[10] saw evaporation-based backward spreading of a virus while performing an infection assay using passive pumping. Lynn et al.[11] used evaporation to their advantage when studying evaporation at the ports as a function of port dimensions. In their paper, they presented a two-step method to predict the liquid evaporation rates of an air–liquid interface over time. First, they determined the shape of the air–liquid meniscus created in a reservoir for a given liquid volume. Second, computational fluid dynamics (CFD) simulations were used to calculate the instantaneous

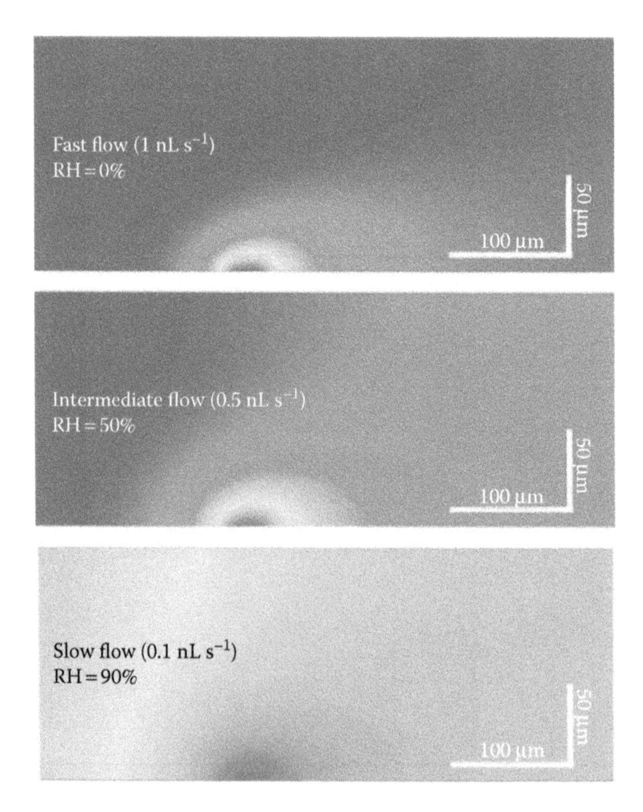

FIGURE 9.6 COMSOL simulation of the concentration of a secreted protein ($D = 10^{-10}$ m^2 s^{-1}) corresponding to different humidity conditions in the environment it is placed in. Lower humidity environments cause higher evaporation rates and stronger flow, effectively washing secreted protein away. RH, relative humidity. (From Berthier, E., Warrick, J., Yu, H., and Beebe, D.J., *Lab Chip*, 8, 860–864, 2008.)

rate of evaporation for the given meniscus shape. The results show that evaporation is a function of the port geometry. This is important as evaporation rates in tubeless microfluidic devices can lead to experimental error. Hence, a user should understand and control for evaporation to avoid unwanted changes in experimental conditions. Microfluidic device design should take into account local evaporation rates to avoid experimental error. Figure 9.7 shows the geometric parameters used to describe the meniscus and the microreservoir for the CFD simulations.[11,12]

Figure 9.8 shows the effect of reservoir liquid volume on evaporation given an expanding reservoir and a contracting reservoir.[11] The result shows that the local evaporation rate is greater for greater reservoir volumes. For a given reservoir volume, the local evaporation rate is greater for an expanding reservoir than for a contracting reservoir. Less liquid volume in a reservoir means gas molecules have to travel larger diffusion distances. Hence, a lower evaporation rate is observed.

Results from Lynn et al.[11] show the evaporation rate as a function of reservoir volume for different channel geometries, assuming constant contact angles and a constant microreservoir bottom diameter, D_1. There is more evaporation for an expanding reservoir and more evaporation for less channel height. Less channel height means a shorter diffusion distance for gas molecules to travel, hence higher evaporation rates. An expanding reservoir has a greater air–liquid surface area (i.e., a larger evaporation area than its contracting counterpart); therefore, evaporation

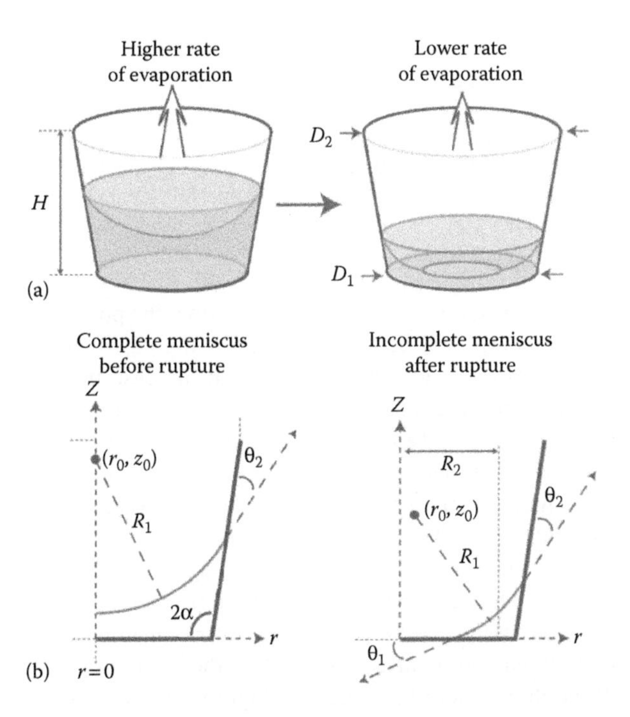

FIGURE 9.7 (a) Complete and incomplete meniscus in a microreservoir. (b) Geometric parameters used to describe the meniscus in a microreservoir. (From Lynn, N.S., Henry, C., and Dandy, D., *Lab Chip*, 9, 1780–1788, 2009.)

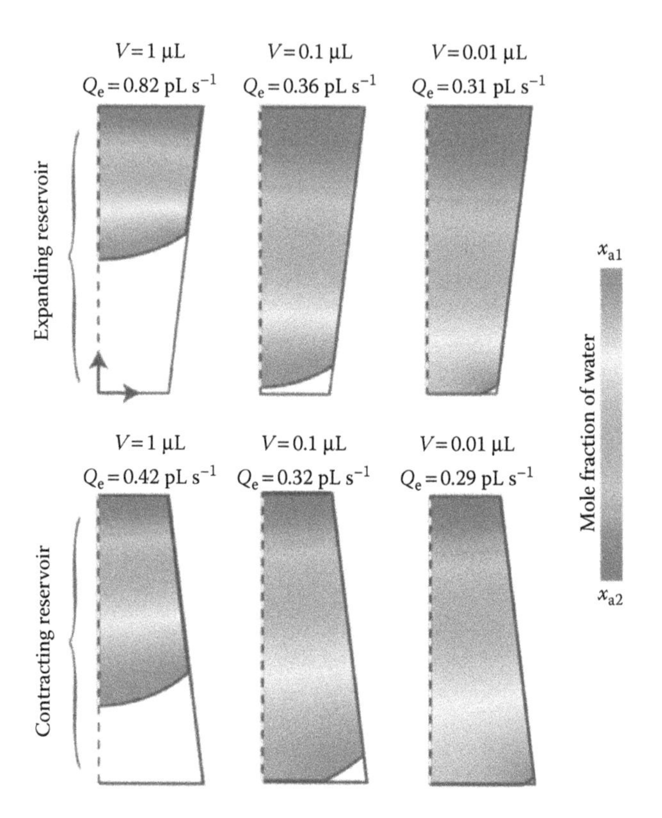

FIGURE 9.8 CFD simulations displaying the mole fraction of water vapor in several reservoirs containing different liquid volumes (V). The rate of evaporation from the air–liquid interface (Q_e) is displayed for each scenario. (From Lynn, N.S., Henry, C., and Dandy, D., *Lab Chip*, 9, 1780–1788, 2009.)

rates for an expanding reservoir are greater. In addition, the position of the meniscus in an expanding reservoir will be higher, reducing the gas diffusion distance and increasing evaporation rates. Results also show an increase in evaporation with a decrease in liquid–reservoir contact angle; this is mainly due to the increase in the air–liquid interface area with the decrease in liquid–reservoir contact angles.

Lynn et al.[11] continued their work by analyzing evaporation-mediated flow in a follow-up paper where capillary forces and evaporation were used as the actuation mechanism for fluid flow.[12] Fluid is introduced into the microfluidic device through an inlet reservoir. Capillary forces move liquid through the hydrophobic microchannel that connects the inlet and outlet reservoirs until liquid at the outlet reservoir forms a meniscus around the corners of the reservoir. The pressure differential arising from this small, curved meniscus situated on the bottom of the outlet reservoir drives liquid through the microfluidic device (Figure 9.9). The system reaches steady state and is able to provide precise flow rates for periods of over an hour. Control over flow rate is achieved by controlling the microfluidic device and outlet reservoir dimensions.

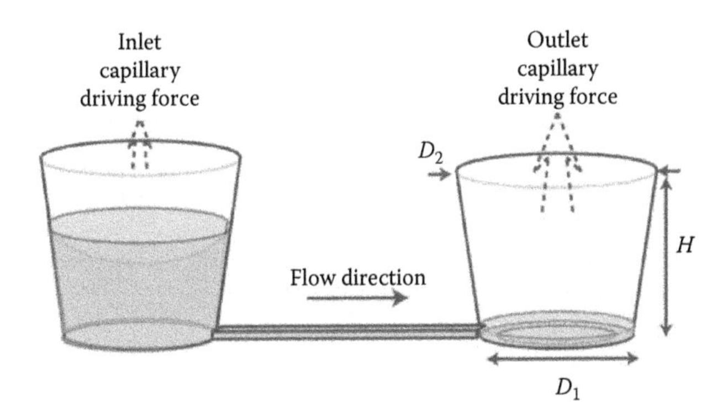

FIGURE 9.9 Diagram of the system described in Lynn et al.[12] (From Lynn, N.S. and Dandy, D., *Lab Chip*, 9, 3422–3429, 2009.)

Surface tension passive pumping as described by Walker and Beebe[4] uses the surface tension inherent in small droplets to drive flow. Evaporation is indeed an important factor in surface tension passive pumping; however, the flow caused by surface tension works in a much faster time scale than the flow obtained by evaporation. While surface tension passive pumping can be analytically described and predicted for flow occurring in seconds to minutes, Lynn et al.[11,12] presented evaporation as a way to control flow for over an hour. Figure 9.10 presents experimental fluid velocity data overlaid on theoretical predictions for two reservoirs, having different volumes of 1.5 μL and 2.0 μL, using coupled capillary/evaporation effects to create fluid flow. It is important to note that surface tension passive pumping[4] is different from capillary flow as defined by the filling of an empty hydrophobic channel due to intermolecular attractive forces between the liquid and the solid surrounding surfaces.

Like the Dandy laboratory, the Khademhosseini laboratory has used evaporation to their advantage. In their work, Du et al.[13] created controllable concentration

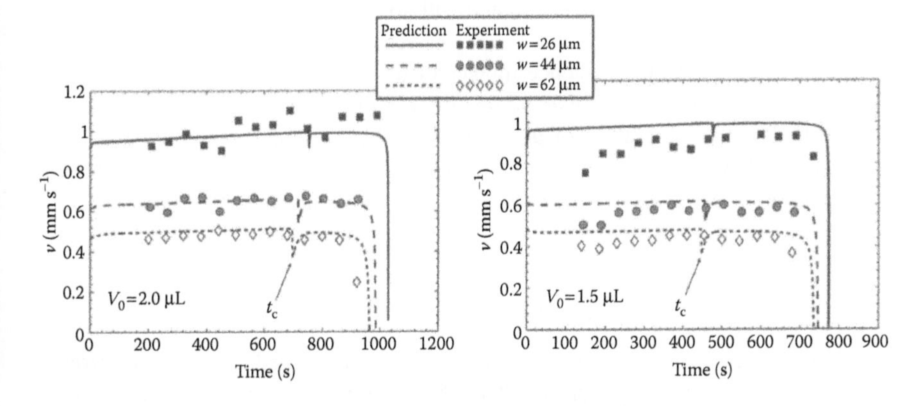

FIGURE 9.10 Experimental and theoretical predictions of average fluid velocity, v (mm s^{-1}), as a function of time (s) using coupled capillary/evaporation effects for different initial inlet reservoir volumes, $V_0 = 2.0$ and 1.5 μL. (From Lynn, N.S. and Dandy, D., *Lab Chip*, 9, 3422–3429, 2009.)

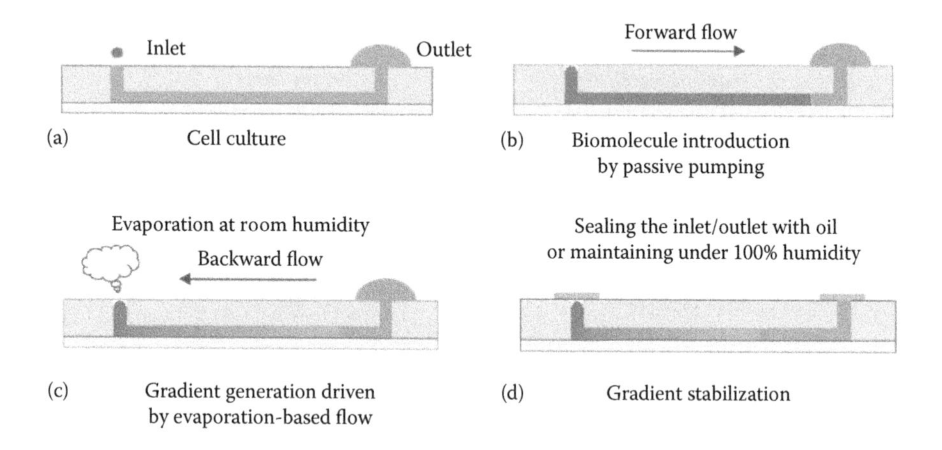

FIGURE 9.11 Schematic of the gradient device developed by Du et al.[13] (a) A microfluidic channel is filled with culture medium. (b) Solution is introduced into the channel by passive pumping. (c) A concentration gradient develops due to evaporation-based backward flow. (d) The gradient profile is stabilized by sealing the inlet and outlet ports. (From Du, Y., Shim, J., Vidula, M., Hancock, M., Lo, E., Chung, B., Borenstein, J., Khabiry, M., Cropek, D., and Khademhosseini, A., *Lab Chip*, 9, 761–767, 2009.)

gradients of molecules in a microfluidic device using a tubeless passive pumping microchannel. The concentration gradient was first established using forward flow induced by passive pumping and then controlled using evaporation-induced backward flow in the same channel. The backward flow component along the microfluidic channel allowed spatial and temporal control of the gradient. Stopping the flow then stabilized the gradient profile (Figure 9.11). To demonstrate the applications of this methodology, a stabilized concentration gradient of a cardiac toxin, alpha-cypermethrin, along the microchannel was used to test the response of HL-1 cardiac cells in the microdevice, which correlated with toxicity data obtained from multiwell plates.

Figure 9.12a shows fluorescence images of forward fluorescein flow into microfluidic channels using a passive pump. Figure 9.12b shows fluorescence images demonstrating the generation of a dynamic gradient of fluorescein solution using evaporation-driven backward flow. Analogous results are shown in Figure 9.13, where the toxicity of a stabilized concentration gradient of alpha-cypermethrin is studied on the HL-1 cells from outlet (5 cm) to inlet (0 cm) after 4 h residence time.[13]

9.4 SAMPLE CONCENTRATION

Following up on the original work with evaporation-mediated sample concentration by Walker and Beebe,[7] Warrick et al.[14] worked with passive pumping to concentrate cells and demonstrated the use of passive pumping to load samples into a cell concentration device (Figure 9.14). The concentrator is used to concentrate a suspension of cells and seed them into a microchannel. Passive pumping was used to drive flow. Unlike the 2002 work that used evaporation to drive flow and concentrate samples

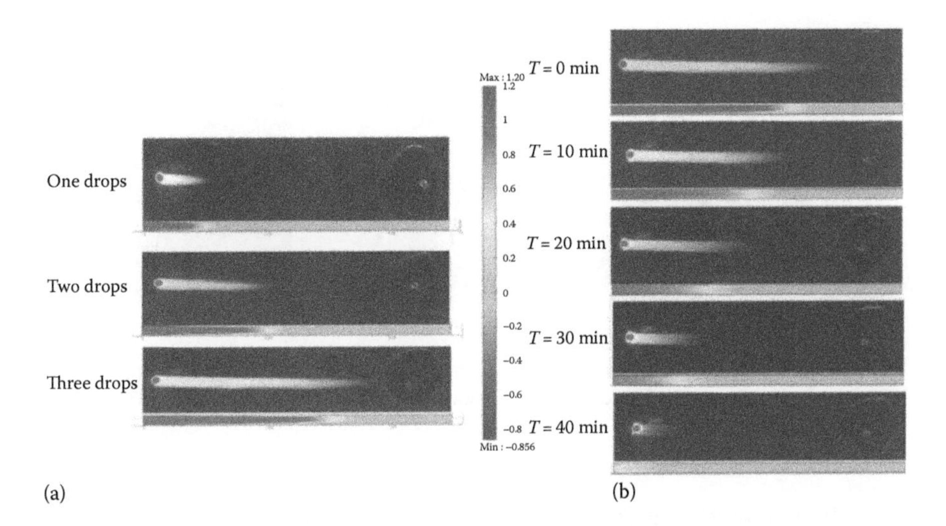

FIGURE 9.12 (a) Forward flow due to passive pumping. (b) Evaporation mediated backward flow. (From Du, Y., Shim, J., Vidula, M., Hancock, M., Lo, E., Chung, B., Borenstein, J., Khabiry, M., Cropek, D., and Khademhosseini, A., *Lab Chip*, 9, 761–767, 2009.)

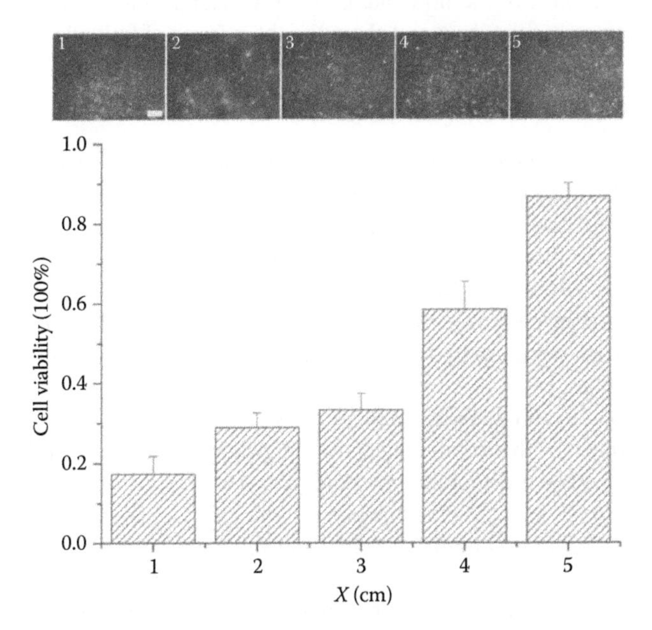

FIGURE 9.13 *Top*: Fluorescent imaging showing the viability of HL-1 cells along the 5 cm length of a microfluidic channel after treatment with the toxin concentration gradient (scale bar: 600 μm). *Bottom*: Quantification of the viability of HL-1 cells along the 5 cm long microfluidic channel after treatment with the toxin concentration gradient. (From Du, Y., Shim, J., Vidula, M., Hancock, M., Lo, E., Chung, B., Borenstein, J., Khabiry, M., Cropek, D., and Khademhosseini, A., *Lab Chip*, 9, 761–767, 2009.)

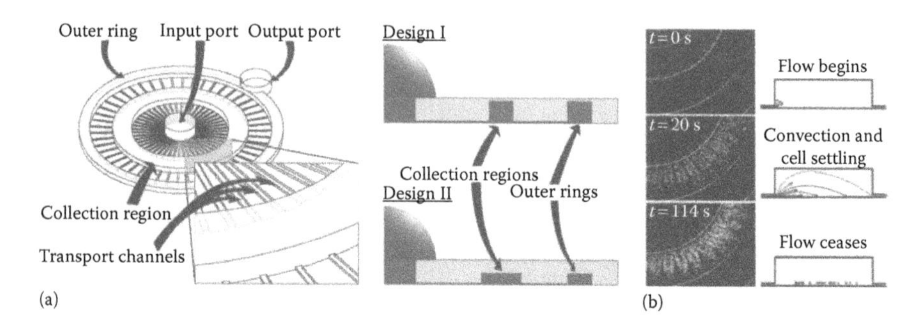

FIGURE 9.14 Passive pumping cell concentration device. (Reprinted adapted from Warrick, J., Casavant, B., Frisk, M., and Beebe, D.J., *Anal. Chem.*, 82, 8320–8326, 2010. With permission.)

at a collection port,[7] Warrick et al. flowed biological cells through channels of different sizes using small, fast channels for flow and larger, slower channels where cells concentrate and adhere to the substrate. While in the substrate, the cells can be cultured and treated. The method can be used for culture and subsequent application of treatment protocols. The authors used passive pumping as a loading mechanism since it allows control of the flow rate used for concentrating samples by changing the wetted diameter at the input and changing the volume of fluid placed at the input.

9.5 CONCENTRATION GRADIENTS

Walker et al.[7] mentioned a bead concentration gradient that formed at the collection port while using evaporation to concentrate samples. Concentration gradients are important for the research community because they allow the study of many biological and pathological processes, such as metastasis, embryogenesis, axon guidance, and wound healing.[15] Microfluidic devices allow researchers to manipulate fluid flow and diffusion profiles to create spatiotemporal biomolecular gradients and allow cell–extracellular environment, cell–cell, cell–matrix, and cell–soluble factor interactions.[15]

Du et al.[13] studied the viability of HL-1 cells along a 5 cm long microfluidic channel after treatment with a toxin concentration gradient (Figure 9.13). In their work, Kim et al.[16] designed a device for chemotaxis based on static diffusion and passive pumping. Chemotaxis is the phenomenon in which cells, bacteria, and other organisms move according to stimuli and chemical gradients in their environment. The 2009 paper describes a chemotaxis device with an *in situ* nanoporous membrane that eliminates the effect of flow for chemotaxis (Figures 9.15 and 9.16).[16] Passive pumping is used to fill the source channel. The nanoporous membrane was used to prevent flow from interfering with diffusion phenomena and ensures that diffusion is the only force that drives the generation of a gradient from the source to the sink channel (Figure 9.15b).[17] Previously, Abhyankar et al.[18] had used a membrane with high fluidic resistance to create a concentration gradient and limit flow due to pressure differentials (i.e., passive pumping) from interfering with gradient.

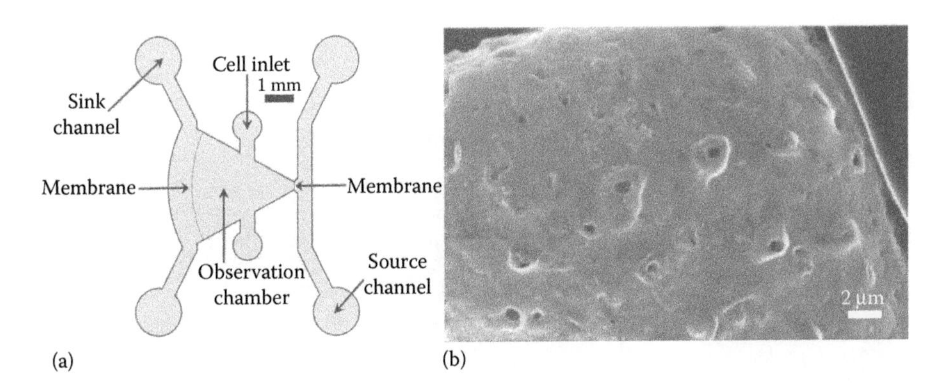

FIGURE 9.15 (a) Passive pumping chemotaxis device.[16] (b) *In situ* nanoporous membrane.[17] (From Kim, D., Lokuta, M., Huttenlocher, A., and Beebe, D.J., *Lab Chip*, 9, 1797–1800, 2009; Kim, D. and Beebe, D.J., *J. Appl. Polym. Sci.*, 110, 1581–1589, 2008.)

Results from fluorescent experiments in which fluorescein was introduced into the device's source channel by passive pumping show no flow into the observation chamber (Figure 9.16). Results demonstrate that a concentration gradient developed in the observation chamber and show neutrophil migration up a chemoattractant concentration gradient (Figure 9.17).

Seidi et al.[19] studied how concentration gradients of 6-hydroxydopamine (6-OHDA) triggered neuronal apoptosis in a pheochromocytoma PC12 neuronal cell line. Mimicking apoptotic cell death as it occurs in Parkinson's disease (PD) in an *in vitro* model is important for developing therapeutic antiapoptotic drugs for treating PD. Such drugs can keep the disease from progressing. The goal of high-throughput screening using such a device is to create an *in vitro* system that achieves the highest rate of apoptosis and the minimum rate of necrosis. The concentration gradient was achieved by passively pumping a small volume of toxin through the inlet of the microchannel. The outlet drop was then aspirated in order to stop flow from the inlet

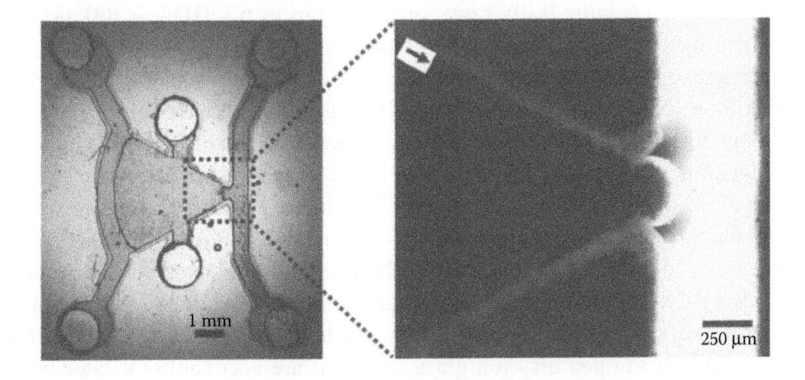

FIGURE 9.16 Nanoporous membrane prevents flow from the sink channel to the observation chamber. (From Kim, D., Lokuta, M., Huttenlocher, A., and Beebe, D.J., *Lab Chip*, 9, 1797–1800, 2009.)

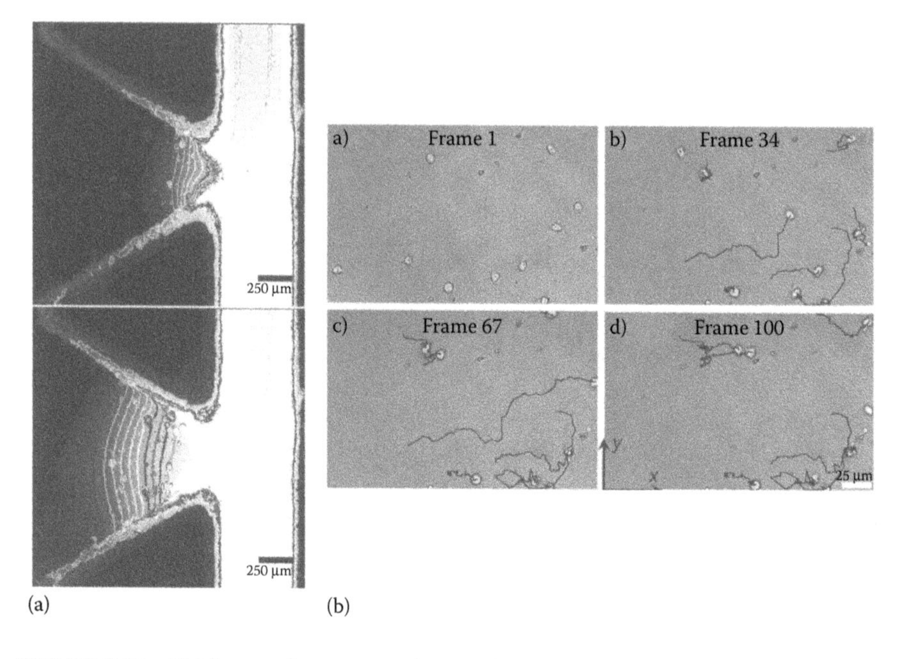

FIGURE 9.17 (a) Fluorescein concentration gradient observed after 5 and 10 min, top and bottom respectively. (b) Neutrophil migration up the concentration gradient at different time steps. (From Kim, D., Lokuta, M., Huttenlocher, A., and Beebe, D.J., *Lab Chip*, 9, 1797–1800, 2009.)

to the outlet. A drop of buffer solution was placed in the outlet, reversing flow from the outlet to the inlet. Liquid was aspirated from the inlet in order to stop flow from the outlet. Both ports are sealed to prevent evaporation and flow (Figure 9.18).

Live/dead staining of PC12 cells cultured over 24 h in microfluidic devices containing 6-OHDA gradients showed higher cell death at higher toxin concentrations within the channel (Figure 9.19). Low concentrations of 6-OHDA (<260 μM) along the gradient resulted in neuronal death mainly induced by apoptosis, while higher concentrations of 6-OHDA resulted in neuronal death mainly induced by necrosis. The maximum percentage of apoptotic cells was achieved at toxin concentrations of less than 300 μM. Toxin concentrations greater than 300 μM resulted in a higher percentage of necrotic events and a decline in the percentage of apoptotic cells.

Hancock et al.[20] used surface tension phenomena in a different way to create flow and make a concentration gradient. They generated centimeters-long gradients of molecules and particles in less than 1 s by using a fluid stripe held in place on a glass slide by a hydrophobic boundary (Figure 9.20). Interestingly, this is not tubeless microfluidics but rather channel-less microfluidics. Gradients produced on the stripe are created in open air on a glass slide and are accessible for later processing and analysis. The flow analysis for this type of system is different from that of surface tension passive pumping as described by Beebe[7] and different from that of surface tension–evaporation coupled flow as described by Dandy[11,12] (Figure 9.21).

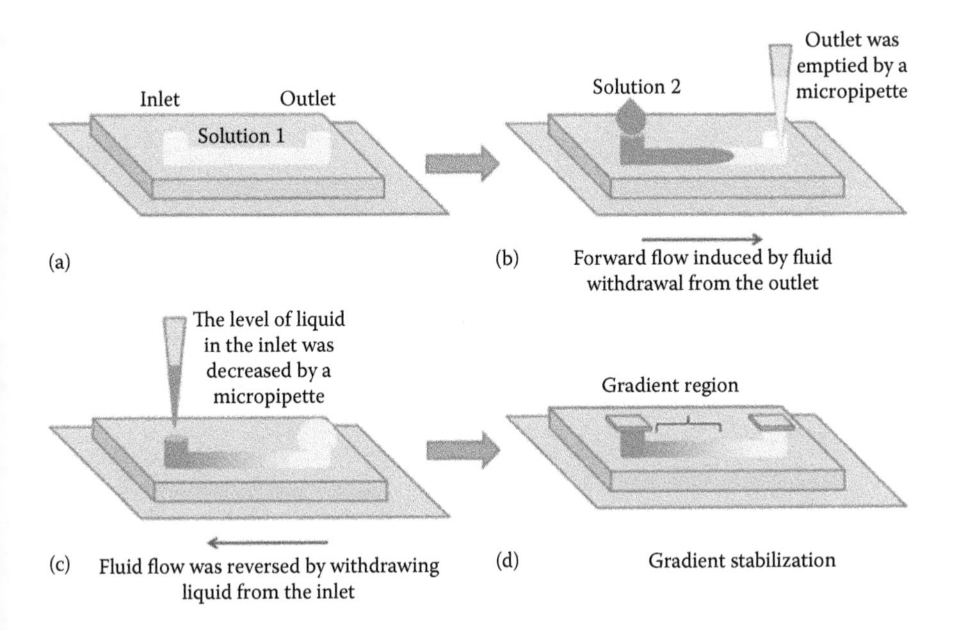

FIGURE 9.18 Microfluidic-based neurotoxin concentration gradient for the generation of an *in vitro* model of PD. (From Seidi, A., Kaji, H., Annabi, N., Ostrovidov, S., Ramalingam, M., and Khademhosseini, A., *Biomicrofluidics*, 5, 022214, 2011.)

9.6 BACKFLOW MECHANISMS

Microscale forces, such as surface tension and evaporation, can be used as an advantage to perform interesting microfluidic experiments. Evaporation has been mostly used to create flow opposite the direction of surface tension forces and to create concentration gradients and sample concentrations. Another microfluidic phenomenon of interest to researchers dealing with surface tension flow is the appearance of a backward-flow mechanism that cannot be explained by evaporation.

In their work, Walker et al.,[7] Lynn et al.,[12] and Du et al.[13] have used evaporation to their advantage to perform a given set of experiments. However, Zhu et al.[10] saw evaporation-induced backward flow play an unsolicited role in their results while investigating virus spread inside a passive pumping channel. The bias in their results was not critical; however, its very existence makes it important to understand any mechanism that can cause backflow in tubeless microfluidic devices.

To this end, Ju et al.[21,22] explained a nonevaporation backward-flow mechanism occurring in passive pumping devices. By ruling out evaporation and surface tension-based inertial interactions, they discovered that rotational flow inside the outlet drop may be a source of inertia that can explain the generation of the backward flow (Figure 9.22a), and outline inlet/outlet volume ratios to prevent the occurrence of the outlet flow (Figure 9.22b). This type of backflow mechanism occurs on a time scale of tens of seconds.

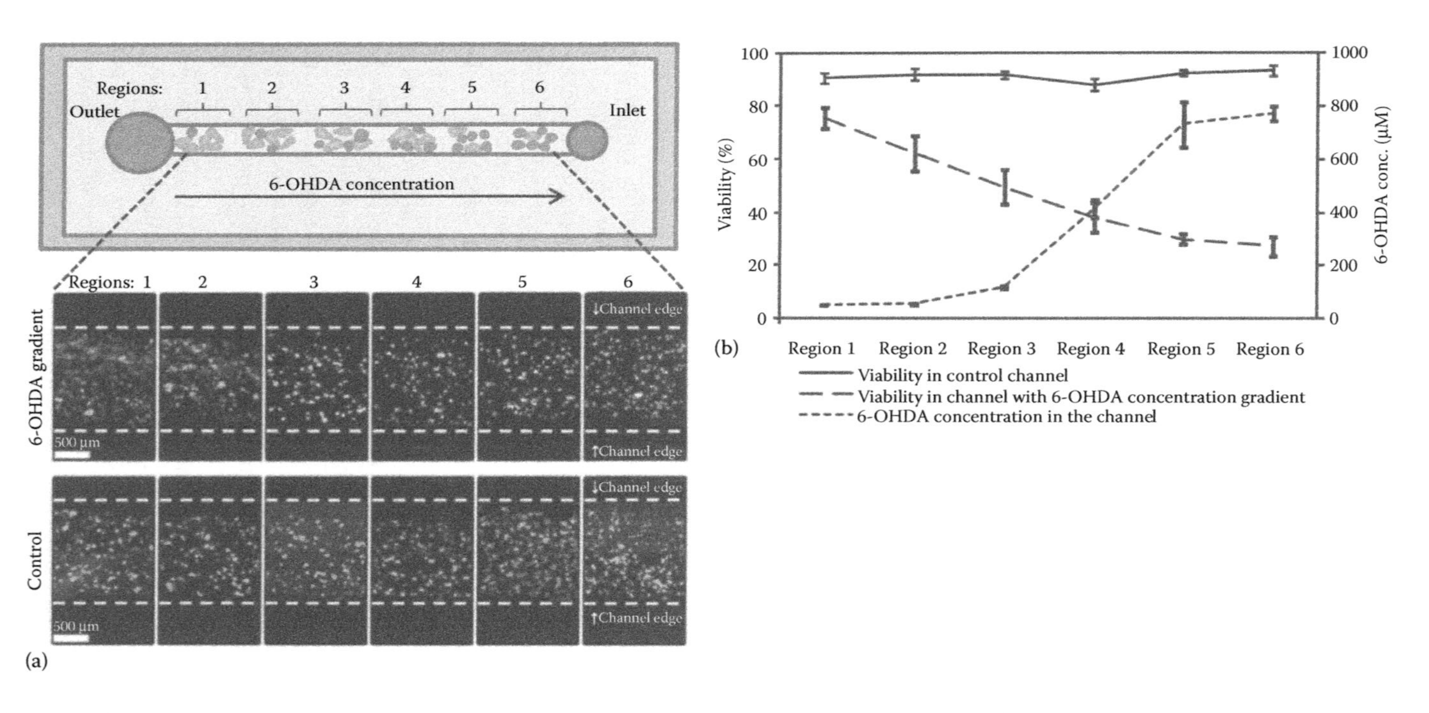

FIGURE 9.19 Microfluidic devices containing 6-OHDA gradients show higher cell death at higher toxin concentrations within the channel. (From Seidi, A., Kaji, H., Annabi, N., Ostrovidov, S., Ramalingam, M., and Khademhosseini, A., *Biomicrofluidics*, 5, 022214, 2011.)

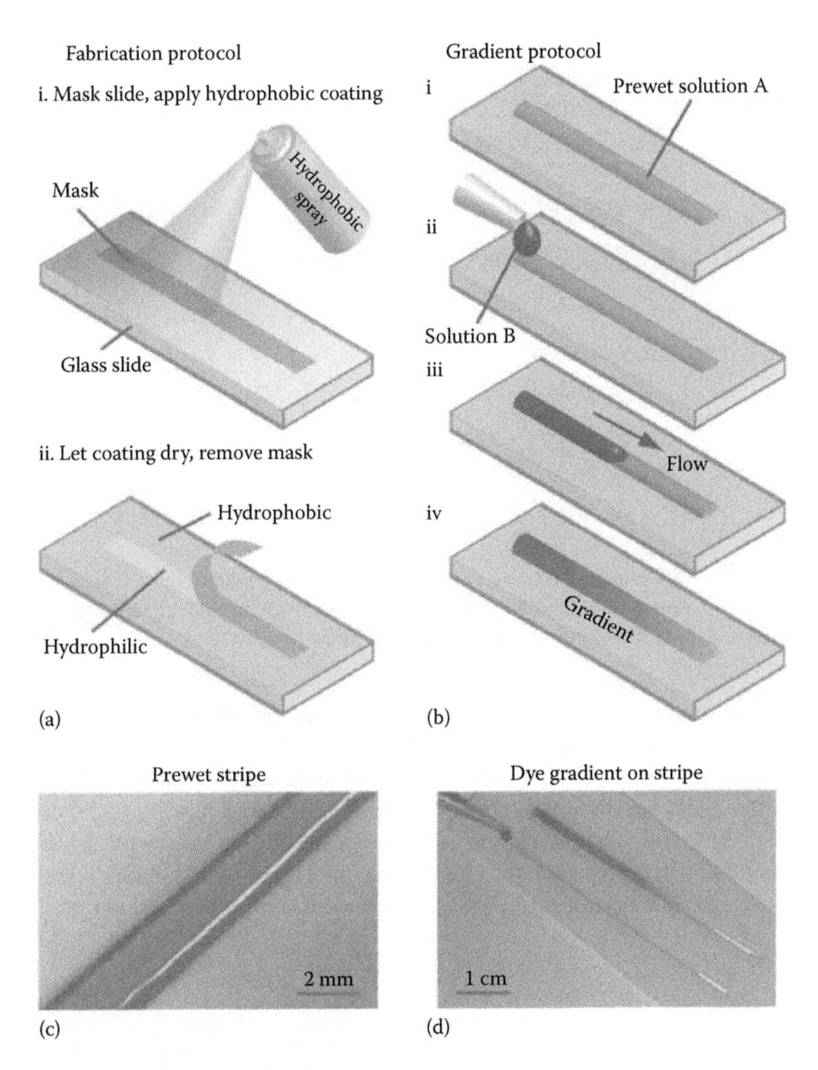

FIGURE 9.20 Generation of surface-tension-driven gradient in a fluid stripe held in place by a hydrophobic surrounding. (From Hancock, M., He, J., Mano, J., and Khademhosseini, A., *Small*, 7, 892–901, 2011.)

Besides evaporation and rotational force at the outlet drop, at least one more backflow mechanism has been discovered. In their work, Resto et al.[23] observed an inertia-based backflow mechanism (Figure 9.23). Unlike the previous mechanisms, this inertia-based backflow occurs in a time scale of milliseconds. This type of backflow is likely caused by the interplay of inertia–surface tension interactions at the inlet drop.

The setup of Resto et al. consisted of using automated micronozzles (nozzle diameter 254 μm) to eject droplets to the inlet port of a passive pumping microfluidic device (Figure 9.24). There is evidence that the inertia imparted by the ejected

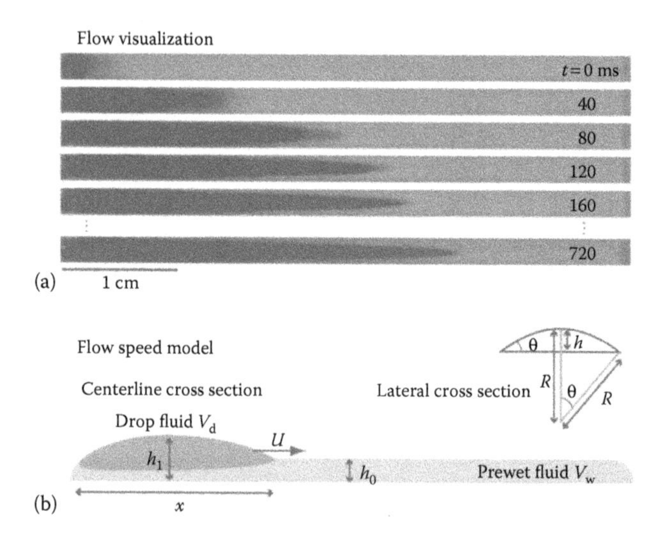

FIGURE 9.21 (a) Consecutive frames of dye flowing across a fluid stripe. (b) Schematic and experimental parameters involved in fluid stripe gradient generation. (From Hancock, M., He, J., Mano, J., and Khademhosseini, A., *Small*, 7, 892–901, 2011.)

droplets on the liquid in the inlet port is nonnegligible. When a droplet hits the inlet port of the device, it immediately collapses. Collapse was observed to occur in approximately 30 ms with their setup.[23] Within these experimental conditions, the drop at the inlet port is seen to rebound immediately after it has finished collapsing (Figure 9.23b). This rebound can be a return-to-equilibrium movement by the inlet drop after the inertia present in the incoming droplet causes the standing liquid at the inlet to overshoot a natural surface tension-dictated equilibrium position.

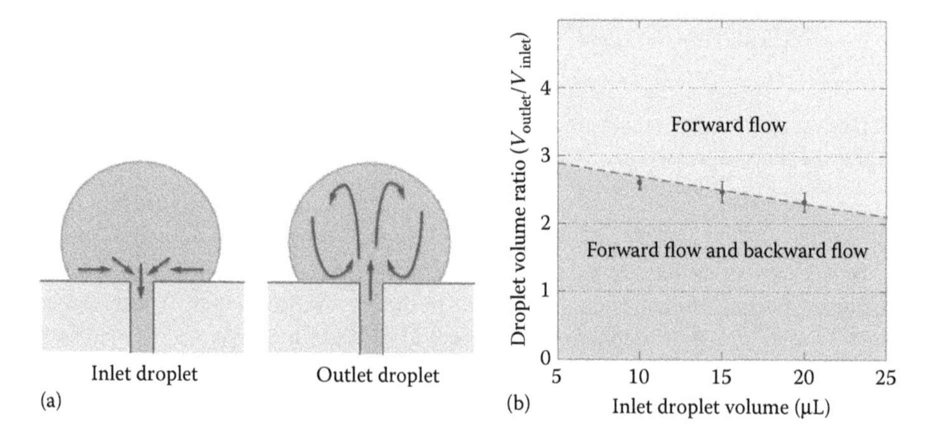

FIGURE 9.22 (a) Rotational flow in the outlet drop. (b) Volume ratios of the two drops should be in the light gray area to prevent backflow. (From Ju, J., Park, J.Y., Kim, K.C., Kim, H., Berthier, E., Beebe, D.J., and Lee, S., *J. Micromech. Microeng.*, 18, 087002, 2008.)

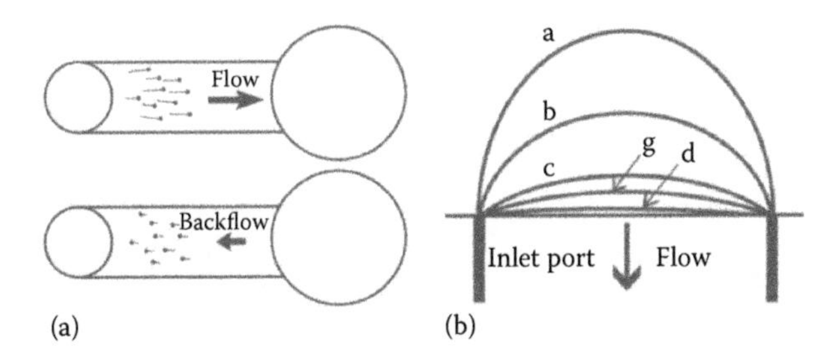

FIGURE 9.23 (a) Backflow observed in red fluorescent beads while they are flowing in a passive pumping channel. (b) Backflow is predicted to be due to inertia–surface tension interactions at the inlet drop. (From Resto, P.J., Mogen, B.J., Berthier, E., and Williams, J.C., *Lab Chip*, 10, 23–26, 2010.)

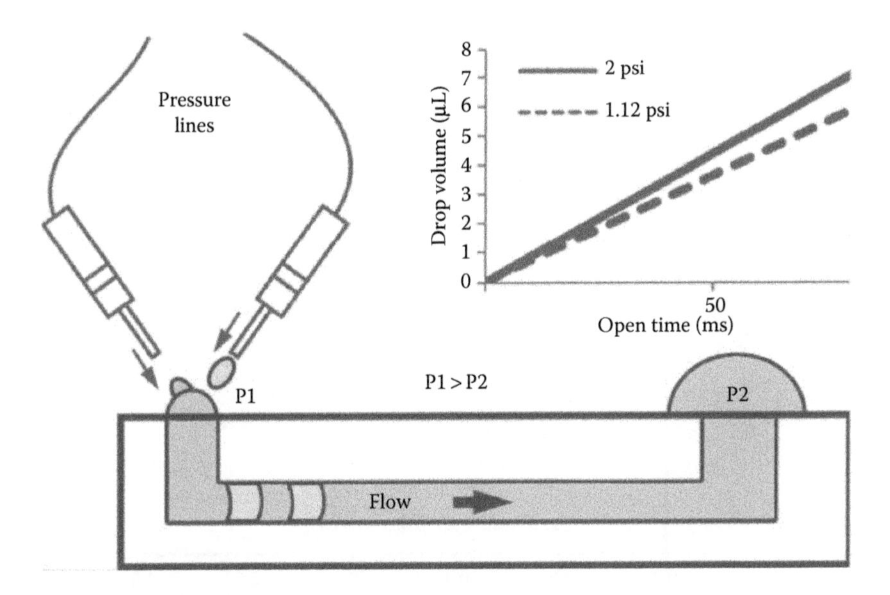

FIGURE 9.24 Microfluidic setup used by Resto et al.[23] (From Resto, P.J., Mogen, B.J., Berthier, E., and Williams, J.C., *Lab Chip*, 10, 23–26, 2010.)

9.7 HIGH THROUGHPUT

Fluid flow in microchannels is used to treat and wash samples for different types of experiments. An advantage of tubeless microfluidic devices is that they can be used for high-throughput applications, most of which currently use standard wells for performing assays. Therefore, an important component of analyzing tubeless microfluidic devices is to understand fluidic wash inside them. In their 2007 work, Warrick et al.[24] analyzed fluidic exchange inside a passive pumping microchannel. They described a linear relationship between intrachannel concentration and the volume of an incoming fluid. One of their relevant conclusions is that it takes five times the channel volume to exchange 95% of the intrachannel liquid (Figure 9.25).

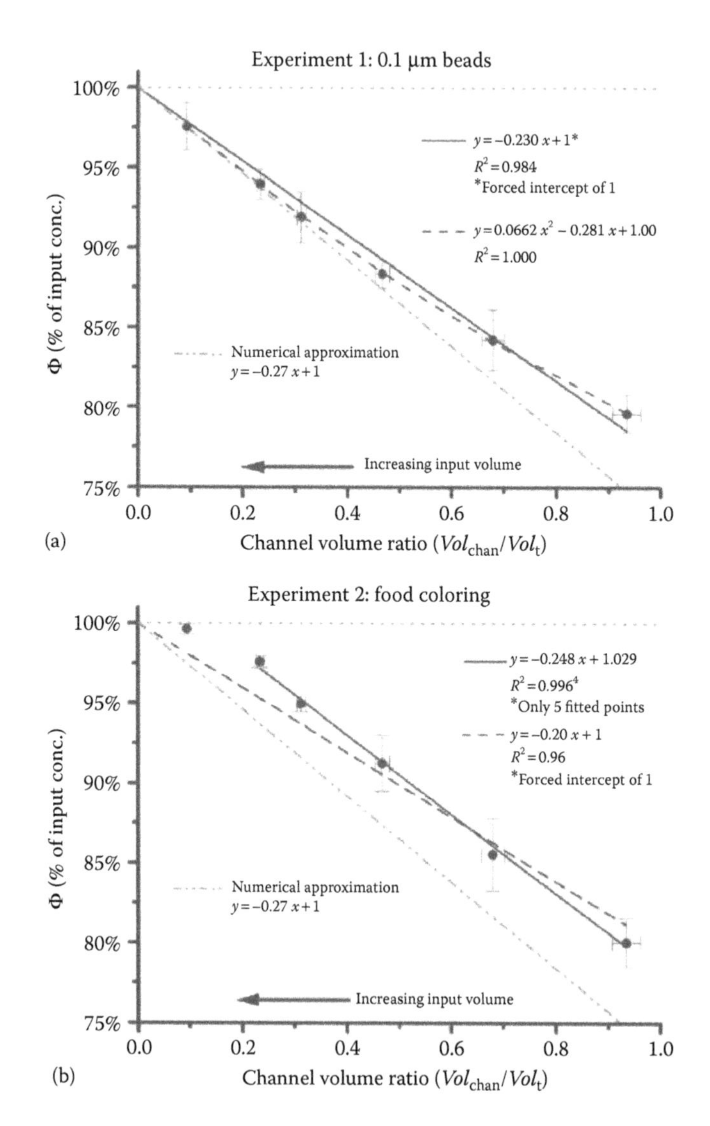

FIGURE 9.25 Fluidic wash inside passive pumping microfluidic devices. (a) Use of fluorescent beads to study washout. (b) Use of food coloring to study washout. (From Warrick, J., Meyvantsson, I., Ju, J., and Beebe, D.J., *Lab Chip*, 7, 316–321, 2007.)

They also proposed a *treat-wait-treat* method for more efficient channel washout, as there are convection and diffusion components to channel washout.

Taking advantage of the simplicity and control of passive pumping, Meyvantsson et al.[25] applied the principles of passive pumping for mammalian cell culture. The authors developed an array of 192 microfluidic channels, each with two access ports, to mimic the microtiter plate standard array (Figure 9.26). They demonstrated the utility of such an array used with an automated liquid handler (ALH) to culture

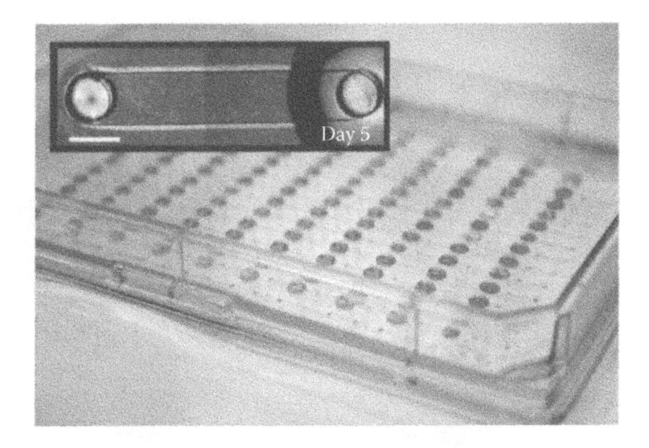

FIGURE 9.26 Microfluidic array composed of 192 independent microfluidic devices. Insert: 3T3 cells cultured in passive pumping channels over a period of 5 days. (From Meyvantsson, I., Warrick, J., Hayes, S., Skoien, A., and Beebe, D.J., *Lab Chip*, 8, 717–724, 2008.)

3T3 cells in passive pumping channels over 5 days with medium changes every 24 h (Figure 9.26).

Meyvantsson et al.[25] showed that passive pumping retains the benefits of laminar flow embodied by microfluidic devices in general. They were able to pattern cells while maintaining fine fluid control inside the device (Figure 9.27). Figure 9.27a and b shows mouse mammary epithelial cells cultured in an eight-channel passive pumping device. They used an eight-channel pipette to simultaneously deliver a drop to each inlet port and maintain an even laminar boundary between the liquid from each inlet. Figure 9.27c and d shows fibroblasts (Hs578Bst) and epithelial cells (HMT-3522 S1) patterned in a three-dimensional (3D) 1rECM gel matrix. Figure 9.27e shows MCF10A patterned in the 3D 1rECM gel. Overall, the techniques outlined in their work show that passive pumping in conjunction with an ALH is capable of performing long-term cell culture. This opens the door for the use of passive pumping in high-throughput biological applications and applications that need cell culture over long periods.

Puccinelli et al.[26] continued this line of work and expanded on the opportunity of using passive pumping for high-throughput assay screening. The authors used an array of passive pumping devices and an ALH to perform In-Cell Western (ICW) assays (Figure 9.28). Their aim was to validate the tubeless microfluidic platform for performing complete cell-based biological assays reliably and repeatedly, from culture to analysis, using automated liquid handling and automated analysis, and develop step-by-step protocols to mimic traditional cell culture by using an ALH (Figure 9.29).

Paguirigan et al.[27] further validated passive pumping devices for use in ICW assays. This time the authors validated high-throughput ICW assays using passive pumping by comparing it directly with macroscale methods. They used transforming growth factor-β-induced epithelial-to-mesenchymal transition of an epithelial

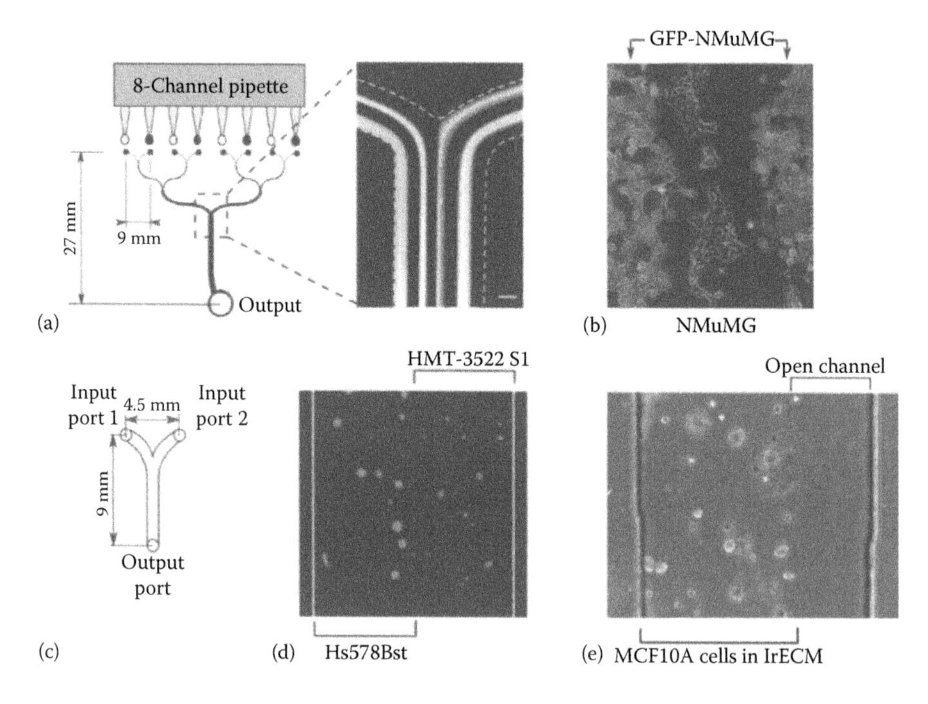

FIGURE 9.27 (a,b) Mouse mammary epithelial cells cultured in an eight-channel passive pumping device. (c,d) Fibroblasts (Hs578Bst) and epithelial cells (HMT-3522 S1) patterned in a 3D 1rECM gel matrix. (e) MCF10A patterned in the 3D 1rECM gel. (From Meyvantsson, I., Warrick, J., Hayes, S., Skoien, A., and Beebe, D.J., *Lab Chip*, 8, 717–724, 2008.)

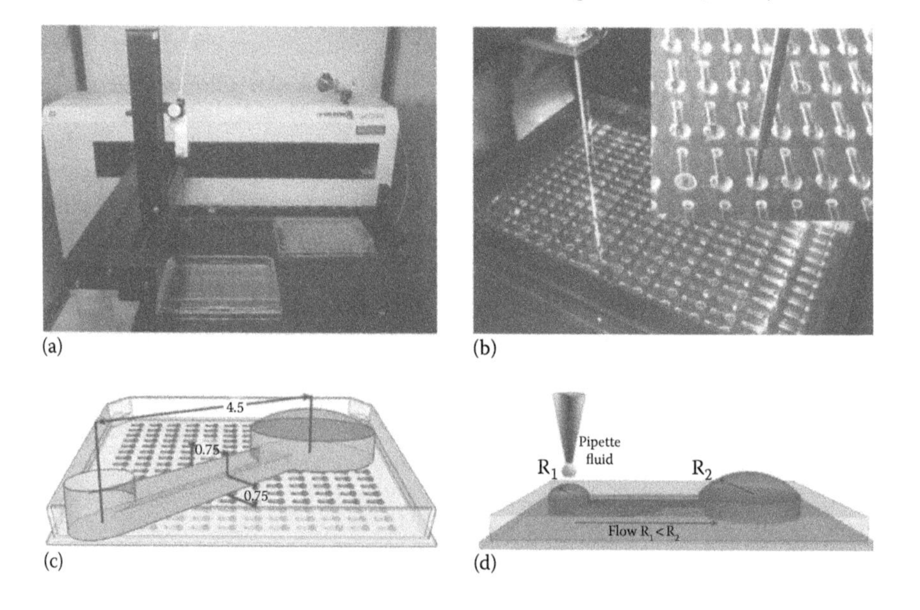

FIGURE 9.28 Passive pumping devices and an ALH used to perform ICW assays. (From Puccinelli, J.P., Su, X., and Beebe, D.J., *J. Assoc. Lab. Autom.*, 15, 25–32, 2010.)

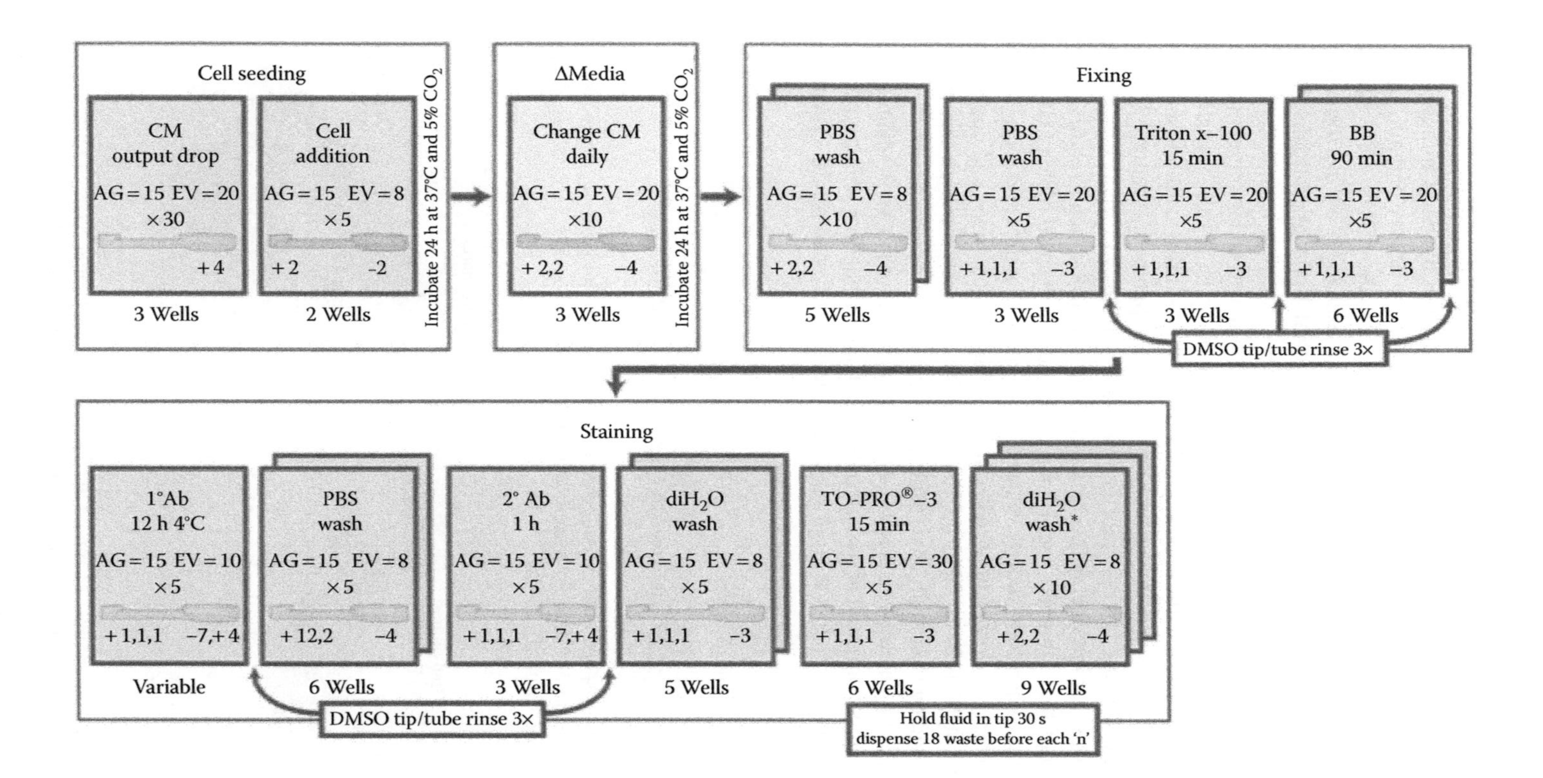

FIGURE 9.29 Step-by-step protocol of robotic manipulations developed by Puccinelli et al.[26] for performing complete cell-based biological assays. AG, air gap; EV, extra volume; n, number of channels; CM, culture medium; PBS, phosphate-buffered saline; PFA, paraformaldehyde; BB, blocking buffer; DMSO, dimethyl sulfoxide; 1°Ab, primary antibody; 2°Ab, secondary antibody; DCV, dispense channel volume; ACV, aspirate channel volume; *Step 2 and 3 as listed, step 1 was identical to TO-PRO-3. (From Puccinelli, J.P., Su, X., and Beebe, D.J., *J. Assoc. Lab. Autom.*, 15, 25–32, 2010.)

cell line as an example to validate the technique as a readout for soluble factor–based assays performed in high-throughput microfluidic channels.

Resto et al.[23] developed a simplified version of an ALH by using an automated micronozzle system to deliver small droplets to the inlet of a passive pumping device (Figure 9.24). Theory states that the smaller the drops, the higher their internal pressure. Therefore, high flow rates and fast fluidic exchanges are performed by delivering small droplets at a high frequency to the inlet of a microfluidic device (Figure 9.30).

Another clever example of passive pumping for automated cell manipulation was demonstrated by Ju et al. in their work using passive pumping in cell programmable assay (CPA) chips.[28] The authors created CPA chips for culturing cells and automating the process of staining using surface tension passive pumping (Figure 9.31). The system was tested using human embryonic kidney (HEK) 293 cells. The rationale for this work was that interfacing robotics and culture control systems require equipment or methods that can make regular use of the device in small biology laboratories challenging. Therefore, they developed these chips to increase the throughput of common laboratory procedures without requiring the machines and equipment needed for most high-throughput approaches—for example, the high-throughput method proposed by Puccinelli et al. using ALHs.

Seidi et al.[19] in their work on creating an *in vitro* model for PD, took advantage of the high-throughput potential of passive pumping. High-throughput screening assays allow researchers to conduct a large number of tests for screening the efficacy of

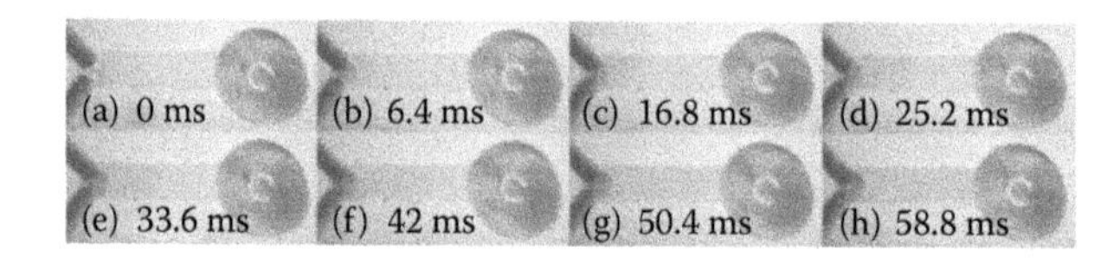

FIGURE 9.30 High-speed fluid exchange inside an open microfluidic device using an automated micronozzle droplet delivery system. (From Resto, P.J., Mogen, B.J., Berthier, E., and Williams, J.C., *Lab Chip*, 10, 23–26, 2010.)

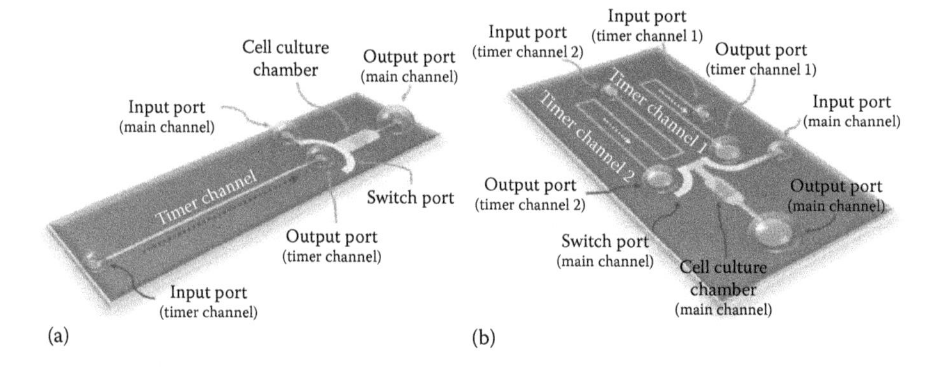

FIGURE 9.31 (a) Single timer channel CPA (sCPA). (b) Dual timer channel CPA (dCPA). (From Ju, J., Warrick, J., and Beebe, D.J., *Lab Chip*, 10, 2071–2076, 2010.)

drug concentrations on cells. Passive pumping allowed Seidi et al. to duplicate experiments typically performed in multiwell plates without the high cost of such assays.

Su et al.[29] took advantage of the high-throughput capabilities of passive pumping devices to investigate the growth and culturing properties of HEK cells transfected with hERG channel protein for the development of a high-throughput platform for screening the cardiac toxicity of drug compounds. The authors performed different biological cell assays and procedures, including traditional western blots and live-cell westerns, to make sure the microchannels were a viable culturing environment for hERG-expressing cells. Patch-clamping experiments were then performed on the microchannel-cultured cells to verify that the hERG ion channels were functioning normally. After the system was validated as a growth environment, the hERG-expressing cells were treated with various drugs to determine the suitability of microchannels for drug toxicity testing. Results show that microfluidic screening is useful and that polystyrene (PS) and cyclo-olefin polymer (COP) microchannels are more appropriate than PDMS devices as they do not absorb small hydrophobic drug molecules. Passive pumping shows its usefulness in eliminating the need for external pumps and tubing and allowing the use of existing biology laboratory equipment for analysis.

9.8 PASSIVE WASHING AND LOGIC CIRCUITRY

A major use of microfluidic devices is to observe and analyze the effect of the microenvironment on cells and in many cases apply and wash away multiple reagents over a period of time. Warrick et al.[24] studied the washing efficiency of passive pumping microchannels and concluded that introducing five times the channel volume will exchange 95% of the original liquid. Meyvantsson et al.[25] kept a consistent laminar boundary layer between adjacent streams of liquid inside a multi-inlet microchannel by using an eight-channel pipette to deliver drops to the inlet simultaneously.

Berthier et al.[30] simplified the maintenance of a consistent laminar pattern within tubeless devices by developing a short-term laminar flow patterning (LFP) technique for multiple liquids using a single pipette (Figure 9.32). LFP allows two or more

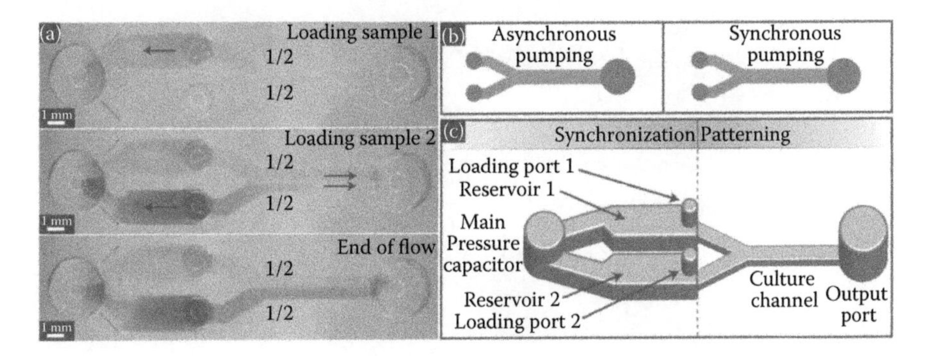

FIGURE 9.32 (a) Loading samples using LFP. (b) Asynchronous versus synchronous pumping in a Y-channel. (c) Fluidic capacitor that enables LFP with passive pumping. (From Berthier, E., Warrick, J., Casavant, B., and Beebe, D.J., *Lab Chip*, 11, 2060, 2011.)

different solutions to flow side-by-side in a channel without convective mixing. The technique can be used to pattern cells and carry out treatments and has the potential to be used for creating chemical gradients. LFP is possible in systems that are able to deliver precise and controllable flow rates through two or more inputs. Passive pumping is capable of performing LFP as long as droplets of identical volumes can be delivered to two or more inlets at exactly the same time, which would require ALHs or other forms of precise liquid ejection systems—for example, a multichannel pipette.

Asynchronous flow in a Y-channel using passive pumping results when there is a difference in the fluid driving pressure at the inlet, due to differences in either the droplet volume or delivery time to the inlet (Figure 9.32b). To prevent asynchronous pumping, Berthier et al.[30] developed a fluidic capacitor designed to allow LFP with passive pumping using nothing more than a manual pipette. The design uses a "synchronization component" that acts as an intermediate storage location for sample fluid and as a single pressure source (Figure 9.32c). The system controls LFP through the channel by allowing liquid placed at each inlet to first rush into the capacitor, creating equal pressures at each inlet; thus, liquid moving through each individual Y-branch has the same driving pressure.

Kim et al.[31] devised a different approach for passive washing using tubeless devices. In their work, they combine passive pumping and a washing valve used as a fluidic regulator to control the merging of two solutions (Figure 9.33a). Essentially, they fill two channels. One is a fluidic resistor; the other is a reaction chamber. The channels are separated by a small constriction. Each channel is filled with a different solution. Eventually the two solutions will meet at the constriction and make contact. When this contact happens, fluid from the fluidic resistor will rush into the reaction chamber, thus washing away the standing liquid (Figure 9.33b).

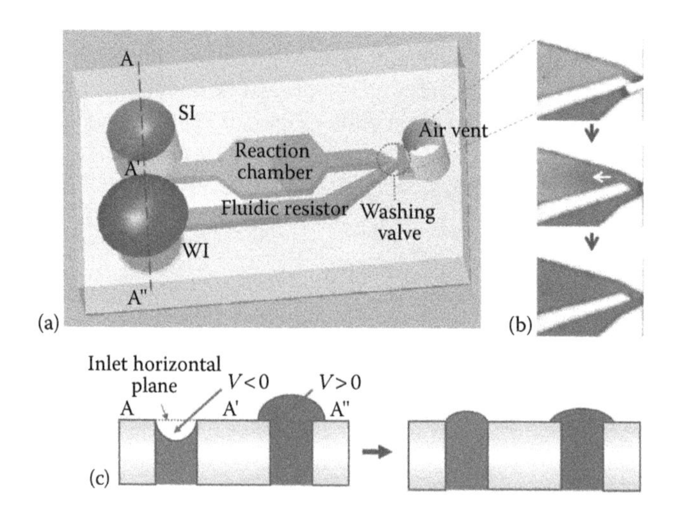

FIGURE 9.33 (a) Schematic of passive washing microfluidic device. (b) Solution washing occurs after contact by two liquids opposing the constriction. (From Kim, B., Kim, S., Yang, S., and Yang, H., *J. Micromech. Microeng.*, 17, N22–N29, 2007.)

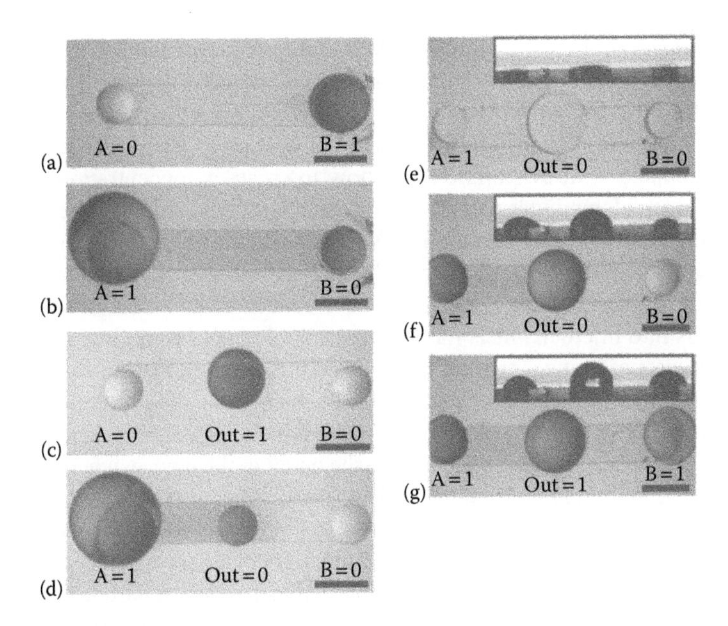

FIGURE 9.34 (a,b) Microfluidic inverter. (c,d) NOR gates. (e–g) AND gates. (From Toepke, M., Abhyankar, V., and Beebe, D.J., *Lab Chip*, 7, 1449–1453, 2007.)

Microfluidic devices are analogous to electrical circuitry. The simplest example is the fluidic resistance inherent to microfluidic devices. Another case is the fluidic capacitor developed by Berthier et al.[30] Toepke et al.[32] went a step further and created microfluidic components analogous to electronic circuit components. This allows passive pumping to be used in multistep, programmable biological protocols without adding complexity to the manufacturing process while limiting the dependence on external equipment.

In microfluidics, fluidic resistance is analogous to electrical resistance, fluid flow is analogous to electrical current, and the pressure differential across a channel is analogous to voltage. Changing the dimensions of a device controls the fluidic resistance. In passive pumping, the pressure differential (i.e., voltage) is controlled by the volume of the drops placed at the inlet and the outlet. Using these concepts, the authors created three classes of logic components: OR/AND, NOR/NAND, and XNOR digital microfluidic gates (Figure 9.34). A programmable timer is constructed by using a high-resistance channel to provide slow flow, acting as a timer, and a second channel to pump the timer output droplet once its volume has achieved a critical size. A slow, perfusion rheostat can be created by using multiple inlet ports with high-resistance channels. This last scheme allows control over the flow rates during passive pumping.

9.9 BIOLOGY

Passive pumping in microfluidics is most commonly used for the study of biological systems. The majority of research in this area aims to control, manipulate, and study the environment of small living things. Meyvantsson et al.[25] showed it is possible to perform cell culture in passive pumping channels. Puccinelli et al.[26] followed this line of work by using an ALH to create cell-handling protocols and demonstrated their work by performing high-throughput ICW assays. Du et al.[13] used passive pumping coupled with evaporation backflow to create a concentration gradient of a cardiac toxin, alpha-cypermethrin, along the microchannel to test the response of HL-1 cardiac cells in the microdevice. Kim et al.[16] made a chemotaxis device that used passive pumping to load a source channel and allow diffusion to create a concentration gradient toward a sink channel. Domenech et al.[33] used passive pumping to load cells into microchannels for studying soluble factor signaling. In cancer and other diseases, paracrine signaling is implicated in disease pathogenesis.[34]

Khnouf et al.[35,36] used passive pumping for high-throughput protein expression. A passive pumping device was filled with the protein expression solution. Nutrient solution was then pumped from the inlet to the outlet of the device, where the reaction takes place. A 5 μL volume from the outlet drop was then mixed with 35 μL of luciferase assay reagent in an individual well of a 384-well microplate. The luminescence signal from the well was correlated with the amount of luciferase present. The method of protein expression in a passive pumping channel was compared with the same method of protein expression in a microcentrifuge tube. The authors concluded that protein expression using nutrient solution in the microfluidic device is slightly higher than that produced by the same experiment in a microcentrifuge tube (Figure 9.35). Protein expression in both scenarios using nutrient solution is much higher than protein expression in a microcentrifuge tube without the use of nutrient solution.

Protein expression yield is a function of the amount of nutrient solution and not the amount of expression solution used in the experiment (Figure 9.36). Use of passive pumping is advantageous as it allows simultaneous production of the same protein under different experimental conditions, simultaneous production of different proteins for high-throughput, high-yield protein expression, and low reagent cost since it consumes a fraction of the reagents used in a commercially available protein expression instrument.

Zhu et al.[10] performed flow-enhanced infection assays in microchannels using passive pumping to inoculate cell monolayers with virus and drive infection spread.

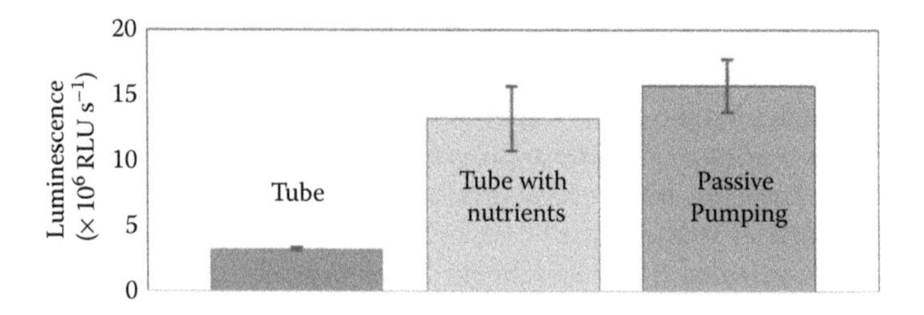

FIGURE 9.35 Amount of protein synthesized between expression in a microcentrifuge tube with and without nutrient solution and in a passive pumping device. (From Khnouf, R., Beebe, D.J., and Fan, Z.H., in *Proceedings of the 12th International Conference on Miniaturized Systems for Chemistry and Life Sciences*, pp. 1740–1742, 12–16 October, San Diego, CA, 2008.)

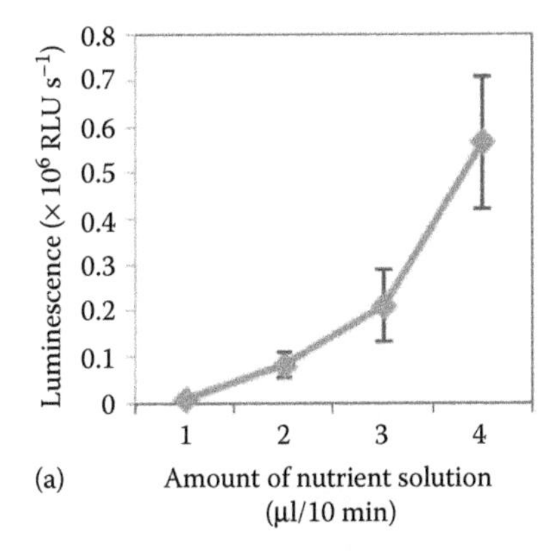

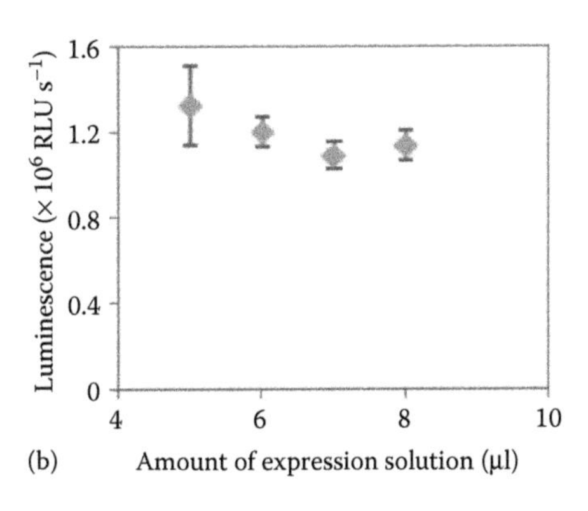

FIGURE 9.36 (a) Protein expression is a function of the amount of nutrient solution supplied. (b) Amount of reaction, in terms of luminescence, as a function of amount of expression solution used. (From Khnouf, R., Beebe, D.J., and Fan, Z.H., in *Proceedings of the 12th International Conference on Miniaturized Systems for Chemistry and Life Sciences*, pp. 1740–1742, 12–16 October, San Diego, CA, 2008.)

Under a lack of flow, virus infections spread locally throughout the microchannels. The experiments measured virus infection levels in terms of plaque-forming units (PFU) and showed higher viral gene expression for higher virus PFU per channel (Figure 9.37).

Under flow conditions, the extent of viral infection spread correlated with the volume of virus added (Figure 9.38b). Similar results are obtained when inocula containing from 300 to 3 million virus particles (i.e., 3×10^2 to 3×10^6 PFU) are added; however, infection spread dropped when the inoculum contained only 30 PFU (Figure 9.38c), suggesting that an infection of anything greater than 300 PFU is enough to saturate the infection capacity of the microfluidic device. A backward spread of infection, opposing the direction of surface tension flow, was observed and blamed on backflow due to evaporation at the inlet port. Cell death decreased accordingly when flow-enhanced infections were treated with increasing levels of antiviral drug (Figure 9.38d). The infection assay using a microfluidic device required only 1/80 of the amount of drug needed in the current gold-standard plaque-reduction assay. Assay sensitivity based on drug concentrations that reduce infection by 50% indicated that the flow-enhanced infection assay was also twice as sensitive as the gold-standard plaque-reduction assay (Figure 9.38e).

Koepsel et al.[37] used passive pumping to create a rapid method for self-assembled monolayer (SAM) cell culture substrates with a functionalized carboxylic acid end that covalently links to a cell adhesion peptide in order to create cell adhesive environments (Figure 9.39). The process uses passive pumping to drive liquid flow. Passive pumping allowed rapid liquid transport and rapid local substrate modification. Use of an array of microchannels allowed for control of local substrate modifications from channel to channel.

Echeverria et al.[38] performed tumor cell migration assays in a 3D extracellular matrix (ECM) inside a passive pumping channel (Figure 9.40). The protocol involved

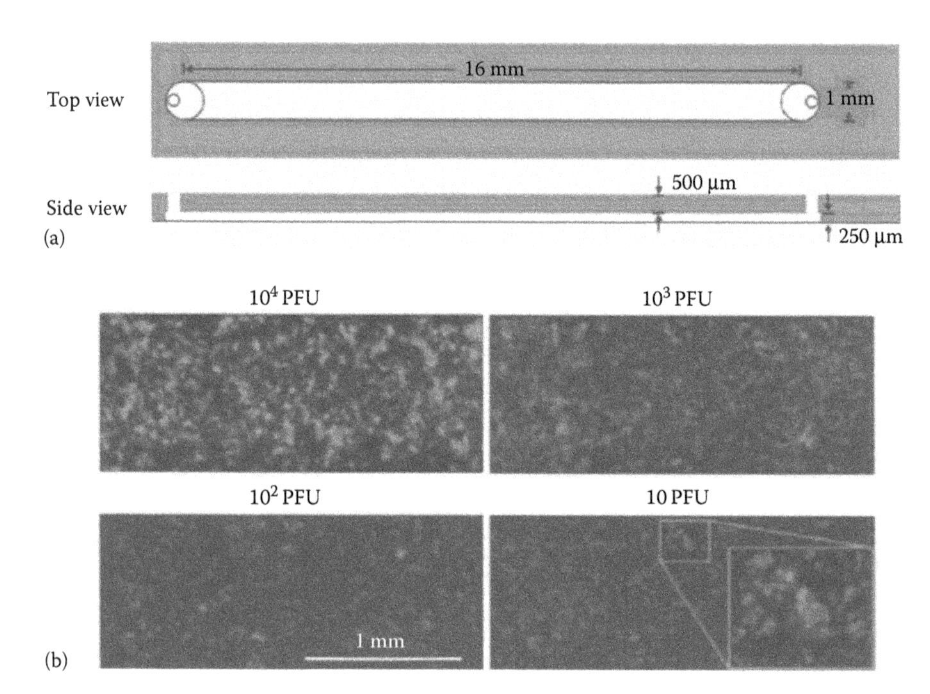

FIGURE 9.37 (a) Microfluidic channel used for cell culture and viral spread. (b) Stained virus (light gray) and stained cells (dark gray) as a function of number of plaque forming units (PFU) per channel. (From Zhu, Y., Warrick, J.W., Haubert, K., Beebe, D.J., and Yin, J., *Biomed. Microdevices*, 11, 565–570, 2009.)

filling the channel with ECM (Matrigel) via the inlet port. The matrix moves into the channel via capillary action. After the channel was filled, the devices were incubated at 37°C for 60 min to allow the matrix to gel. Tumor cells were then added to the output port and incubated for 5 days at 37°C and 5% CO_2 (Figure 9.41). The medium was changed daily by replacing the medium at the output port. The fluid content inside the channel can be exchanged without disturbing the gel by using surface tension passive pumping. Experiments were then performed to study the effect of inhibitors on PC3-M cell invasion. Results show a decrease in the invasion distance for cells under inhibitor treatment. Overall, the paper demonstrated the use of a passive pumping device in extracting quantitative data on a compound's effect on tumor cell migration through 3D collagen.

Mohammed et al.[39] developed a microfluidic brain slice device (μBSD) that integrates passive pumping technology with a standard brain slice perfusion chamber (RC-26GPL, Warner Instruments) in order to deliver fluid to regions within the brain slice perfusion chamber. They bonded the standard brain slice perfusion chamber to a custom-made PDMS microfluidic device having four independent channels, each of which uses passive pumping to drive flow (Figure 9.42). This applies microfabrication techniques and the advantages of microfluidic principles to commercially available, *in vitro* neurobiology equipment in order to perform microscale electrophysiology research. Caicedo et al.[40] followed up this work with a simulation to

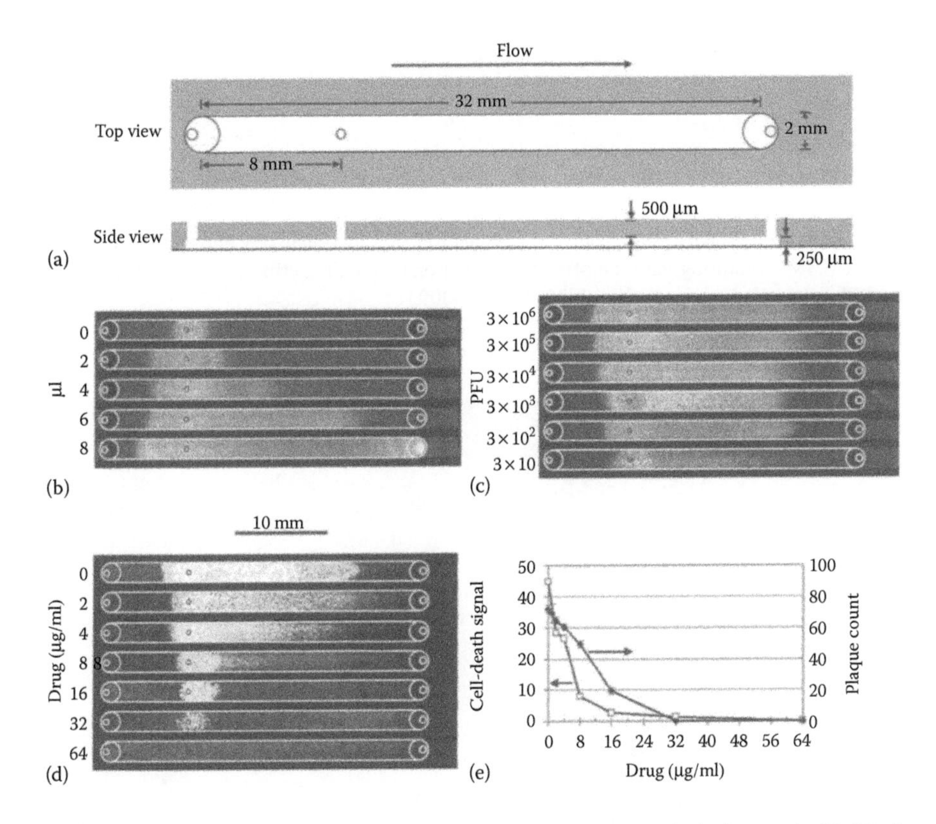

FIGURE 9.38 (a) Microfluidic channel used for cell culture and viral spread. (b) Viral infection spread correlated with the volume of virus added. (c) Infection of greater than 300 PFU is enough to saturate the infection capacity of the microfluidic device. (d) Cell death decreased accordingly when flow-enhanced infections were treated with increasing levels of antiviral drug. (From Zhu, Y., Warrick, J.W., Haubert, K., Beebe, D.J., and Yin, J., *Biomed. Microdevices*, 11, 565–570, 2009.)

characterize the flows through the μBSD by using multiphysics simulations of injections into a porous matrix and optimize the spacing of ports.

Seidi et al.[19] developed an *in vitro* model for PD using passive pumping. Their results show that low concentrations of 6-OHDA (<260 μM) along the gradient resulted in neuronal death mainly induced by apoptosis, while higher concentrations of 6-OHDA induced neuronal death mainly by necrosis (Figure 9.43). The maximum percentage of apoptotic cells was achieved at toxin concentrations of less than 300 μM. Toxin concentrations greater than 300 μM resulted in a higher percentage of necrotic events and a decline in the percentage of apoptotic cells. The accuracy of the gradient screening system was compared with a parallel experiment in a multiwell system. Using both systems, the highest rate of cellular apoptosis was achieved with a 6-OHDA concentration of ~260 μM. Given their low use of reagents and ability to create gradients, passive pumping gradient systems are useful in experiments for a range of drug screening applications.

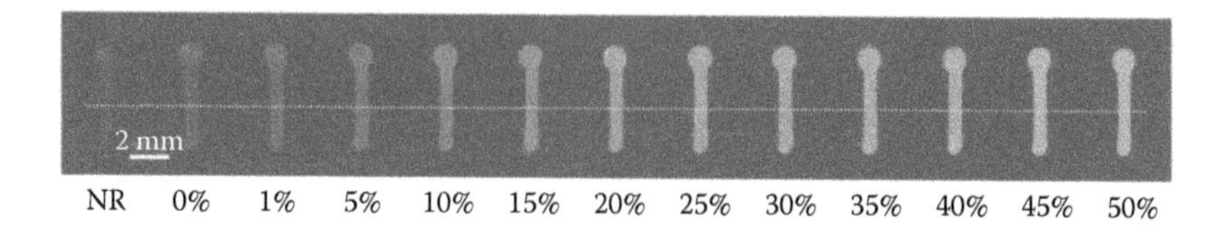

FIGURE 9.39 Image result from the conjugation of functionalized fluorescein to a patterned SAM containing varied mol% of the functionalized alkanethiolate. (From Koepsel, J. and Murphy, W., *Langmuir*, 25, 12825–12834, 2009.)

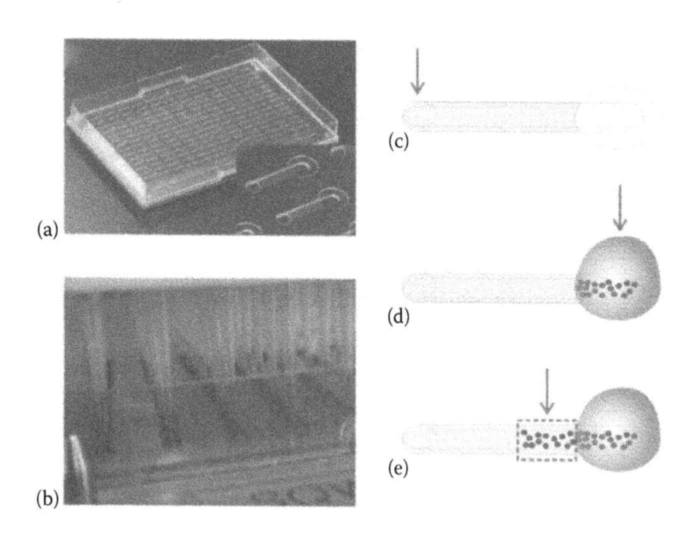

FIGURE 9.40 Protocol for 3D tumor cell migration. (a) Passive pumping channels used for experiments. (b) Automated liquid handler for automated cell culture. (c) Channel is loaded with Matrigel through the inlet port. (d) Tumor cells are loaded through the outlet port. (e) Tumor cells migrating toward inlet port. (From Echeverria, V., Meyvantsson, I., Skoien, A., Worzella, T., Lamers, C., and Hayes, S., *J. Biomol. Screen.*, 15, 1144–1151, 2010.)

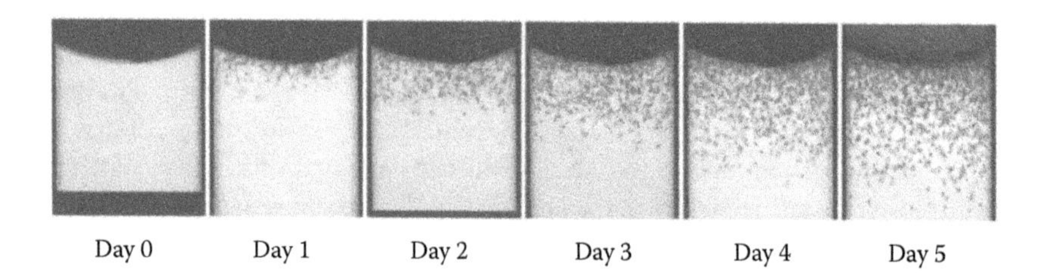

FIGURE 9.41 Tumor cells migrating from the output toward the input through the matrix over the course of 5 days. (From Echeverria, V., Meyvantsson, I., Skoien, A., Worzella, T., Lamers, C., and Hayes, S., *J. Biomol. Screen.*, 15, 1144–1151, 2010.)

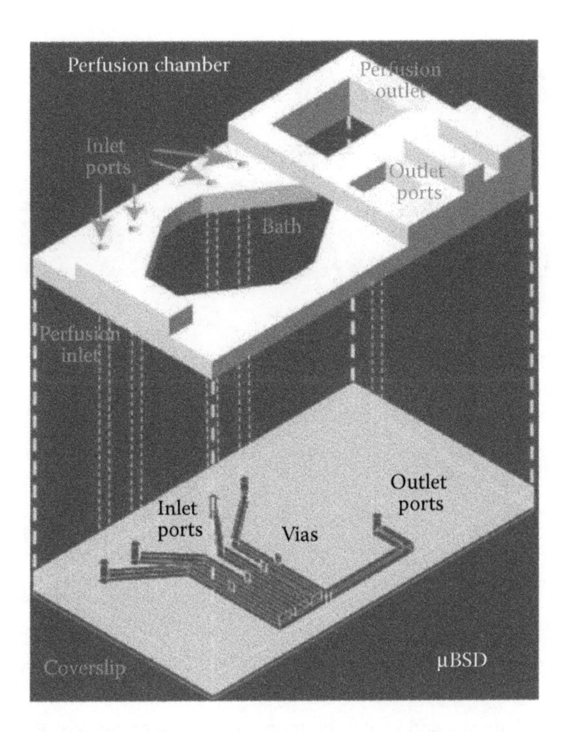

FIGURE 9.42 Microfluidic brain slice device (μBSD) integrates passive pumping technology with a standard brain slice perfusion chamber (RC-26GPL, Warner Instruments). (From Mohammed, J., Caicedo, H., Fall, C., and Eddington, D.T., *Lab Chip*, 8, 1048, 2008.)

Su et al.[29] investigated the growth and culturing properties of HEK cells transfected with hERG channel protein. Their goal was to develop a high-throughput platform for screening the cardiac toxicity of drug compounds. Using passive pumping, they demonstrated microchannel cell culture, drug treatment, protein analysis, and immunostaining of normal and hERG-transfected HEK cells. Their results provided the groundwork for future microfluidic-based drug screening assays for hERG-related cardiotoxicity.

Sung et al.[34] made an *in vitro* microfluidic model to study the transition of ductal carcinoma *in situ* (DCIS) to invasive ductal carcinoma (IDC). Use of a compartmentalized microchannel allowed seeding of epithelial cells next to stromal fibroblasts and the analysis of paracrine interaction as a function of distance; in other words, the authors were able to study interaction at and far from the interface. This is an added functionality not available in traditional coculture systems. This system is also compatible with existing high-throughput systems. The experiments utilize a Y-channel passive pumping device with two inlets and one outlet. The inlets are loaded simultaneously using a multipipette to ensure controlled loading of each type of cell to its respective side of the Y-channel. Alternatively, it would be possible to use the microfluidic capacitor developed by Berthier et al.[30] A spacer gel and a channel with three and four inputs were used to study morphological changes via soluble factors by limiting contact between cells but still allowing diffusion of molecules. Results show

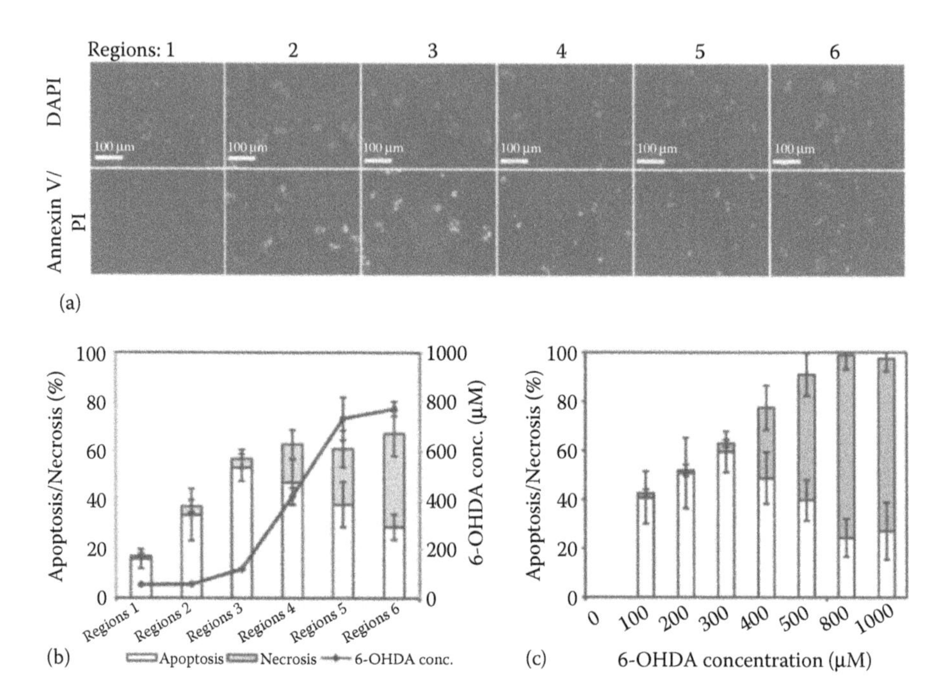

FIGURE 9.43 (a) Cells stained to detect apoptosis and necrosis. (b) Ratio of apoptosis to necrosis as a function of higher concentrations of 6-OHDA, using passive pumping. (c) Ratio of apoptosis to necrosis as a function of higher concentrations of 6-OHDA, using a multiwell system. (From Seidi, A., Kaji, H., Annabi, N., Ostrovidov, S., Ramalingam, M., and Khademhosseini, A., *Biomicrofluidics*, 5, 022214, 2011.)

morphological changes in mammary epithelial cells (MCF-DCIS) when human mammary fibroblasts (HMFs) were cultured some distance (0.5–1.5 mm) from the MCF-DCIS cells, suggesting interaction of soluble factors at the beginning of the transition from DCIS to IDC. Greater gap lengths with the spacer gel resulted in less cellular morphological changes. Cell–cell contact with HMFs allowed the MCF-DCIS cells to finally complete the transition to invasion.

PDMS absorbs small hydrophobic molecules, leaches un-cross-linked oligomers into solution, and allows evaporation through the bulk material.[25,33] For these reasons, PDMS has been shown to alter the biochemical microenvironment in cell cultures. This can bias results in biological experiments in *in vitro* experimentation. Young et al.[41] used a streamlined hot-embossing process to create microfluidic devices out of PS, the most commonly used material for *in vitro* cell-based research (Figure 9.44).

9.10 INTEGRATED ELECTRODES

Another approach to studying and manipulating cells in microfluidic devices is to integrate electrodes into the devices. This contrasts with biochemical analysis, which involves assaying and staining to obtain results from an experiment. Ju et al.[42] used passive pumping as the fluid actuation mechanism for an cellular electrofusion

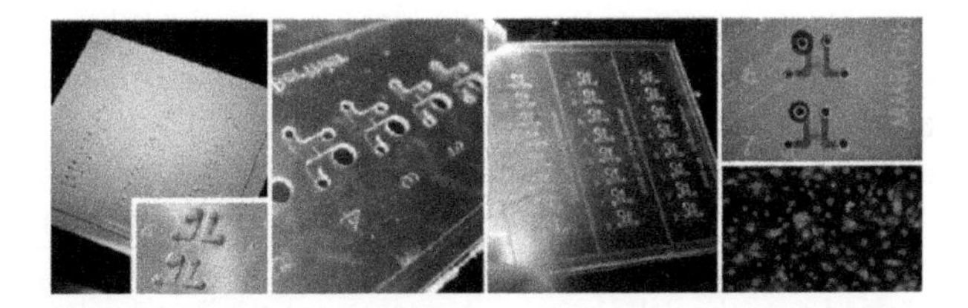

FIGURE 9.44 Polystyrene (PS) microfluidic devices made with a hot-embossing process. (From Young, E.W.K., Berthier, E., Guckenberger, D.J., Sackmann, E., Lamers, C., Meyvantsson, I., Huttenlocher, A., and Beebe, D.J., *Anal. Chem.*, 83, 1408–1417, 2011.)

chip (Figure 9.45). Cell fusion is an important cellular process that occurs during muscle and bone cell differentiation, during embryogenesis, and during morphogenesis. Studying fusion *in vitro* is important for understanding the properties of cells. The advantage of using passive pumping is that it allows for the slow delivery of cells without the need for an expensive pump, and it reduces dead volume.

The process involved filling the electrofusion chip with a hypertonic fusion solution using a micropipette. Then, 30 μL of hypertonic fusion solution was dropped into the outlet port while a mix of protoplasts and hypertonic solution was dropped into the inlet port. Passive pumping allowed the mixed protoplasts to flow into the microchannel. The flow is very slow in the fusion chamber, so an operator can monitor the movement of cells with a microscope. An ac field applied to the microelectrodes to form

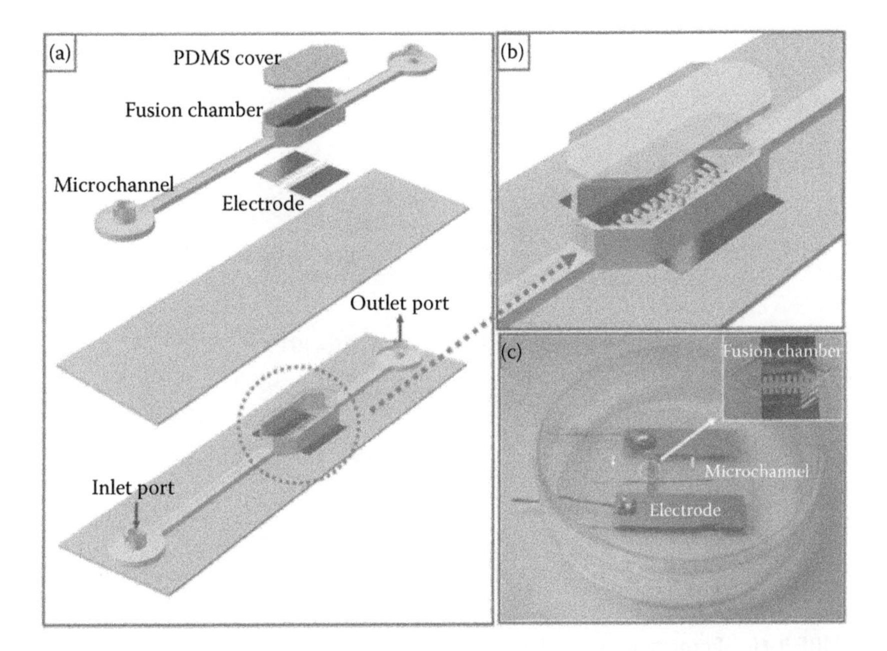

FIGURE 9.45 (a) Schematic of a 3D electrofusion chip used by Ju et al.[42] (b) Region of cell fusion and integrated electrode. (c) Photograph of electrofusion chip. (From Ju, J., Ko, J.-M., Cha, H.-C., Park, J.Y., Im, C.-H., and Lee, S.-H., *J. Micromech. Microeng.*, 19, 015004, 2009.)

pearl chains between the electrodes was followed by a short, high-voltage dc pulse to fuse the cells. Finally, culture medium was flowed into the microchannel via the inlet port to wash away the hypertonic fusion solution.

The formation of the pearl chains depended on the pressure differential across the channel (Figure 9.46). A lower pressure differential between the inlet and outlet droplets allowed for better pearl chain formation. After electrofusion, fresh culture medium was flowed over the cells to provide nutrient to the fused cells. The delivery of the culture medium through passive pumping did not affect the fused cells, demonstrating the cell and medium transport capabilities of such a flow system. After electrofusion, the cells can be collected using a micropipette. Otherwise, passive pumping allows the fused cells to be cultured inside the microdevice just by replacing the cell medium.

Mohanty et al.[43] used passive pumping as the flow actuation mechanism in an electrophoresis device for the injection and extraction of DNA (Figure 9.47). The use of passive pumping, along with simple microelectromechanical systems (MEMS) techniques, allows for a simple electrophoresis device suitable for research laboratories that operate on a budget.

Sample loading and extraction occurs via passive pumping (Figure 9.48). The first step is to fill the microelectrophoresis channel with separation buffer, Fluid A. A drop of sample, Fluid B, is placed in Port 1. This drop causes volume displacement into the separation channel that moves toward Port 2.

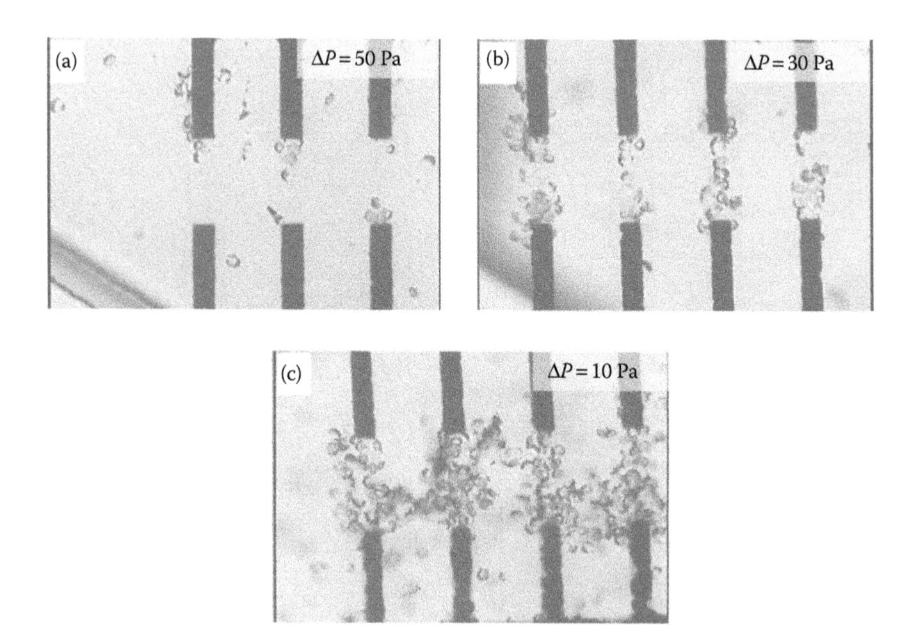

FIGURE 9.46 Formation of pearl chains as a function of the pressure differential (ΔP) across the microdevice. (a–c) Pearl chain formation at $\Delta P = 50$, 30, and 10 Pa, respectively. (From Ju, J., Ko, J.-M., Cha, H.-C., Park, J.Y., Im, C.-H., and Lee, S.-H., *J. Micromech. Microeng.*, 19, 015004, 2009.)

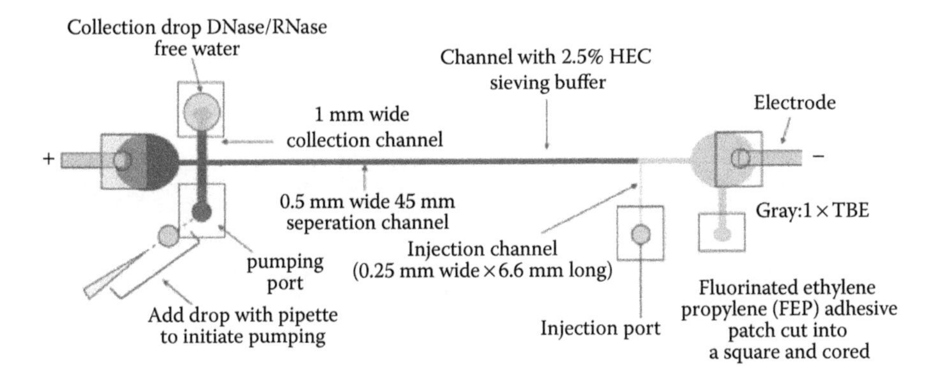

FIGURE 9.47 Device used for microcapillary electrophoresis. (From Mohanty, S.K., Warrick, J., Gorski, J., and Beebe, D.J., *Electrophoresis*, 30, 1470–1481, 2009.)

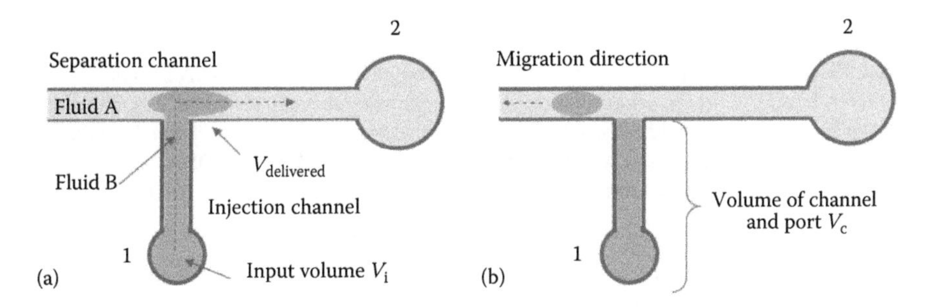

FIGURE 9.48 Sample loading into a microelectrophoresis device. (a) DNA is loaded into Port 1 and flows toward Port 2. (b) A electric field is applied driving the DNA down the separation channel. (From Mohanty, S.K., Warrick, J., Gorski, J., and Beebe, D.J., *Electrophoresis*, 30, 1470–1481, 2009.)

An electric field is applied to the sample that causes the sample to move down the separation channel (Figure 9.49). A band of sample DNA forms after some travel through the separation channel (Figure 9.49f). Passive pumping is again used to collect the sample DNA band (Figure 9.50). When the DNA band reaches the collection channel, a large drop is placed at collection Port 2 and a small drop at collection Port 1. Passive pumping is used to move the DNA band into collection Port 2. After flowing the sample DNA band into collection Port 2, a micropipette is used to remove the sample off the chip. The authors report a DNA capture efficiency of between 77% and 85% depending on the separation channel dimensions used.

Chen et al.[44] studied passive pumping flow by measuring changes in chronoamperometric current using an electrochemical detector (Figure 9.51). The authors measured flow rate as a function of viscosity, as a function of the pressure difference between the inlet and the outlet generated by surface tension, and as a function of the gravity exerted by the height difference between the inlet and the outlet. Results show that flow rate increases as viscosity decreases, flow rate increases as the

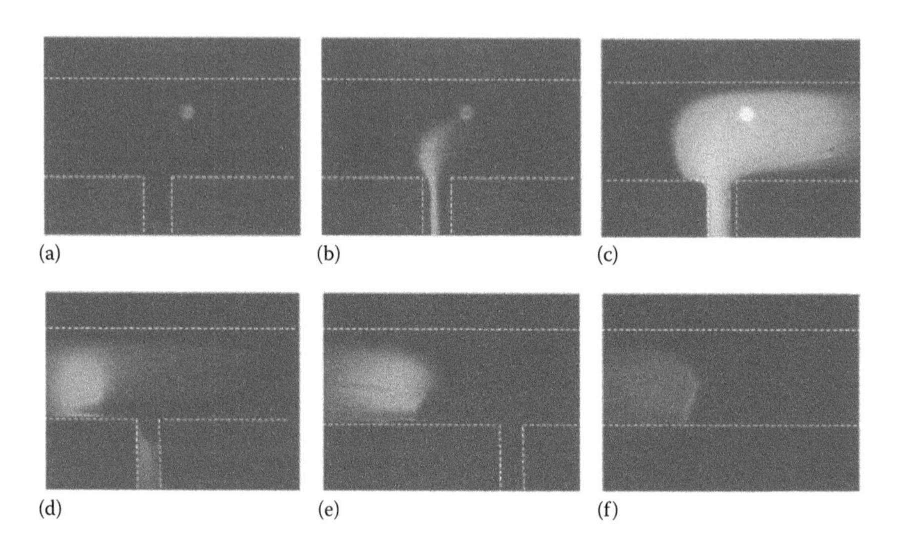

FIGURE 9.49 (a) Empty T junction, with the image taken before the DNA sample arrived. (b) DNA begins to fill the separation chamber. (c) Sample moves toward Port 2 (to the *right*). (d) An applied electric field causes the DNA to move down the separation channel. (e) Sample as it stacks while moving down the separation channel. (f) Band of DNA forms. (From Mohanty, S.K., Warrick, J., Gorski, J., and Beebe, D.J., *Electrophoresis*, 30, 1470–1481, 2009.)

pressure gradient increases, and flow rate increases as the pressure generated by the height difference between the inlet and the outlet increases.

To the best of our knowledge, this paper by Chen et al.[44] has been the only one to mention electrical measurement of the change in current as a function of the liquid exchange between two solutions. The authors exchanged the background electrolyte for a solution of 2 μL of hexacyanoferrate(II) solution at different concentrations (Figure 9.52). Results show a linear relationship between current measurements and the addition of the hexacyanoferrate(II) solution. Hence, a electrochemical sensor inside a passive pumping device can be used to detect the real-time fluidic exchange between a resting liquid and an incoming liquid.

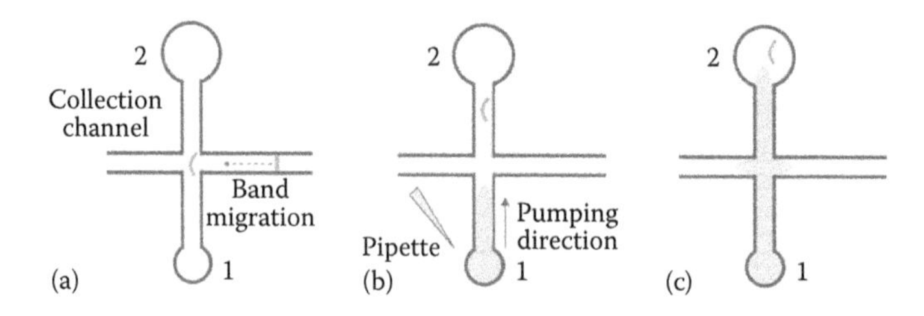

FIGURE 9.50 Sample collection. (a) DNA band migrating toward the collection channel. (b) Passive pumping used to move the DNA toward collection Port 2. (c) DNA band ready for collection. (From Mohanty, S.K., Warrick, J., Gorski, J., and Beebe, D.J., *Electrophoresis*, 30, 1470–1481, 2009.)

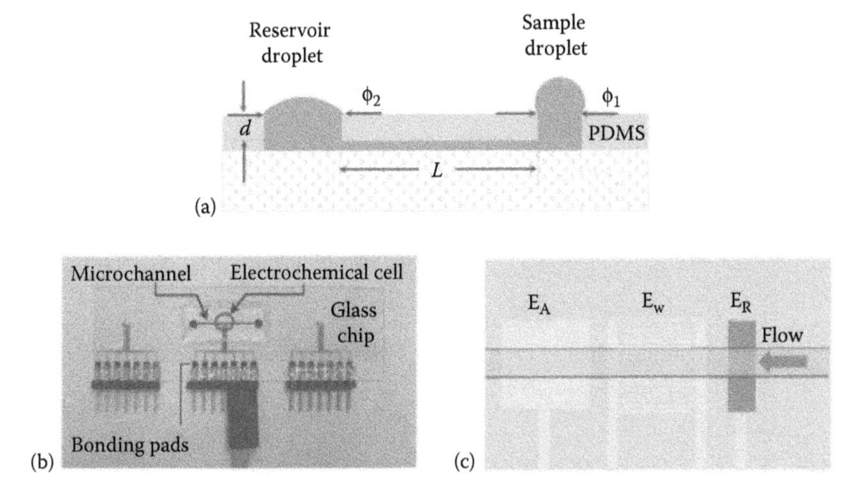

FIGURE 9.51 Electrochemical detector inside the passive pumping device. (From Chen, I. and Lindner, E., *Anal. Chem.*, 81, 9955–9960, 2009.)

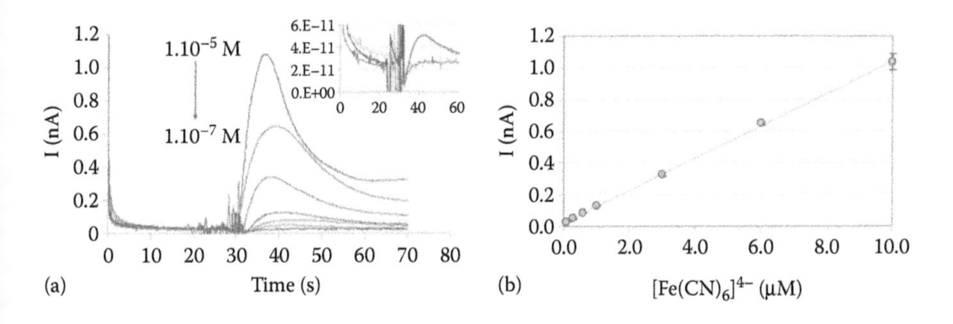

FIGURE 9.52 (a) Current–time measurements recorded with the passive-pump-driven device following the introduction of 2 μL drops of hexacyanoferrate(II) at different concentrations. (b) Correlation between the peak currents and the injected sample concentration. (From Chen, I. and Lindner, E., *Anal. Chem.*, 81, 9955–9960, 2009.)

Another unique application of passive pumping was the conversion of a passive pumping device into a Coulter counter. A Coulter counter is an apparatus for counting and sizing particles. A microfluidic Coulter counter is appealing because it would bring the benefits of microfluidics to an already popular technique. These benefits include reduced sample volume, simplified operation, high throughput, portability, and low unit cost. Many biomedical applications would benefit from such a microfluidic particle counter. In their work, McPherson et al.[45] developed such a counter, using passive pumping as the actuation mechanism for a Coulter counter device (Figure 9.53). They used hydrodynamic focusing to align the particles and maximize counting efficiency. Their results show that a microfluidic particle counter based on passive pumping is as effective as one based on a syringe pump.

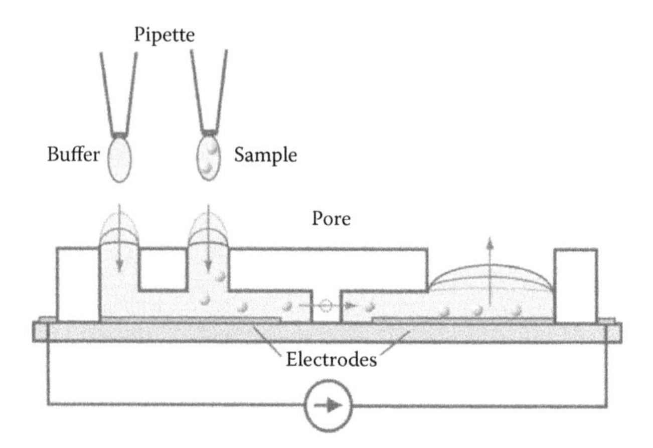

FIGURE 9.53 Passive pumping particle counter. (From Mcpherson, A. and Walker, G., *Microfluid. Nanofluid.*, 9, 897–904, 2010.)

9.11 OTHER TYPES OF TUBELESS DEVICES

The Wheeler laboratory at the University of Toronto developed a different type of tubeless device. Their design used a digital microfluidic setup to manipulate nanoliter-sized droplets on an open surface that contains an array of electrodes[46] (Figure 9.54). The authors used this system to carry out complete mammalian cell culture and obtain dose–response toxicity data for cells suspended in 150 nL droplets. Cells are seeded and grown on the digital microfluidic device by using an ECM protein coating on adhesion pads where the cells grow. Medium is exchanged by driving a droplet of new medium across the adhesion pad, thereby exchanging a portion of the old medium for new medium. Cell passaging is done by driving a droplet of trypsin over the cells, detaching the cells from the adhesion pads, and moving the suspended cells to a different adhesion pad on the electrode surface. This type of system allows for automated *in vitro* cell culture and assaying.

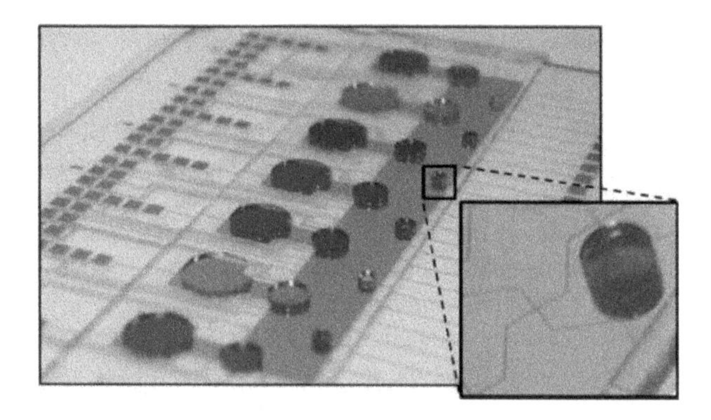

FIGURE 9.54 Digital microfluidic system for automated mammalian cell culture. (From Barbulovic-Nad, I., Au, S., and Wheeler, A.R., *Lab Chip*, 10, 1536–1542, 2010.)

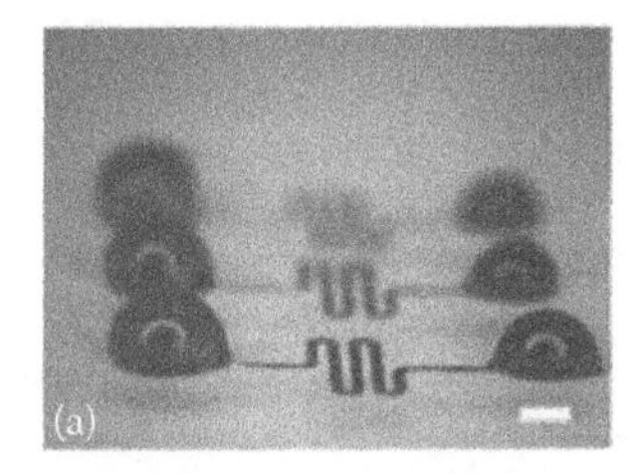

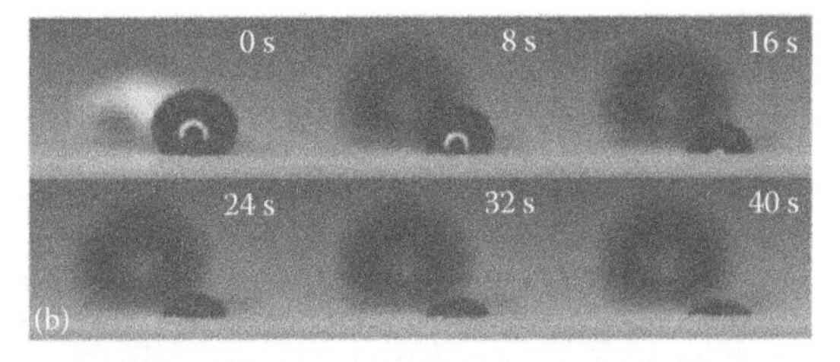

FIGURE 9.55 Microfluidic channels made by SU-8 superhydrophobic nanocomposite. (From Hong, L. and Pan, T., *Microfluid. Nanofluid.*, 10, 991–997, 2011.)

Other laboratories have developed surface microfluidic devices by using hydrophobicity to contain liquid in a channel. Hancock et al.[20] created a surface microfluidic stripe held in place on a glass slide by a hydrophobic boundary (Figure 9.20). They used this type of surface tension flow to generate centimeters-long gradients of molecules and particles (Figure 9.20).

Pan et al. developed another way of making surface microfluidic devices. In their works, the authors created microfluidic channels by containing liquid in hydrophilic channels surrounded by a superhydrophobic substrate. In one case, the superhydrophobic substrate was created by coating SU-8 with a photopatternable superhydrophobic nanocomposite (PSN)[47,48] (Figure 9.55). The PSN was composed of inert polytetrafluoroethylene (PTFE) nanoparticles in the SU-8 matrix. In another case, superhydrophobic surface layers were formed by creating high-density nanofibrous structures on a PDMS matrix using a CO_2 laser. The combination of nanostructures and the molecular properties of the laser-treated PDMS surfaces results in surface superhydrophobicity.[49] Among other proposed uses, the authors showed the usability of this technology by performing surface tension passive pumping on a channel-less device.

The analogous flow rate equation based on inlet drop height is given by $(4\gamma H/H^2+r^2)-P_0=(3\mu l/wh^3 2)(H^2+r^2)(\mathrm{d}H/\mathrm{d}t)$. When a superhydrophobic surrounding is used, the Laplace pressure inside a droplet sitting in a hydrophilic circle is dictated by $\Delta P=2\gamma C=(4h\gamma/r^2+h^2)$, where h is the droplet height, r is the radius of the circular boundary, and γ is the surface tension.

The droplet with higher internal pressure will flow down the pressure gradient until the pressure drop across the channel is leveled. Unlike with surface tension passive pumping, the droplet with higher internal pressure using the current superhydrophobic scheme is not always coincident with a smaller liquid volume. Because of this, it is possible to create bidirectional flow by changing the volume of the droplets placed at different ends of the microchannels. Although not obvious, bidirectional flow is also possible with surface tension passive pumping, as described by Walker et al.,[4] by adding a larger droplet at the receiving end.

REFERENCES

1. D. J. Beebe, G. Mensing and G. Walker, Physics and applications of microfluidics in biology. *Annu. Rev. Biomed. Eng.*, 4, 261–286, 2002.
2. C. H. Ahn, J. Choi, G. Beaucage, J. H. Nevin, J. Lee, A. Puntambekar and J. Y. Lee, Disposible smart lab on a chip for point-of-care diagnostics. *Proc. IEEE*, 92(1), 154–173, 2004.

3. D. Laser and J. Santiago, A review of micropumps. *J. Micromech. Microeng.*, 14, R35–R64, 2004.
4. G. Walker and D. J. Beebe, A passive pumping method for microfluidic devices. *Lab Chip*, 2, 131–134, 2002.
5. E. Berthier and D. J. Beebe, Flow rate analysis of a surface tension driven passive micropump. *Lab Chip*, 7, 1475–1478, 2007.
6. I. Chen, E. Eckstein and E. Lindner, Computation of transient flow rates in passive pumping micro-fluidic systems. *Lab Chip*, 9, 107–114, 2009.
7. G. M. Walker and D. J. Beebe, An evaporation-based microfluidic sample concentration method. *Lab Chip*, 2, 57–61, 2002.
8. E. Berthier, J. Warrick, H. Yu and D. J. Beebe, Managing evaporation for more robust microscale assays, Part 1: Volume loss in high throughput assays. *Lab Chip*, 8, 852–859, 2008.
9. E. Berthier, J. Warrick, H. Yu and D. J. Beebe, Managing evaporation for more robust microscale assays. Part 2: Characterization of convection and diffusion for cell biology. *Lab Chip*, 8, 860–864, 2008.
10. Y. Zhu, J. W. Warrick, K. Haubert, D. J. Beebe and J. Yin, Infection on a chip: A microscale platform for simple and sensitive cell-based virus assays. *Biomed. Microdevices*, 11(3), 565–570, 2009.
11. N. S. Lynn, C. Henry and D. Dandy, Evaporation from microreservoirs. *Lab Chip*, 9, 1780–1788, 2009.
12. N. S. Lynn and D. Dandy, Passive microfluidic pumping using coupled capillary/evaporation effects. *Lab Chip*, 9, 3422–3429, 2009.
13. Y. Du, J. Shim, M. Vidula, M. Hancock, E. Lo, B. Chung, J. Borenstein, M. Khabiry, D. Cropek and A. Khademhosseini, Rapid generation of spatially and temporally controllable long range concentration gradients in a microfluidic device. *Lab Chip*, 9, 761–767, 2009.
14. J. Warrick, B. Casavant, M. Frisk and D. J. Beebe, A microfluidic cell concentrator. *Anal. Chem.*, 82, 8320–8326, 2010.
15. B. Chung and J. Choo, Microfluidic gradient platforms for controlling cellular behavior. *Electrophoresis*, 31, 3014–3027, 2010.
16. D. Kim, M. Lokuta, A. Huttenlocher and D. J. Beebe, Selective and tunable gradient device for cell culture and chemotaxis study. *Lab Chip*, 9(12), 1797–1800, 2009.
17. D. Kim and D. J. Beebe, Interfacial formation of porous membranes with poly(ethylene glycol) in a microfluidic environment. *J. Appl. Polym. Sci.*, 110, 1581–1589, 2008.
18. V. Abhyankar, M. Lokuta, A. Huttenlocher and D. J. Beebe, Characterization of a membrane-based gradient generator for use in cell-signaling studies. *Lab Chip*, 6, 389–393, 2006.
19. A. Seidi, H. Kaji, N. Annabi, S. Ostrovidov, M. Ramalingam and A. Khademhosseini, A microfluidic-based neurotoxin concentration gradient for the generation of an *in vitro* model of Parkinson's disease. *Biomicrofluidics*, 5, 022214, 2011.
20. M. Hancock, J. He, J. Mano and A. Khademhosseini, Surface-tension-driven gradient generation in a fluid stripe for bench-top and microwell applications. *Small*, 7(7), 892–901, 2011.
21. J. Ju, J. Y. Park, K. C. Kim, H. Kim, E. Berthier, D. J. Beebe and S. Lee, Backward flow in a surface tension driven micropump. *J. Micromech. Microeng.*, 18, 087002, 2008.
22. J. Ju, J. Park, E. Berthier, D. J. Beebe and S. Lee, Mathematical and experimental study on backward flow in a surface tension driven micropump. In *Proceedings of the 12th International Conference on Miniaturized Systems for Chemistry and Life Sciences*, 12–16 October, San Diego, CA, 2008.
23. P. J. Resto, B. J. Mogen, E. Berthier and J. C. Williams, An automated microdroplet passive pumping platform for high-speed and packeted microfluidic flow applications. *Lab Chip*, 10, 23–26, 2010.

24. J. Warrick, I. Meyvantsson, J. Ju and D. J. Beebe, High-throughput microfluidics: Improved sample treatment and washing over standard wells. *Lab Chip*, 7, 316–321, 2007.
25. I. Meyvantsson, J. Warrick, S. Hayes, A. Skoien and D. J. Beebe, Automated cell culture in high density tubeless microfluidic device arrays. *Lab Chip*, 8, 717–724, 2008.
26. J. P. Puccinelli, X. Su and D. J. Beebe, Automated high-throughput microchannel assays for cell biology: Operational optimization and characterization *J. Assoc. Lab. Autom.*, 15, 25–32, 2010.
27. A. L. Paguirigan, J. P. Puccinelli, X. Su and D. J. Beebe, Expanding the available assays: Adapting and validating in-cell Westerns in microfluidic devices for cell-based biology. *Assay Drug Dev. Techn.*, 8(5), 591–601, 2010.
28. J. Ju, J. Warrick and D. J. Beebe, A cell programmable assay (CPA) chip. *Lab Chip*, 10, 2071–2076, 2010.
29. X. Su, E. Young, H. Underkofler, T. Kamp, C. January and D. J. Beebe, Microfluidic cell culture and its application in high-throughput drug screening: Cardiotoxicity assay for hERG channels. *J. Biomol. Screen.*, 16, 101–111, 2011.
30. E. Berthier, J. Warrick, B. Casavant and D. J. Beebe, Pipette-friendly laminar flow patterning for cell-based assays. *Lab Chip*, 11, 2060–2065, 2011.
31. B. Kim, S. Kim, S. Yang and H. Yang, Passive washing using inlet-pressure difference and a washing valve. *J. Micromech. Microeng.*, 17, N22–N29, 2007.
32. M. Toepke, V. Abhyankar and D. J. Beebe, Microfluidic logic gates and timers. *Lab Chip*, 7, 1449–1453, 2007.
33. M. Domenech, H. Yu, J. Warrick, N. Badders, I. Meyvantsson, C. Alexander and D. J. Beebe, Cellular observations enabled by microculture: Paracrine signaling and population demographics. *Integr. Biol.*, 1, 267–274, 2009.
34. K. Sung, N. Yang, C. Pehlke, P. Keely, K. Eliceiri, A. Friedl and D. J. Beebe, Transition to invasion in breast cancer: A microfluidic in vitro model enables examination of spatial and temporal effects. *Integr. Biol.*, 3, 439–450, 2011.
35. R. Khnouf, D. J. Beebe and Z. H. Fan, Protein expression in array devices with passive pumping. In *Proceedings of the 12th International Conference on Miniaturized Systems for Chemistry and Life Sciences*, pp. 1740–1742, 12–16 October, San Diego, CA, 2008.
36. R. Khnouf, D. J. Beebe and Z. H. Fan, Cell-free protein expression in a microchannel array with passive pumping. *Lab Chip*, 9, 56–61, 2009.
37. J. Koepsel and W. Murphy, Patterning discrete stem cell culture environments via localized SAM replacement. *Langmuir*, 25(21), 12825–12834, 2009.
38. V. Echeverria, I. Meyvantsson, A. Skoien, T. Worzella, C. Lamers and S. Hayes, An automated high-content assay for tumor cell migration through 3-dimensional matrices. *J. Biomol. Screen.*, 15(9), 1144–1151, 2010.
39. J. Mohammed, H. Caicedo, C. Fall and D. T. Eddington, Microfluidic add-on for standard electrophysiology chambers. *Lab Chip*, 8, 1048–1055, 2008.
40. H. Caicedo, M. Hernandez, C. Fall and D. T. Eddington, Multiphysics simulation of a microfluidic perfusion chamber for brain slice physiology. *Biomed. Microdevices*, 12, 761–767, 2010.
41. E. W. K. Young, E. Berthier, D. J. Guckenberger, E. Sackmann, C. Lamers, I. Meyvantsson, A. Huttenlocher and D. J. Beebe, Rapid prototyping of arrayed microfluidic systems in polystyrene for cell-based assays. *Anal. Chem.*, 83, 1408–1417, 2011.
42. J. Ju, J.-M. Ko, H.-C. Cha, J. Y. Park, C.-H. Im and S.-H. Lee, An electrofusion chip with a cell delivery system driven by surface tension. *J. Micromech. Microeng.*, 19, 015004, 2009.
43. S. K. Mohanty, J. Warrick, J. Gorski and D. J. Beebe, An accessible micro-capillary electrophoresis device using surface- tension-driven flow. *Electrophoresis*, 30, 1470–1481, 2009.

44. I. Chen and E. Lindner, Lab-on-chip FIA system without an external pump and valves and integrated with an in line electrochemical detector. *Anal. Chem.*, 81(24), 9955–9960, 2009.
45. A. Mcpherson and G. Walker, A microfluidic passive pumping Coulter counter. *Microfluid. Nanofluid.*, 9, 897–904, 2010.
46. I. Barbulovic-Nad, S. Au and A. R. Wheeler, A microfluidic platform for complete mammalian cell culture. *Lab Chip*, 10, 1536–1542, 2010.
47. L. Hong and T. Pan, Surface microfluidics fabricated by superhydrophobic nanocomposite photoresist. In *Proceedings of the IEEE 23rd International Conference on Micro Electro Mechanical Systems MEMS*, pp. 420–423, 24–28 January, Wanchai, Hong Kong, 2010.
48. L. Hong and T. Pan, Surface microfluidics fabricated by photopatternable superhydrophobic nanocomposite. *Microfluid. Nanofluid.*, 10, 991–997, 2011.
49. S. Xing, R. S. Harake and T. Pan, Droplet-driven transports on superhydrophobic-patterned surface microfluidics. *Lab Chip*, 11, 3642–3648, 2011.

Index

A

Active microfluidic droplet generator (μDG), 173
Active pumps, 222
A/D converter, 145–146
Adherent cells, physical properties of, 126
Affinity-based electrochemical DNA sensors, 24
  label-based, 35
  label-free, 35
  performance measures, 24
Affordable, sensitive, specific, user-friendly, rapid and robust, equipment-free, and deliverable to users (ASSURED), 156
AFM, *see* Atomic force microscopy (AFM)
Agarose droplet-based SCGA technology, 188–189, 197
  agarose droplet generation, 190
  cell culture and preparation, 189–190
  emulsion PCR, 190–191
  fluorescence imaging, 191
  for genetic detection and multilocus sequencing, of single mammalian cells, 188–189
    genome purification, 191–193
    single-cell encapsulation, 191–192
  method, defined, 188
  multiplexed single-cell droplet PCR, 193
    high-throughput digital genetic analysis of cancer cells, 193–194
  single-cell sequencing, multiplex, 191, 195–196
  single-genome purification, 190
Agarose droplet encapsulation, 196
ALH, *see* Automated liquid handler (ALH)
Alkanethiol self-assembled monolayer (SAM), 34
Allan deviation, 146–147
Aluminum (Al) electrode, 36
3-Aminopropyl triethoxysilane (APTES), 134–135
Apertures, 105–106
  increasing range of, effect of, 107–109
APTES, *see* 3-aminopropyl triethoxysilane (APTES)
ASSURED, *see* Affordable, sensitive, specific, user-friendly, rapid and robust, equipment-free, and deliverable to users (ASSURED)
Atomic force microscopy (AFM), 88–89, 122
Attogram resolution, 115
Au electrode, 27, 32, 35
Au thin-film electrode, 36
Automated liquid handler (ALH), 240–241

B

Backfilling, 34
Bacterial pathogen, single-cell analysis of, 180
Bayesian BSS methods, 9–10
Bayesian minimum mean square error (MMSE) estimator, 10
Bayesian source separation method for mixtures, 14–15
Bayes' rule, 9
BCL-2/IgH translocation t(14;18), in blood cancers, 188
Biological analyses, 172–173; *see also* Droplet-based microfluidics, for bioanalysis
  constraint encountered, 201
  complicated active operations, 173
  sample selection and preparation, 202
Biomedical research, 21
Biotin–streptavidin chemistry, 37
Blind source separation (BSS), 4, 8
  application, 11–15
  Bayesian BSS methods, 9–10
  ICA methods, 8–9
  nonlinear BSS models, 10
  problem description, 7–8
  scale ambiguity and, 15–16
  strategies to perform, 8
Brain slice perfusion chamber, 250, 252
BSS, *see* Blind source separation (BSS)

C

Cantilever resonator, 113
Celera Genomics, 21
Cell-based biological assays, 241, 243
Cell fusion process, 255
Cell mass, 103
Cell programmable assay (CPA) chips, 244
Cells, measuring physical properties of, 102–103
  adherent, 126
  deformability
    decoupling, 123–124
    single-cell, 122–123
    viscoelastic models, 120–122
  density
    average, 117
    single-cell, 118–120

future opportunities for, 126–128
growth, 115–117
exponential model, 116–117
single cell, 116
hydrodynamic stretching, 124–126
mass, 112–117
single cell, 115
resonance frequency, dependence of, 113
size, 104
decoupling, 123–124
sorting *vs.* measuring, 104–105, 126–127
volume, 103, 105–112
mass-based measurement, 120
Cellular electrofusion
3D electrofusion chip, schematic of, 255
passive pumping, as fluid actuation mechanism, 254
advantages, 255–256
in electrophoresis device, with MEMS, 256–257
pearl chains formation, between electrodes, 256
process, 255–256
Cellular heterogeneity
role in biological functions, 172
in tumors, 196
Centrifugal pumps, 222
Charge-coupled device (CCD) camera, 21
Chemical heating, use of, 166
Chemical modification, of DNA targets, 35
Chemotaxis
defined, 232
device, 232–233
Circulating tumor cells (CTCs), 172
Clustering analysis, 7
CMOS biochip, 38–65
ADC circuit components
bandwidth and stability, 48–49
current sources and switches, 50–51
digital counter and control circuitry, 51
integration amplifier, 46
integration capacitor, 49
thermal and flicker noise, 47–48
track-and-latch comparator, 49–50
two-stage operational amplifier (op amp), 46–47
chip packaging and experimental setup, 66
control amplifier circuits, 51–53
CV measurements of bulk redox species, 68–71
electrical characterization, 66–68
electrode–electrolyte interfaces, 45–46
fabrication and postprocessing, 61–65
Al etching, 62–63
improved, 63, 65
initial electrode, 62–63
integrated WE and CE, destruction of, 64
Ti/Au layers, 63–64
sensor detection limit and noise analysis, 53–61
ADC integration effect, on detection limit, 59–60
ADC quantization noise power, 57
amplifier thermal noise, 56–57
average output noise power, at end of integration period, 55–56
flicker noise power at integrator output, 57
as function of WE area, 59
input-referred flicker noise of integration amplifier, 57
integrator output voltage, 54–55
noise voltage at integrator output, 56
open-loop frequency response, 56
output-referred noise due to control amplifier, 56
quantization-noise-free detection limit, 58–59
quantization-noise-free SNR, 58
sensor dynamic range, 60–61
shot noise from redox current, 55
SNR of integrator output, 54
square voltage (power) at integrator output, 54
target coverage detection limit, 57–59
thermal noise current with PSD, 56
system architecture
dual-slope ADC, 42–45, 54
potentiostat control amplifiers, 45
working electrode array, 45
system requirements
bandwidth, 41
input currents, 40–41
silicon area of sensor interface electronics, 41–42
working electrodes (WE), 41
CMOS electrochemical DNA sensors, 36–38
capacitance-based and label-free, 37–38
CMOS microarray, for DNA polymerization detection, 38
CombiMatrix DNA microarray platform, 36–37
construction of portable, 36
field-effect transistor–based DNA sensing, 36
immobilization of DNA probes at electrode sites, 36
Infineon CMOS microarrays, 37
Toshiba CMOS microarrays, 37
CMOS-integrated ion-sensitive FET (ISFET) arrays, 36
Coculture systems, 253
CombiMatrix DNA microarray platform, 36–37
"Complementary DNA" (cDNA) arrays, 25

Complementary metal-oxide-semiconductor (CMOS)
biochip (*see* CMOS biochip)
CMOS electrochemical DNA sensors (*see* CMOS electrochemical DNA sensors)
Complex full-back-end CMOS process, 37
Concentration gradients; *see also* Tubeless microfluidic devices
fluorescent experiments, 233–234
goal of high-throughput screening, 233–234
greater toxin concentrations, 234
importance, 232
microfluidic-based neurotoxin gradient, 233–235
with 6-OHDA gradients233, 234
surface tension phenomena, to create flow and, 234
viability of HL-1 cells, 231–232
Constriction-based microfluidic chip, 123
Continuous-time current-to-voltage converter, 37
COP microchannels, *see* Cyclo-olefin polymer (COP) microchannels
Coulter counter, 105–106, 259–260
CPA chips, *see* Cell programmable assay (CPA) chips
Cyclic voltammetry (CV), 32–34
Cyclo-olefin polymer (COP) microchannels, 245
Cystic fibrosis, 35

D

Damping coefficient, 120
Darcy's law, 160
Decoupling
of cell deformability and size, 123–124
of physical properties, 127–128
Deformability of cells
decoupling, 123–124
by hydrodynamic stretching, 124–126, 125
single-cell, 122–123
viscoelastic models, 120–122
Deformation of fibroblast, 122
Deoxynucleotide triphosphates (dNTPs), 38
Dideoxy chain termination techniques, 21
Digital microfluidic gates, 247
DNA microarrays, 21–22
DNA sequencing technology, 21
Dominant forces, acting at microscale, 221
Driver mutations, leading to carcinogenesis, 172
Droplet-based digital microfluidics, 172, 197
advantages, 196
droplet generation, by flow focusing and cross-flow shearing, 173
microscale multiphase flow dynamics, 172
overcoming limitations, 173
Droplet-based microfluidics, for bioanalysis, 201
application in chemical and biological research, 202
attractive platform, for small-volume, 201
challenges addressed, 201
limiting reagent dilution caused by diffusion, 201
minimizing cross-contamination, 202
surface-related adsorptive losses, 202
Taylor dispersion, 201
detection strategies, employed *in situ*
electrochemical detection, and advantages, 208–209
in-droplet fluorescence detection, for analyzing droplets contents, 208
in-droplet Raman spectroscopy, 208
microcoil NMR probe, for high-throughput analysis, 209
nuclear magnetic resonance (NMR), 209
surface-enhanced Raman spectroscopy (SERS), 208
droplet contents detection, limited to optical methods, 202
droplet extraction, and content analysis
contents extraction, into channel for CE separation, 209
corona treatment to hydrophilize PDMS chip portion, 210
droplet-based PDMS micro-fluidic assembly, 210–211
electricity-based method to control, 210
matrix-assisted laser desorption ionization mass spectrometry (MALDI-MS), 209
surface modification method, for stable interface, 209–210
using MS, 209
droplet fusion, reagents combination and mixing, 205
active fusion methods, 206
flowing reagent in microchannel, two laminar streams, 205–206
in-channel droplet fusion, passive and active methods, 206
method employing two pneumatic valves, integrated at double-T intersection, 207
pillar-induced droplet merging device, 206
surface-induced droplet fusion method, 206
using picoinjector to add reagents to droplets, 206
valve-based droplet generation, 206–207
droplet incubation, method for, 208
droplets generation, approaches, 203
droplet volume and generation frequency, 203

on-demand droplet generation, 203–205
promising tool, understanding biological processes, 215
proteomic analysis, enhanced LC/MS-based (*see* Proteomic analysis, enhanced LC/MS-based)
provide reliable quantitative analysis, 202
single-cell chemical analysis, 212, 215
droplet-based platforms
combining with, LC, CE, and MS, 202
to interrogate, complex biological systems, 173
"Dropspots" device, 208
D-shaped fiber-based biosensor, 87–90
cladding region in, 92–93
fabrication, 91–92
optical microscope photographs, of light scattered from, 88, 90
photoresist on the surface, 92–93
sensitivities, 90
SMF-based, 92–94
high extinction ratio for the r-leaky mode, 96
high-order leaky mode, 96
PLL layer, 96–98
surface LPG, 95–96
temperature sensitivity and resonant wavelength shift, 95
wavelength shift of r-mode in, 97–98
TE and TM modes, 94–95
transmission characteristics, 90–91, 95
Dual-slope ADC, 42–45, 54
Dynal beads, 143
Dynamic micropumps, 222

E

Earth's magnetic field, 139
EDC, *see* 1-ethyl-3-(3-dimethylaminopropyl) carbodiimide (EDC)
Electrical characterization, of active CMOS biochip, 68–71; *see also* CMOS biochip
differential nonlinearity (DNL) values for ADCs, 67
of dual-slope ADC, 66–67
integral nonlinearity (INL) values for ADCs, 67
Randles–Sevcik equation, 69
SNR and signal-to-noise-and-distortion ratio (SNDR) of the ADC, 67
WE surface, cleaning of, 69
Electrochemical detection, to collect droplets information, 208–209
Electrochemical DNA sensing, 22, 27–29
affinity-based electrochemical DNA sensors, 24
performance measures, 24
CMOS platform, 23
real-time DNA detection, 28–29
review, 35
sensing methodology, 27–28
Electrochemistry, principles of
cyclic voltammetry (CV), 32–34
electrochemical reactions at equilibrium, 30–31
electrode–electrolyte interface, 29–30
electrode materials, 32
potentiostat operation, 31–32
Electroosmotic flow, 110
Electroosmotic pumps, 222
Emulsion polymerase chain reaction (ePCR), 172
genetic targets, counting, 172
for high-throughput single-molecule amplification, 172
limitations, 172–173
Enzyme DNA polymerase, 38
Enzyme-linked immunosorbent assay (ELISA), 131–132
advantages, 133–134
vs. immunomagnetic assay, 132–134
limitations, 132–133
steps, 132–133
ePCR, *see* Emulsion polymerase chain reaction (ePCR)
eSensor DNA chip, 35
1-Ethyl-3-(3-dimethylaminopropyl)carbodiimide (EDC), 135
Eukaryotic cells, 102

F

FACS, *see* Fluorescence-activated cell sorter (FACS)
Faradaic processes, 55
Ferrocene, 35
Fiber grating–based biosensors, 87, 98
Fiber-optic biosensors, 87
Fibroblast, full model of, 121
Field-effect transistor–based DNA sensing, 36
Flex board method, 136
Flow-enhanced infection assays
in microchannels using passive pumping, 248–249
assay sensitivity, based on drug concentrations, 249, 251
backward spread of infection, 249
decreased cell death, when treated with increased antiviral drug, 249, 251
under lack of flow and flow conditions, viral infection spread, 249–251
Fluorescence-activated cell sorter (FACS), 106, 127

Fluorescence-based DNA microarrays, 22, 25–27
  applications, 25
  probe and target sequence, 26
    hybridization process equation, 26–27
    probe–target surface binding, equation of, 26
Fluorescence detection, 208
Fluorescent dyes, 26
Fluorophore-labeled ssDNA "target" molecules, 21
Fractional occupancy, 26

G

Genetic biomarkers, 172
Genome sequencing, 21
Genomic assays, 21
Giant magnetoresistance (GMR) sensors, 140, 146
GMR-based detection system, *see* Giant magnetoresistance (GMR) sensors
Gold-standard diagnostic assays, 155
Gold-standard plaque-reduction assay, 249
Gradient-based learning rule, 12

H

Hall sensor, 135–136, 141–142, 146
  crucial challenges, 141
  magnetization-based Hall effect detection, 143–144
Hemochromatosis, 35
He-Ne laser, 88
HGP, *see* Human Genome Project (HGP)
Hoechst oxidation, 37
Horseradish peroxidase (HRP), 37
Human Genome Project (HGP), 21
Human papillomavirus, 35
Hybridization process
  backfilling and, 34–35
  CMOS electrochemical DNA detection system for, 38
  multiplexed and specific DNA sensing, 78–80
  in real time, 28–29, 37
    DNA probe–target hybridization, 76–77
  of surface-bound probe, 26
Hydrodynamic stretching of cells, 124–126
Hydrodynamic stretching technique, 126

I

ICA, *see* Independent component analysis (ICA)
IMA, *see* Magnetic immunoassays (IMA)
Immunomagnetic detector chip, in microfluidic system, 135–136
In-Cell Western (ICW) assays
  array of passive pumping devices and ALH, 241–242
  platform validation, for complete cell-based biological assays, 241, 243
  validation using passive pumping, 241, 244
Independent component analysis (ICA), 8–9
  electrode's slope, 13
  and valences, 11–13
In-droplet Raman spectroscopy, 208
Infineon CMOS microarrays, 37
Integrated electrochemical measurement, of DNA probe surface coverage
  biochip preparation and experimental procedure, 71–72
  measurement results, 72
Integrated MOSFETs, 36
International Human Genome Sequencing Consortium, 21
In vitro microfluidic model, to study transition of
  ductal carcinoma *in situ* (DCIS), to invasive ductal carcinoma (IDC), 253–255
    cell–cell contact with HMFs, 254
    interaction of soluble factors, at beginning of transition, 254
  Y-channel passive pumping device, 253
Ion-selective electrode (ISE), 4–5
  ISE array (ISEA) dataset, 14, 16
Ion-sensitive field-effect transistor (ISFET), 5, 11

K

Kelvin body, 121–122

L

Label-based electrochemical DNA sensors, 35
  basic DNA hybridization detection, 74–75
  chip surface preparation, 74
  design of an active CMOS biochip for, 39–40
  DNA oligonucleotide sequences, 73
  DNA target concentration series, 75–76
  genetic mutation detection, 78
  multiplexed and specific DNA sensing, 78–80
  real-time, 76–77
  surface heating effects, 80
Label-free DNA sensors, 35
  for detecting changes in capacitance, 37–38
Lab-on-a-chip platforms
  advantages, 201, 221
  for biomarker screening, 135–137
Laminar flow patterning (LFP) technique, 245
Large label effects, 136–138
  advantages, 137
  assay accuracy and, 138
  assays utilizing microbeads, 137
  noise problem, 137
Linear-quadratic models, 12

Linear-sweep voltammetry, 37
Long-period gratings (LPGs)-based biosensors, 87
Low-cost Coulter counters, 109–111
Lucas–Washburn prediction, 159
Lucas–Washburn relation, 158

M

Magnetic bead microrheometry, 122
Magnetic field at the sensor, 138–139
Magnetic immunoassays (IMA)
  assay protocols, microfluidic and packaging technologies for, 134–138
    gate microfluidic control system, 135
    Hall sensor, 135–136
    immobilizing capture antibodies, 134–135
    large label effects, 136–138
    microfluidic integration and assay automation, 135–136
  integrated detection for
    Hall effect detection platform, 141–142
    optical detection of magnetic labels, 141
    state-of-the-art methods and limitations, 138–142
  sensitive detection platform, construction of, 142–143
Magnetic relaxation, detectors based on, 144–147
  Allan deviation measurements, 146–147
Magnetization-based Hall effect detection, 143–144
Magnetohydrodynamic pumps, 222
Mammalian cells; *see also* Agarose droplet-based SCGA technology
  agarose droplet-based SCGA technology for genetic detection and multilocus sequencing in single, 188
  digital microfluidic system, for automated culture in, 260
  droplet-based genetic analyses, 187
  microfluidic agarose encapsulation, and SCGA of, 191–193
  single-cell multiplex gene detection, 187–188
    agarose droplet generation, 190
    cell culture and preparation, 189–190
    emulsion PCR, 190–191
    fluorescence imaging, 191
    method, defined, 188
    multiplexed single-cell droplet PCR, 193
    single-cell sequencing, multiplex, 187–188, 191, 195–196
    single-genome purification, 190
MAP, *see* Maximum *a posteriori* (MAP) estimation
Markov chain Monte Carlo methods, 14
Mass-based measurement of cell volume, 120
Maximum *a posteriori* (MAP) estimation, 10
Maxwell body, 121
MCH, *see* Mercaptohexanol (MCH)
MEMS, *see* Microelectromechanical systems (MEMS)
Mercaptohexanol (MCH), 34
Microarray-based tests, 22
  affinity-based electrochemical DNA sensors, 24
  commercial microarray systems, 27
  electrochemical DNA sensing, 22, 27–29
    CMOS platform, 23
  fluorescence-based DNA microarrays, 25–27
  instrumentation cost associated with, 22
  laboratory-based clinical applications, 22
Microcapillary electrophoresis, passive pumping use in
  device used, 256–257
  electrical measurement, of current change, 258–259
  electric field applied to sample, causes movement, 257–258
  electrochemical detector use, 257, 259
  flow rate measurement, as function
    of gravity exerted, 257
    of pressure difference, between inlet/outlet, 257
    of viscosity, 257
  sample collection process, 257–258
  sample loading, and extraction, 256–257
Microdroplets, 202
Microelectromechanical systems (MEMS), 102, 256
  resonators, 113
Microfluidic agarose encapsulation
  agarose matrix, protects genomic DNA from physical damage, 193
  of cells in droplets, roles served, 188–189
  enables reproducible single-cell DNA extraction and isolation, 191–192
  key benefit, 193
Microfluidic brain slice device (μBSD), 250, 251, 253
Microfluidic Coulter counters, 106–107
  design and operation of a flow-focusing, 108
  four-aperture, 109
  high-conductivity fluid and low-conductivity fluid, 108–109
  hydrodynamic focusing and, 107–109
  volume measurements to aperture geometry and cell shape/orientation, 112
Microfluidic devices
  allows fluid flow and diffusion profiles manipulation, 232
  analogous to electrical circuitry, 247
  components, analogous to electronic circuit components, 247
  control fluid flow, 222
  electrodes integration into, 254–257

and fluidic resistance, 247
hERG-related cardiotoxicity, microfluidic-based drug screening assays for, 245, 253
logic components, classes of, 247
passive, advantages, 222
polystyrene (PS), made with hot-embossing process, 254–255
protein expression, 248–249
tubeless (*see* Tubeless microfluidic devices)
usage, 245
to wash away multiple reagents, 245
Microfluidic droplet technology
droplet generation techniques
conventional, based on T-junction or flow focusing, 204
integration with multiplex PCR, 180
single-cell analysis, advantageous for, 187
Microfluidic emulsion generator array (MEGA) systems
assembly, 174
as core droplet-based digital microfluidic platform, 196–197
design, 174–175
for detection of extremely low frequency events, in vast population, 175, 176
droplet generation, 174–175
homemade manifold module used, 175–176
fabrication, 174
multilayer devices, consisting, 175
uniform droplets, generation of, 175–176
using 96-channel
assess encapsulation performance, 176
droplet generation, characterization of, 175–176
PCR amplification, 175–176
Microfluidic particle counter, 259–260
Microfluidics, 102
advantages, 201
passive pumping device
mathematical model characterizing, two phase inlet drop collapse, 223
study of interplay between convection and diffusion inside, 225
used for study of biological systems, 247–248
protein expression using nutrient solution in, 248–249
technologies
advantages, 221
goal, 221
usage, 172
Micropipette aspiration, 122
Microscale devices, 222
Microscale forces, 235
Multilayer perceptron (MLP) neural networks, 6
Multiplex single-cell PCR, 186
Multisizer 4, 106
Mutual information, 9

N

Nicolsky–Eisenman (NE) equation, 5–6
Nonevaporation backward-flow mechanism, 235
Nonlinear BSS models, 10
Nuclear magnetic resonance (NMR), 209
Nucleotide sequence, 25

O

On-chip sorters, 104
On-demand droplet generation, approaches
droplet volume, depends on parameters, 205
generation process, of individual droplet, 204
pneumatic valve-assisted, 203–204
valve integrated system for, 204

P

Paper-based diagnostic devices, 156–157
Paper microfluidics tools, 158–162
to control flow rates of fluids in paper networks, 158–161
for fluidic resistance, 160–161
off-switch for fluid flow, 161
on-switch for fluid flow, 161
Para-aminophenol (p-AP), 37
Passive pumping; *see also* Tubeless microfluidic devices
with ALH, capable of performing long-term cell culture, 240–241
devices, hERG-transfected HEK cells for growth and culturing properties investigation, 245, 253
gradient systems, 251
in high-throughput biological applications, 241
in vitro model for PD using, 233–235, 244–245, 251, 254
Passive pumping particle counter, 259–260
Pathogen detection, 180
PBS, *see* Phosphate buffer saline (PBS)
2PDN, *see* Two-dimensional paper networks (2PDN)
Phosphate buffer saline (PBS), 134–135
Photolithographic techniques, 25
Photomultiplier tube (PMT), 21
Photopatternable superhydrophobic nanocomposite (PSN), 261
"Physical fingerprinting" of cells, 128
Pinched-flow fractionation, 104–105
Plasmodium falciparum malaria-infected erythrocytes, 118, 122

Point-of-care (POC) genetic diagnostic applications, 22
Point-of-need/point-of-care diagnostic devices, 133
Polydimethylsiloxane (PDMS), 135
altering biochemical microenvironment, in cell cultures, 254
valve-based method, for droplet generation, 179
Poly-L-lysine (PLL), 96–98
Polystyrene (PS) microchannels, 245
Polytetrafluoroethylene (PTFE) nanoparticles, 261
Post-nonlinear (PNL) models, 10
Potentiometric sensors, 4–6
interference phenomenon in, 5–6
ion-selective electrode (ISE), 4–5
NE equation, 5–6
PrA, *see* Protein A (PrA)
Programmable droplet generation, 179–180; *see also* Droplet-based microfluidics, for bioanalysis
adjusting pumping period, control droplet generation, 179
effects of
flow conditions on, 176, 178
mechanical pump actuation, minor role, 177
oil flow rate, 176–179
pulsatile on-chip pumping, 176–178, 180
pumping conditions on, 178
piezoelectric actuators, and electrowetting for, 173
pneumatic valves fabricated by soft lithography used for, 173
Prokaryotic cells, 102
Protein A (PrA), 135
Proteomic analysis, enhanced LC/MS-based
"bottom-up" proteomics workflow, 211–212
integrated microfluidic platform, to combine myoglobin and pepsin, 212–214
peptide fragments identification, 215
sequence of apomyoglobin, 215
MS-based proteomics studies, 211
schematic, 213
"top-down" proteomics approach, 211–212
Pt microelectrode sites, 37

Q

Quartz block, 87

R

Reciprocating pumps, 222
Redox-label-based electrochemical DNA sensors, 35
Reynolds numbers, 104, 107–108
Root-mean-square roughness values, 88, 90
Rotary pumps, 222

S

Sanger sequencing, 21
Self-organizing maps (SOM), 7
Single-cell analysis, microfluidic droplet technology for, 187
Single-cell chemical analysis, by using droplet-based microfluidics, 212, 215
Single-cell/copy genetic analysis (SCGA) technology, 173, 180
Single-cell genetic analysis (SCGA), multiplex
cell culture, 180–181
detection limit, improvement, 187
displays tolerance to PCR inhibition, 186
dynamic range, for *E. coli* O157 detection, 185–186
feasibility, for large-scale genetic analysis of mammalian cells, 186–187
and genotyping with high speed with sensitivity, 197
higher sensitivity, required for zero-pathogen-tolerance policy, 186–187
to improve detection sensitivity, 183–184
limit of detection (LOD), for *E. coli* O157 detection, 184–186
nonspecific amplification, minimize, 187
PCR preparation, and beads, 181
bead recovery, 182
flow cytometry quantitation, 182
PCR reaction, steps, 181–182
perform analysis, at elevated cell density, 183–184
process involved, 180–181
single-cell pathogen detection, at dilute regime, 182–183
for ultrasensitive pathogen detection, 197
Single cells, measuring physical properties of, 102–103
deformability, 122–123
density, 118–120
growth, 116
mass, 115
Single-genome extraction, 188
Single micron-sized cells, 103
Single-mode fiber (SMF), 87–88; *see also* SMF-based D-shaped fiber-based biosensor
Single-nucleotide polymorphism (SNP) analysis, 22
Single-stranded DNA (ssDNA)
"probe" molecules, 21
target molecules, 26
Smart sensor arrays (SSAs), 3
based on supervised signal processing methods, 4

chemical sensor arrays, 6–7
limitations, 4
in potentiometric sensors, 4–6
signal processing techniques in, 6
supervised qualitative chemical analysis, 6–7
unsupervised qualitative chemical analysis, 6–7
SMF, *see* Single-mode fiber (SMF)
SMF-based D-shaped fiber-based biosensor, 92–94
high extinction ratio for the r-leaky mode, 96
high-order leaky mode, 96
PLL layer, 96–98
temperature sensitivity and resonant wavelength shift, 95
wavelength shift of r-mode in, 97–98
SMR, *see* Suspended microchannel resonator (SMR)
SOM, *see* Self-organizing maps (SOM)
Spring constant, 120
Stokes shift, 26
Stretched fibroblast, 121
Superhydrophobic substrate, 261
Support vector machines (SVM), 6
Surface-enhanced Raman spectroscopy (SERS), 208
Surface-hybridized DNA targets, 35
Surface microfluidic devices, 260–261
Surface plasmon resonance (SPR) phenomenon, 87
Surface superhydrophobicity, 261
Surface tension passive pumping; *see also* Tubeless microfluidic devices
actuation mechanism, 223
on channel-less device, 261
characterization of flow, small to large drop, 222–223
evaporation, an important factor in, 229
first described by, 223, 229
passive and tubeless, 222
use of PDMS microfluidic device, 224
2D surface topography, 88
Suspended microchannel resonator (SMR), 113–115
SVM, *see* Support vector machines (SVM)

T

Tetramethylbenzidine (TMB), 37
Thiol-metal chemisorption, 34
Time-dependent magnetization of a magnetic domain, 144–145
T-junction droplet generator, 204
TMB, *see* Tetramethylbenzidine (TMB)
Toshiba CMOS microarrays, 37
Total internal reflection (TIR) condition, 88
Transimpedance amplifier, field-effect (ISFET) sensor, 38
Treat-wait-treat method, 240
Tubeless microfluidic devices
advantage, 222, 239
backward-flow mechanism
evaporation-induced, 235
inertia-based, inertia–surface tension interactions, 237–239
microfluidic setup, 237, 239
rotational flow inside the outlet drop as source of inertia, 235, 238
concentration gradients
importance, 232
microfluidic-based neurotoxin, for generation of an *in vitro* model of PD, 233–235, 251
of molecules, controllable, 230
viability of HL-1 cells, after treatment with toxin, 231–232
evaporation
and capillary forces, used as actuation mechanism for fluid flow, 228–229
cause flow from large drop to small drop, 225
effect of reservoir liquid volume on, 227–228
electroosmotic flow, 222
evaporation-induced backward flow, 230–231, 235
evaporation-mediated sample concentration, 230
importance and issue, 224
interplay between convection and diffusion, 225–226
less invasive method, 225
methods to mitigate, invasive, 224–225
passive pumping–based assay, 225
rates in, lead to experimental error, 227
two-step method, to predict liquid evaporation rates, 226–227
used to create flow opposite, direction of surface tension forces, 235
volume loss associated with, 226
ways to mitigate convection–diffusion effects include, 226
as way to control flow, 229
flow rates and liquid velocities estimation, 222
fluorescence images, generation of dynamic gradient, 230–231
fluorescent experiments, 233–234
generation of surface-tension-driven gradient, in fluid stripe, 234, 237
goal of high-throughput screening, 233–234
ICW assays, 241–242
to validate platform, for complete cell-based biological assays, 241, 243

LFP technique, 245–246
microfluidic channels SU-8, made by superhydrophobic nanocomposite, 261
micronozzle droplet delivery system, 244
with 6-OHDA gradients, higher cell deaths at higher toxin concentrations, 234, 236
neuronal death mainly induced by apoptosis, and necrosis, 234, 236, 251, 254
triggered neuronal apoptosis in PC12 neuronal cell, 233
passive pumping
with ALH for mammalian cell culture, pattern cells, 240–242
for automated cell manipulation, 244
channel used, for high-throughput protein expression, 248
combine with washing valve, fluidic regulator, 246
to concentrate cells, device, 230, 232
developing platform for screening cardiac toxicity of drug compounds, 245
forward flow due to, 230–231
high-throughput screening assays, 241, 244–245
in vitro model for PD using, 244–245, 251
with LFP, fluidic capacitor, 245–246
microchannels, flow-enhanced infection assays, 248–251
microchannels, washing efficiency of, 245
in microfluidics, used for study of biological systems, 247–248
for screening cardiac toxicity of drug compounds, 245
self-assembled monolayer (SAM) cell culture, 249, 252
simplified version of ALH, 239, 244
study fusion *in vitro,* advantages of using, 255
use advantageous, in protein expression, 248–249
using in cell programmable assay (CPA) chips, 244
passive washing, schematic of, 246
surface microfluidic devices, by using hydrophobicity, 237, 261
surface tension passive pumping, 222–223
system for automated mammalian cell culture, 260
treat-wait-treat method, channel washout, 240
understanding fluidic wash inside, passive pumping, 239–240
used for high-throughput applications, 239
Tumor cell migration assays, 249–250, 252
Two-dimensional paper networks (2PDN)
applications
chemical signal amplification, 164–166
sample dilution and mixing, 162–163
small molecule extraction, 163
assays for, 155–156
for automated multistep sample processing, 157–158
complementary advances, 166
diagnostic devices, 156–157
paper microfluidics tools, 158–162
to control flow rates of fluids in paper networks, 158–161
for fluidic resistance, 160–161
off-switch for fluid flow, 161
on-switch for fluid flow, 161

V

Valve-based methods, droplet formation, 173
Valve integrated system, droplet generation, 204

W

Water, importance of, 103–104
Wheeler laboratory at the University of Toronto, 260
"Working" electrodes (WEs), 22

Y

Young–Laplace equation, 110